Applied Nonparametric Statistical Methods

OTHER STATISTICS TEXTS FROM CHAPMAN & HALL

Practical Statistics for Medical Research
Douglas Altman
The Analysis of Time Series –An Introduction
C. Chatfield
Problem Solving: A Statistician's Guide
C. Chatfield
Statistics for Technology
C. Chatfield
Introduction to Multivariate Analysis
C. Chatfield and A. J. Collins
Applied Statistics: Principles and Examples
D. R. Cox and E. J. Snell
An Introduction to Statistical Modelling
A. J. Dobson
Introduction to Optimization Methods and their Applications in Statistics
B. S. Everitt
Multivariate Statistics – A Practical Approach
B. Flury and H. Riedwyl
Readings in Decision Analysis
S. French
Multivariate Analysis of Variance and Repeated Measures
D. J. Hand and C. C. Taylor
Multivariate Statistical Methods – a primer
Bryan F. Manly
Elements of Simulation
B. J. T. Morgan
Probability Methods and Measurement
Anthony O'Hagan
Essential Statistics
D. G. Rees
Foundations of Statistics
D. G. Rees
Decision Analysis: A Bayesian Approach
J. Q. Smith
Applied Statistics: A Handbook of BMDP Analyses
E. J. Snell
Applied Statistics: A Handbook of GENSTAT Analyses
E. J. Snell and H. R. Simpson
Elementary Applications of Probability Theory
H. C. Tuckwell
Intermediate Statistical Methods
G. B. Wetherill
Statistical Process Control: Theory and practice
G. B. Wetherill and D. W. Brown
Statistics in Research and Development
Second edition
R. Caulcutt
Modelling Binary Data
D. Collett
Statistical Analysis of Reliability Data
M. J. Crowder, A. C. Kimber, T. J. Sweeting and R. L. Smith

Further information of the complete range of Chapman & Hall *statistics books is available from the publishers.*

Applied Nonparametric Statistical Methods

Second edition

P. Sprent

Emeritus Professor of Statistics
University of Dundee
Scotland

Published by Chapman & Hall, 2–6 Boundary Row, London SE1 8HN

Chapman & Hall, 2–6 Boundary Row, London SE1 8HN, UK

Blackie Academic & Professional, Wester Cleddens Road, Bishopbriggs, Glasgow G64 2NZ, UK

Chapman & Hall Inc., 29 West 35th Street, New York NY10001, USA

Chapman & Hall Japan, Thomson Publishing Japan, Hirakawacho Nemoto Building, 6F, 1–7–11 Hirakawa-cho, Chiyoda-ku, Tokyo 102, Japan

Chapman & Hall Australia, Thomas Nelson Australia, 102 Dodds Street, South Melbourne, Victoria 3205, Australia

Chapman & Hall India, R. Seshadri, 32 Second Main Road, CIT East, Madras 600 035, India

First edition 1989
Reprinted 1990
Second edition 1993

Printed in Great Britain by Clays Ltd, Bungay, Suffolk

ISBN 0 412 44980 3

A catalogue record for this book is available from the British Library

Library of Congress Cataloging-in-Publication data available

Contents

Preface

The first edition of this book, published in 1989, provided a practical introduction to statistical techniques known as **nonparametric methods**. This extended version has the same aim but differs in two major respects: firstly it takes account of the practical implications of developments since 1990 in computer software specifically designed to exploit the versatility of nonparametric methods, and secondly it emphasizes many links between procedures which at first sight appear to be tackling very different problems. The revision has provided an opportunity to introduce additional material especially in the area of categorical data analysis.

Many research workers, industrialists and business men carry out their own basic statistical analyses. This book does not aim to turn non-statisticians into statisticians, but it tries to indicate the strengths and weaknesses of nonparametric methods, as well as warning of potential pitfalls.

Many standard and relatively straightforward nonparametric procedures are dealt with in Chapters 1 to 8 using examples to explain assumptions and demonstrate techniques; theory is kept to a minimum. We show how basic problems are tackled and attempt to clear up common misapprehensions so as to help students of statistics meeting the methods for the first time as well as workers in other fields faced with data needing simple but informative analysis. Chapters 1 to 6 cover similar ground to the corresponding chapters in the first edition but there is some additional material and many new examples. Treatment of correlation and regression has been extended to two chapters, 7 and 8, and the old Chapter 8 on categorical data has been replaced by a new and extended treatment of this topic in Chapters 9 and 10. It is in these chapters that the many links between apparently diverse procedures introduced earlier in the book are most clearly demonstrated. Chapters 11 and 12 give an updated coverage of material in Chapters 9 and 10 of the first edition.

I have retained features of the earlier version which received favourable comment from reviewers and correspondents. As well as the major changes indicated I have included brief summaries at the end of Chapters 2 to 11 to guide the reader with a specific problem more quickly to relevant details. Most examples conclude with notes about appropriate computer software under the heading *Computational aspects*. I have listed in Section 12.2 a number of techniques not dealt with in this book which may be of interest to readers with particular problems, giving references to texts where detailed accounts of these may be found.

When appropriate I have used real (or at least realistic) data in examples and exercises; some were specially obtained for this book, some extracted from larger published data sets with sources indicated. References to these source will often show that the original data have been analysed with different objectives and using more advanced techniques (parametric or nonparametric). Most important of all, I have made use of a number of data sets (some unpublished) kindly sent to me by practising statisticians who have generously provided me with background information and allowed me to use such data as I saw fit. The sources of such material are indicated in the text and my thanks go to Thomas E. Bradstreet (Merck and Company, USA), Timothy P. Davis (Ford Motor Company, UK), Byron Morgan and his colleagues Ian Joliffe and James Matcham (University of Kent, UK), Nigel Smeeton (UMDS, Guy's campus, UK) and Chris Theobald (University of Edinburgh, UK), for providing such data. In many cases they sent me very extensive data sets and only a small fraction has been used for illustrative purposes. Nigel Smeeton also drew my attention to the biblical reference in Section 1.1.3.

My warmest thanks also go to Cyrus Mehta and Nitin Patel (Cytel Corporation, USA) and to Nigel Smeeton for their critical reading and helful comments on selected draft chapters and especially to Giles E. Thomas of the University of Dundee, UK, for a detailed assessment of the completed manuscript.

P. Sprent
August 1992

1

Introducing nonparametric methods

1.1 BASIC STATISTICS

Only a basic knowledge of statistics is needed for sensible use of simple nonparametric methods.

The reader without previous statistical training should use in parallel, and refer as necessary to, an introductory text such as the straightforward *Statistics Without Tears* (Rowntree, 1981) or the fuller *Foundations of Statistics* (Rees, 1987); the more sophisticated but still basic approach in Chapters 1 to 8 of *Statistics for Technology* (Chatfield, 1983) may appeal to those happy with mathematical notation. There are many other good introductory texts; some emphasize applications in areas such as medicine, biology, agriculture, the social sciences, etc., which may appeal to workers interested in the particular fields.

We summarize a few relevant statistical concepts in an appendix where an A precedes each section number; references to these are given in the form 'see Section A6', etc. Some tables for nonparametric procedures are given at the end of the book, identified by roman numerals and referred to in the text as 'Table I', 'Table II', etc.; modern computer packages for nonparametric methods (Section 1.5) make such tables less important than they used to be.

In this chapter we survey some fundamentals.

1.1.1 Parametric and nonparametric methods

Families of probability distributions depend on quantities called parameters. The well-known **normal** or **Gaussian distribution** family has two, usually denoted by μ and σ^2, specifying the mean and variance respectively. Normal distributions are often relevant to continuous data such as measurements.

Binomial distributions, associated with certain counts, have two parameters n, p, where n is the total number of observations and p is the probability that a specified event occurs at each observation. Subject to certain conditions, the total number of occurrences, r, $0 \leq r \leq n$, of that event in the n observations has a binomial distribution. Other families include the uniform (or rectangular), multinomial, Poisson, exponential, double exponential, gamma, beta and Weibull distributions. This list is not exhaustive and do not worry if you are not familiar with all of them.

Given a set of independent observations (called a random sample) from a normal distribution, we often want to infer something about the unknown parameters. The sample mean provides a point estimate of the mean μ (sometimes called, in statistical jargon, the mean of the **population**). In this situation the *t*-test (Section A4) is appropriate to decide whether to accept an *a priori* hypothesized value μ_0 for the population mean. More usefully, we may determine what is called a **confidence interval** for the 'true' population mean: this is an interval for which we have, in a sense to be described in Section 1.4.1, reasonable confidence that it contains the true but unknown mean μ.

Similarly, for a binomial distribution if we observe r favourable outcomes in a 'sample' of n observations, then r/n is a point estimate of p and we may test whether this estimate supports an *a priori* hypothesized value p_0, say, or obtain a confidence interval for the true value of p, the probability of success at each independent observation.

The normal distribution is strictly relevant to continuous data such as measurements, but it often works quite well if the measurement scale is coarse (e.g. for examination marks recorded to the nearest integer). More importantly, it is useful for approximations in a wide range of circumstances. The binomial distribution is relevant to counts of 'favourable' events. These are often called *successes*, but the distribution is applicable to many dichotomous outcome situations. For example, the number of male children in a family of size n has a binomial distribution with $p=\frac{1}{2}$ (approximately, see Section 1.1.3); the number of 'sixes' recorded in 10 casts of a fair die has a binomial distribution with $n=10$ and $p=\frac{1}{6}$.

It is often reasonable to assume that observations come from a particular family of distributions. Moreover, experience, backed by theory, suggests that, for many measurements, inferences based on the assumption that observations form a random sample from some normal distribution may not be misleading even if the normality assumption is incorrect; but this is not always true.

Inferences about parameter values for a specific family of distributions, or other distributional characteristics that imply specific values of these parameters, are **parametric** inferences. An assumption that samples come from a specific family of distributions may be inappropriate. For example, we may not have examination marks for individual candidates but only know the numbers in banded and ordered grades designated Grade A, Grade B, Grade C, ... Given the numbers of candidates in each grade for two different schools, we may use such information to see if it provides evidence of a difference in performance between schools that might be attributable to unequal teaching standards or to an ability of one school to attract more able pupils. This type of inference is **nonparametric**.

Even when we have precise measurements it may be obvious we cannot assume a normal distribution because normality implies certain properties of symmetry and spread. It may be clear that a distribution cannot be normal simply by looking at the mean, standard deviation and range of values. For example, if all observations are zero or positive and the standard deviation exceeds the mean, our sample is certainly not one from a normal distribution. There are well-known distributions that are flatter than the normal (e.g. the continuous uniform, or rectangular, distribution) or skew (e.g. the exponential and gamma distributions). In practice we are often able to say little more than that our sample appears to come from a distribution that is skew, or very peaked, or very flat, etc. Here nonparametric, or as they are sometimes called, **distribution-free** methods, are again appropriate.

Some writers regard 'distribution-free' and 'nonparametric' as synonymous: this ignores subtle distinctions that need not worry us here. Paradoxically, many tests that are universally regarded as nonparametric or distribution-free do involve parameters and distributions (often the familiar normal or binomial distributions). This is because the tags 'nonparametric' and 'distribution-free' apply **not** to the distribution of the test statistics, but to the fact that the methods can be applied to samples that come from populations having any of a wide class of distributions which need only be specified in broad terms, e.g. as being continuous, symmetric, identical, differing only in median or mean, etc. They need not belong to specified families such as the normal, uniform, exponential, etc. Logically, the term 'distribution-free' may then be more appropriate but 'nonparametric' is well established in popular usage although some writers, e.g. Marriott (1990), plead for the use of *distribution-free*. There is a grey area between what is clearly a distribution-free situation and parametric inference. Some of the association tests described in Chapters 9 and 10 fall in this region.

1.1.2 When do we need nonparametric methods?

Even when a parametric test does not depend too critically on an assumption that our sample comes from a distribution in a particular family, if there is doubt then a nonparametric test needing weaker assumptions is preferable. Nonparametric methods are often the only ones available for data that simply specify order, ranks or counts in various categories.

Weaker assumptions do not mean that nonparametric methods are assumption-free. In most statistical problems what can be deduced depends upon what assumptions can validly be made. An example illustrates this.

Example 1.1

A contract for the manufacture of metal rods specifies (among other requirements) that not more than 2.5% of those supplied should have a cross-sectional diameter exceeding 30 mm. Some quality control methods used to see if this condition is met are based on counts of the proportion of defectives (i.e. rods exceeding 30 mm diameter) in samples. Such tests extend to assessing whether two machines produce the same proportion of defectives. Even if we conclude that two machines produce the same proportion of defectives, this does not mean that the distribution of diameters is the same for each.

For example, suppose that diameters of items produced from the first machine have a normal distribution with mean 27 mm and standard deviation 1.53 mm; then conventional normal distribution theory tells us that 2.5% of all items produced will have a diameter exceeding 30 mm. This is because, for any normal distribution, 2.5% of all items have a diameter at least 1.96 standard deviations above the mean; in this case, 2.5% exceed $27+1.96\times1.53=30$. If the second machine produces items with diameters that are uniformly distributed between 20.25 and 30.25 mm (i.e. with mean diameter 25.25 mm) it is easy to see that once again 2.5% would have diameters exceeding 30 mm (since any interval of 0.25 mm between 20.25 and 30.25 mm contains a proportion $^1/_{40}=0.025$, i.e. 2.5%, of all production). Here we have the same proportion of defectives in two populations, yet each has a different mean and their distributions do not even belong to the same family. The assumption that diameters have a uniform distribution is unlikely to arise in practice, but we consider a more realistic situation involving different means in Exercise 1.1. It is clear from this illustration that we cannot use the sample numbers of defectives to determine whether the mean diameters of items produced by each machine are equal.

On the other hand, if we assume that the distributions of diameters of items from the two machines come from populations that differ, if at all, only in their mean values, then if we know the proportion of defectives in samples of 200 from each machine, we could test whether the means can reasonably be supposed to be identical. The test would not be very efficient; we would do better to measure actual diameters of each item in much smaller samples, and dependent upon the distributional assumptions one makes, use an appropriate parametric or nonparametric test. We say more about hypothesis tests in Section 1.3.

Means and medians (Section A1.2) are common measures of location. Statistical tests and estimation procedures for measurement data are often concerned with location. Does a sample come from a population with some pre-specified mean or median? Can we assert that two samples come from populations whose means differ by at least 10 units? Given a sample, what is an appropriate estimate of the population mean or median? How good an estimate is it? Increasingly, and particularly in the context of industrial quality control, interest is being focused also on variability or dispersion. Consumers not only demand that their computers or cars are of high quality (give a good average performance) but that they do so consistently. Sales of goods often depend upon personal recommendations and qualified endorsements that warn of niggling faults on delivery are not good publicity. There are parametric and nonparametric methods for assessing variability.

Some nonparametric techniques require only minimal information. We may test whether it is reasonable to assume weights of items have some prespecified median θ, say, if we know only how many items in a sample of n have weights greater than θ. If it were difficult, expensive, or impossible to get exact observations, but easy to determine numbers above (or below) θ, this nonparametric approach may be very cost effective.

Example 1.2

In medical studies sometimes we can only monitor the progress of patients for a limited time after treatment; in practice often anything from 12 months to 5 or 6 years. Dinse (1982) gives (as part of a larger investigation) data for survival times (weeks) for 10 patients with symptomatic lymphocytic non-Hodgkins lymphoma. The precise survival time is not known for one patient who was alive after 362 weeks. The observation *time to death* is said to be *censored* for that patient. The survival times in weeks were

49, 58, 75, 110, 112, 132, 151, 276, 281, 362*

The asterisk denotes the censored observation.

We may ask if these data are consistent with a median survival time of 200 weeks. A population median of 200 implies that in a random sample about half the observations should be below 200 and about half should be above. If we score an observation above 200 as a 'plus' and one below as a 'minus' then, since any one observation in a random sample is equally likely to be above or below the population median, it follows that the number of plus scores will have a binomial distribution with $p=\frac{1}{2}$ and $n=10$ (the sample size) *if the population median is indeed 200*. This is analogous to the the distribution of the number of heads when a fair coin is tossed 10 times. In our data only three survival times exceed 200 (i.e. three are 'plus' scores). We show in Section 1.3.1 that this is not sufficient evidence to reject the hypothesis that the median is 200 if we assume our data are a random sample from some population of patients with the disease.

The test proposed above is called the **sign** test. The result is not very interesting in itself. It would be more useful to be able to say that our observations imply that it is reasonable to assert that the median survival time for patients with this disease is between 75 and 275 weeks or something of that sort. This is what confidence intervals (Section 1.4.1) are about. In his paper Dinse was interested, among other things, in whether the median survial times differed between symptomatic and asymptomatic cases; he used this sample and another of 28 asymptomatic cases to compare the survival time distributions in more detail. In this other sample 12 of the 28 observations were censored at values of 300 or more. We show in Example 5.9 that we reject the hypothesis of equal medians for symptomatic and asymptomatic cases. These data were also considered by Kimber (1990).

Hypothesis testing has many limitations. Non-rejection of the hypothesis that the population median is 200 in Example 1.2 does not **prove** the hypothesis is correct; it only means that currently we have insufficient evidence to reject it. We discuss this more formally in Section 1.3.

1.1.3. A historical note

The first chapter of the Book of Daniel records that on the orders of Nebuchadnezzar certain favoured children of Israel were to be specially fed on the king's meat and wine for three years. Reluctant to defile himself with such luxuries, Daniel pleaded that he and three of his brethren be fed instead on pulse for 10 days. After that time the four were declared 'fairer and fatter in flesh than all of the children which did eat the portion of the king's meat.' This evidence was taken on common sense grounds as 'proof' of the superiority of a diet of pulse. In Example 1.4 we illustrate how we test such evidence more formally to justify the common sense conclusion arrived at in the Old Testament. Although the biblical analysis was informal it contains the germ of a nonparametric test.

Arbuthnot (1710) observed that in each year from 1629 to 1710 the number of males christened in London exceeded the number of females. He regarded this as strong evidence that the probabilities of any birth being male or female were not *exactly* equal, a discrepancy he attributed, with male chauvinism, to 'divine providence'. A sign test (slightly modified) is appropriate for these data and indicates we should reject the hypothesis $p=\frac{1}{2}$; the situation is akin to observing 82 heads in 82 consecutive tosses of a coin. If this happened would you believe the coin was 'fair'?

Karl Pearson (1900) proposed the chi-squared goodness-of-fit test applicable to any discrete distribution, and Spearman (1904) proposed a rank correlation coefficient (see Section 7.1.3) that bears his name, but systematic study of nonparametric inference dates only from the 1930s.

Research at that time was stimulated by attempts to show that even if an assumption of normality often stretched credulity, then at least in some cases making it would not greatly alter valid conclusions. This inspired work by R.A. Fisher, E.J.G. Pitman and B.L. Welch on **randomization** or **permutation** tests which were then too time consuming for general use, a problem now overcome by appropriate statistical program packages.

At the same time there was a growing realization that observational data consisting simply of preferences or rankings could be used to make inferences in a way that required little computational effort.

A few years later F. Wilcoxon and others showed that even if we have precise numerical data, we sometimes lost little useful information by ranking them in increasing order of magnitude and basing analyses on computationally simple procedures using these ranks. Indeed, if assumptions of normality are not justified, analyses based on ranks or on some transformation of them may be the most efficient available. Nonparametric methods then became practical tools to use either when data were by nature simply ordinal (ranks or preferences) as distinct from precise measurements (interval or scalar); or as a reasonably efficient method that reduced computation even when full numerical data were available, but those data could easily be replaced by ranks. At the time hypothesis testing was usually easy; unfortunately the more important interval estimation (Section 1.4) was less easy.

In parallel with the above, techniques relevant to counts were developed. Counts often represent the numbers of items in categories which may be either **ordered**, e.g. examination grades; or **nominal**, e.g. Roman Catholic, Church of England, Presbyterian, atheist, etc. Unlike ordered categories, nominal ones cannot be placed in ascending or descending order of magnitude. A concise summary of the properties of different data categories is given by Siegel and Castellan (1988, Section 3.3).

In the pre-computer era simplicity only applied to basic nonparametric procedures and these lacked the flexibility of much 'linear model' and 'least squares' theory that are cornerstones of normal distribution parametric inference. Many newer advanced and flexible nonparametric methods are tedious only in that they require repeated applications of simple calculations, a task for which computers are admirably suited and easily programmed.

The dramatic post-war development of nonparametric methods is described by Noether (1984). Some idea of the volume of literature is given in a nonparametric bibliography compiled by Singer (1979) who only included work relevant to applications in psychology. Kendall and Gibbons (1990) gives over 350 references for the one topic of rank correlation.

Computers have revolutionized our approach to data analysis and statistical inference (see e.g. Durbin, 1987). Pious hopes that data will fit a restricted mathematical model with few parameters and emphasis on simplifying concepts such as linearity have been replaced by the use of robust methods and by exploratory data analyses to investigate different potential models, areas where nonparametric methods sometimes have a central role. Generalized linear models described by McCullagh and Nelder (1989) at the theoretical level and by Dobson (1990) at the practical level often blend parametric and nonparametric approaches.

Nonparametric methods are in no sense a preferred method of analysis in all situations. A strength is their applicability in situations where there is insufficient theory or data to specify, or test compatibility with, highly specific models.

The most important practical development since the first edition of this book has been the marketing of computer software (Section 1.5) to carry out permutation tests. Results may be compared with those given by asymptotic theory, which, in the past, was often used in situations where validity was dubious.

1.2 SAMPLES AND POPULATIONS

Much statistical inference assumes observations are a random sample from some population. Specification of the population may be imprecise. If we select 20 books at random from 10,000 volumes in a library, and record the number of pages in each book in the sample, then inferences made from the sample about the mean or median number of pages per book apply strictly to the population of books in that library. If the library covers a wide range of fiction, non-fiction and reference works it is reasonable to assume that any inference applies to a wider population of books; perhaps to all books published in the United Kingdom or the United States or wherever the library is situated. If the books in the library are all English language books, inferences may not apply to books in Chinese or Russian.

We seldom deal with strictly random samples from a clearly specified population. More commonly it may be reasonable to assume observations represent a sample with the essential properties of a random sample from a vaguely-specified population. For example, if a new diet is tested on pigs and we have measurements of weight increments for 20 pigs at one agricultural experimental station we might assume that these are something like a random sample from all pigs of that breed raised under similar conditions. Inferences might apply widely if the experimental station adopted common farming practices and if these were fairly uniform wherever that breed of pig was raised. This would not be so if the experimental station adopted different husbandry practices from those used on most pig farms.

The abstract notion of sampling from an infinite population (implicit in normal distribution theory) often works well in practice, but is never completely true!

There are situations where the sample is essentially the whole population. For example, at the early stages of testing a new drug for treating a rare disease there may be just, say, nine patients available for test. A common procedure is to allocate, say, four of these patients at random to receive this drug. The remaining five are untreated (or treated with a drug already in use). Because the four patients are selected at random from the nine available, if the drug has no effect (or is no better than one currently in use) it is unlikely that a later examination would indicate that the four patients receiving the new drug had responded better than any of the others. This is possible, but it has a low probability, which we can calculate on the assumption that the new drug is ineffective or no better than the old drug (see Example 1.4, Section 1.3.2). Clearly, if the drug is beneficial the probability of better responses among those treated with it is increased.

We shall often draw attention to the validity and scope of inferences in the light of the nature of our data. For example, the set of 10 survival times of patients with symptomatic lymphocytic non-Hodgkins lymphoma given in Example 1.2 came from a study conducted by the Eastern Co-operative Oncology Group in the USA and represented all patients so afflicted that were available for that study. In making inferences about the median or other characteristics of the survival time distribution, it is reasonable to assume these inferences are valid for all such patients receiving similar treatment and alike in other relevant characteristics; e.g. with a similar age distribution. The patients in this study were all male, so clearly it would be unwise to infer, without further evidence, that the survival times for females would have the same distribution.

Fortunately the same nonparametric inference procedures are usually valid whether we are sampling from an infinite population, a finite population or when our sample units are the entire relevant population. How widely any inferences will be valid differs for each situation. A full discussion of the implications of this general applicability is given for several specific tests by Lehmann (1975, Chapters 1–4).

1.3 HYPOTHESIS TESTS

Estimation (Section 1.4) is very often the appropriate aim of a statistical analysis, but because estimation procedures are most easily explained in terms of testing a range of hypotheses, we must understand testing.

We assume some familiarity with simple parametric hypothesis tests such as the *t*-test and chi-squared test, but review some fundamentals in the light of altered practices made possible by computer software. Some of these changes are especially relevant to nonparametric inference.

Consider a *t*-test of hypotheses about an unknown mean, μ, of a normal population based on a random sample of n observations from that population. We specify a **null hypothesis**, H_0, that μ takes a specific value μ_0. We continue to accept this hypothesis unless there is strong evidence that we should reject it in favour of the alternative hypothesis H_1, that μ takes some other value, i.e. $\mu \neq \mu_0$. This is concisely summarized as:

$$\text{Test } H_0 : \mu = \mu_0 \text{ against } H_1 : \mu \neq \mu_0 \tag{1.1}$$

A test is based on a function of the sample values called a **statistic**. Here the relevant statistic, t, is calculated by formula (A4.1) in Section A4.1. The classic procedure is to compare the magnitude, without regard to sign, of t, written $|t|$, with a value t_α, so chosen that when H_0 is true

$$\Pr(|t| \geq t_\alpha) = \alpha \tag{1.2}$$

We call α the **significance** level. Historically it was customary to choose $\alpha=0.05$, 0.01 or 0.001. These probabilities are often expressed as percentages, i.e. 5%, 1% or 0.1% and the corresponding levels are often called significant, highly significant and very highly significant. For consistency in this book we specify significance levels as percentages. We reject H_0 at a $100\alpha\%$ significance level if our observed t is such that $|t| \geq t_\alpha$.

The rationale behind the test (1.1) is that if H_0 is true then values of t near zero are more likely than large values of t, either positive or negative. Large values of $|t|$ have a greater probability of occurence under H_1. It follows from (1.2) that if we perform a large number of such tests on a sequence of independent random samples when H_0 is true we shall incorrectly reject H_0 in a proportion α of these; e.g. if $\alpha=0.05$ we would reject the null hypothesis when it were true in the long run in 1 in 20 tests, i.e. in 5% of all tests.

Rejection of H_0 when it is true is an **error of the first kind**, or Type I error. When we choose α we fix the probability of making such an error. The set of t values such that $|t| \geq t_\alpha$ in (1.2) defines a **critical** or **rejection** region and α is the **size** of this region. Using a critical region of size α implies we accept H_0 if $|t| < t_\alpha$. If we follow this rule we sometimes accept H_0 when in fact H_1 is true. Accepting H_0 when H_1 is true is an **error of the second kind**, or Type II error. Let β denote the probability of a Type II error. The probability of a Type II error depends on the precise value of μ. A moment's reflection shows that the further μ is from μ_0, the more likely we are to get large values of $|t|$, i.e. values in the critical region, so that β decreases as $|\mu-\mu_0|$ increases. If we decrease α (say from 0.05 to 0.01) our critical region becomes smaller, so that for a given μ we increase β because the region in which we accept H_0 is larger. If we increase the sample size, n, we decrease β for a given μ and α. Thus β depends on the true value of μ (over which we have no control) and the value of n and α (over which we may have some control).

A good test is one for which β is small, or alternatively $1-\beta$, which is called the **power** of the test, is large. For samples from a normal distribution and all choices of n, α and for any μ, the t-test is more powerful than any other test for (1.1).

The test in (1.1) is called a **two-tail** test. The reason is clear from Figure 1.1a. This shows a typical t-distribution, that with 12 degrees of freedom. The distribution is symmetric about zero and from standard t-tables we find the critical region of size $\alpha=0.05$ consists of tail regions $t \geq 2.18$ and $t \leq -2.18$. These are shaded in Figure 1.1a and each has associated probability 0.025.

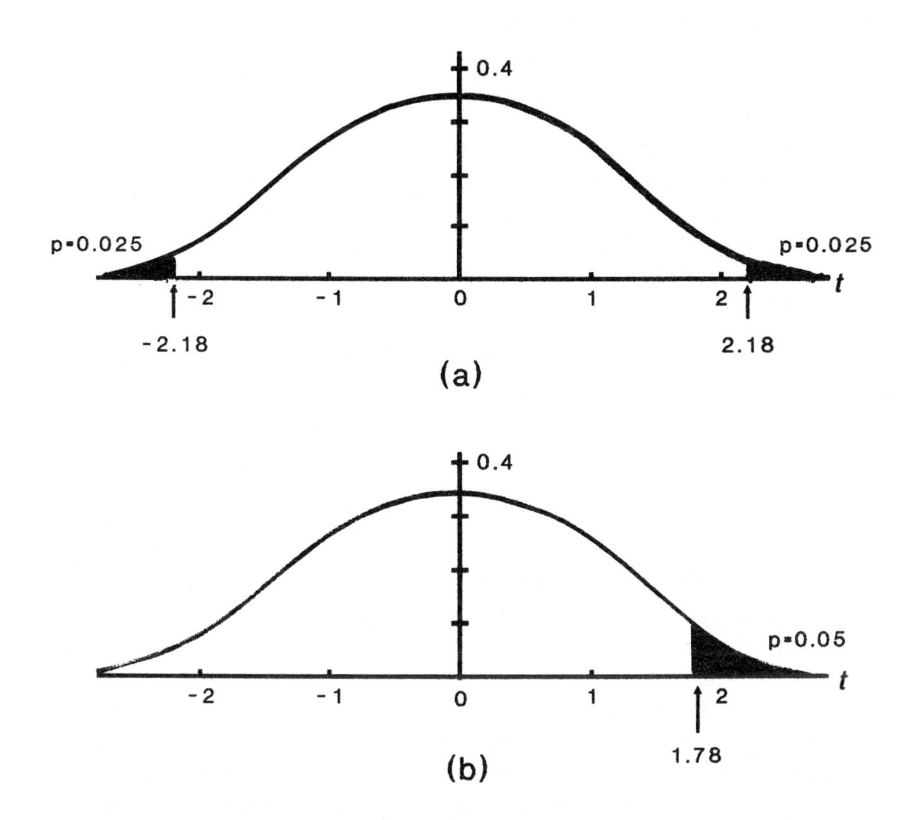

Figure 1.1. The probability density function for the t-distribution with 12 degrees of freedom. The shaded tail regions in (a) each have an associated probability $p=0.025$ and jointly provide a critical region of size $\alpha=0.05$ for rejection of H_0 in a two-tail test. In (b) the corresponding upper-tail critical region for a one-tail test has associated probability $p=0.05$ and defines the critical region of size $\alpha=0.05$.

Specification of H_0 and H_1 is determined by the logic of a problem. Two other common choices are

(i) Test $H_0 : \mu=\mu_0$ against $H_1 : \mu>\mu_0$ (1.3)

(ii) Test $H_0 : \mu\leq\mu_0$ against $H_1 : \mu>\mu_0$ (1.4)

These both lead to a *one-tail* (here *right* or *upper-tail*) test, since in each case large *positive* values of t favour H_1, whereas small positive values or *any* negative value indicate that H_0 is more likely to hold. For the t-test with 12 degrees of freedom and $\alpha=0.05$ tables indicate that the relevant critical region is associated with $t\geq1.78$, a region shaded in Figure 1.1b.

The modification to a one-tail test if the inequalities in (1.3) and (1.4) are reversed is obvious. The critical region becomes the left, or lower, tail rather than the upper tail. For example, if the amount of a specified impurity in 1000 g ingots of zinc produced by a standard process is known to be normally distributed with a mean of 1.75 g and it is hoped that a steam treatment will remove some of this impurity we might steam treat a sample of 15 ingots and determine the amount of impurity left in each ingot. If the steam used is free from the impurity the treatment cannot increase the level and either it is ineffective or it reduces the impurity. It is therefore appropriate to test

$$H_0 : \mu=1.75 \text{ against } H_1 : \mu<1.75$$

In a situation where ingots have an unknown mean impurity level, but a batch is acceptable if for specified n, α we accept $H_0 : \mu\leq 1.75$ but not otherwise, the appropriate test would then be

$$H_0 : \mu\leq 1.75 \text{ against } H_1 : \mu>1.75$$

Adherence to the historical choices $\alpha=0.05$, 0.01 or 0.001 is largely dictated by available tables and the selected significance level will depend partly on the consequences in terms of cost, etc., that arise when we make a Type I error.

If we restrict α to the two values 0.05, 0.01, then, for a two-tail t-test with 12 degrees of freedom the minimum values of $|t|$ for significance at these levels are respectively 2.18 and 3.05. If we observe $t=2.78$ we would only claim significance at the 5% level. In fact $\Pr(|t|\geq 2.78)=0.0167$, so we are entitled to claim significance at the $100\alpha=1.67\%$ level, rather than the more conservative 5%. Many computer packages give the calculated t value, say t_0, for any data and also $\Pr(t\geq t_0)$ if t_0 is positive and $\Pr(t\leq t_0)$ if t_0 is negative. These are the tail probabilities relevant to one-tail tests. We double these probabilities to get the relevant two-tail probability. Significant results for which we reject H_0 are associated with low values for these tail probabilities (conventionally less than approximately 0.05). Thus if we find that our test statistic t_0 is such that $\Pr(t\geq t_0)=0.027$ we may properly speak of significance at the 0.027 or 2.7% level in a right-tail test. One word of caution; if we are performing a test of the hypotheses in (1.3) or (1.4) and observe a negative t value such that, say, $\Pr(t\leq t_0)=0.035$ we *do not reject* H_0, since the low probability is associated with the left-tail and therefore certainly does **not** support the hypothesis $\mu>\mu_0$. Some programs give $\Pr(|t|\geq t_0)$ if a two-tail test is specified. The symmetry of the t distribution means that the critical region components in a two-tail test are symmetric about the mean value zero. The doubling of one-tail probabilities to give the corresponding two-tail test probability or significance level applies in other tests such as the F-test for equality of variance based on samples from two normal populations, but in these cases the two subregions are not symmetric about the mean. However, in many applications where relevant statistics have a chi-squared or F-distribution a one-tail (upper-tail) test is appropriate.

Knowing actual tail probabilities for a given value of a test statistic such as t removes the need to confine significance levels to those available in tables and having the actual probability of making a Type I error helps a decision maker who often wants to weigh up a number of factors before deciding whether to accept or reject a hypothesis. These may include, for example, economic factors, the well-being of, or danger to, those affected by the decision, particularly in a medical context, or there may be safety considerations or legal requirements if toxic substances are involved.

It is important to realise that a statistically significant result may be of little or no practical importance. It is a common misconception that a low tail probability (often reported as a 'p' value in published reports) is an indicator of practical importance. We shall see why this is not necessarily true in Section 1.4.2. Knowing actual tail probabilities is, however, useful when comparing the performance of different tests.

1.3.1 Some nonparametric test statistics

Some statistics used for nonparametric tests have familiar distributions such as the normal, t, F or chi-squared distribution. However, in some situations our statistics have discontinuous distributions and this raises a further problem that must be understood, although it has limited practical impact except for very small samples.

Example 1.2 (continued)

We wanted to test the hypothesis that the median θ of the population of survival times was 200 against the alternative that it had some other value,

$$\text{i.e. to test } H_0 : \theta = 200 \text{ against } H_1 : \theta \neq 200 \tag{1.5}$$

We proposed a two-tail test based on the number of sample values exceeding 200 (recording each as a 'plus'), indicating that under H_0 the number of plus signs has a binomial distribution with $p=½$ and $n=10$. The probability of observing r plus signs in 10 observations when $p=½$ is given by the binomial formula

$$\Pr(X=r) = {}^{10}C_r(½)^{10}$$

The values of these probabilities, P, for each value of r between 0 and 10, correct to 4 decimal places, are

r	0	1	2	3	4	5	6	7	8	9	10
P	0.0010	0.0098	0.0439	0.1172	0.2051	0.2461	0.2051	0.1172	0.0439	0.0098	0.0010

The probability of 3 or less plus signs is 0.1172+0.0439+0.0098+0.0010 =0.1719. The decision is clearly to accept H_0, since the probability of getting 3 or less plus signs or 7 or more plus signs (the 'tail' probabilities for our observed statistic) is 2×0.1719=0.3438, implying that departures from the expected number of plus signs, 5, as large or larger than that observed will occur in slightly more than one-third of all samples when H_0 is true.

In the classical approach to hypothesis testing we might seek a critical region of size 0.05, or in each tail a region with associated probability 0.025, relevant to the test (1.5). Clearly if we choose the values 0, 1, 9, 10 as our critical region, because of symmetry the associated probability is 2×(0.0010+0.0098)=0.0216. If we include the additional points 2 and 8 in our critical region the associated probability is 0.1095. Unlike the continuous t-statistic, for which it is possible to have a critical region of any desired size, we find discontinuities in the 'size' of the critical region as we add points to it. Our statistic – the number of plus signs – has a discrete distribution. This means there is no direct way of obtaining a critical region of exact size 0.05; we must choose between regions of size 0.0216 or 0.1095. A common recommendation is to choose the largest possible region for which the associated probability is less than our pre-specified α (here 0.05) and to refer to this as a critical region of **nominal** size 0.05, but **actual** or **exact** or **true** size (in this example) 0.0216.

A device called a randomization procedure exists for making decisions so that in the long run we make an error of the first kind in repeated testing with probability at or close to some chosen nominal level, rather than at an actual level. In practice we are primarily interested in what happens in our one test, so it seems better, when we know them, to use actual levels, rather than worry unduly about nominal levels. There is, however, a case for forming a tail probability by allocating only one half of the probability that the statistic equals the observed value to the 'tail' when determining the size of the 'critical' region. This approach has many advocates; we do not use it in this book, but if it is used this should be done consistently. Modern computer programs calculate actual (often calling them exact) tail probabilities for many parametric and nonparametric tests, but pragmatism means we cannot ignore the established concept of nominal levels entirely, for if we have no appropriate program, we must use tables where information is usually only given for nominal levels like α=0.05 or 0.01, thus forcing us to quote these nominal rather than the actual levels of significance. Nominal and actual levels may differ appreciably, especially in small samples.

Differences between nominal and actual levels for one-tail tests also arise. Good computer programs for nonparametric tests quote the probability that a value greater than or equal to the test statistic will be attained if this probability is less than ½, otherwise they give the probability that a value less than or equal to the test statistic is obtained. This can be regarded as the probability of errors of the first kind in a one-tail test if we decide to reject at a significance level equal to that probability. In practice we only reject H_0 if this 'tail' probability is sufficiently small (and in the appropriate tail!). In general, we recommend doubling a one-tail probability to obtain the actual significance level for a two-tail test. If the test statistic has a symmetric distribution this is equivalent to considering equal deviations from the mean value of the statistic. However, in the case where the statistic does not have a symmetric distribution, taking tails equidistant from the mean will not be equivalent to doubling a one-tail probability. This is analogous to the situation with the F-test commented upon in Section 1.3.

Example 1.3 shows a further difficulty that sometimes arises due to discontinuity in nonparametric testing; namely, that if we adhere to the convention that we only reject H_0 if $\alpha \leq 0.05$ (or at any rate a value not very much greater than this), we may never reject because no outcome implies significance. This problem arises mainly when we try to make inferences from very small samples.

Example 1.3

Suppose that a test statistic, S, say, has a binomial distribution under H_0 with some fixed n and $p=\frac{3}{4}$. For the case $n=10$ the probabilities for the various values r of the statistic S, where r takes integral values between 0 and 10 are given by

$$\Pr(S=r) = {}^{10}C_r(\tfrac{3}{4})^r(\tfrac{1}{4})^{10-r}$$

The relevant probabilities are

r	0	1	2	3	4	5	6	7	8	9	10
P	0.0000	0.0000	0.0004	0.0031	0.0162	0.0584	0.1460	0.2503	0.2816	0.1877	0.0563

We cannot find a critical region for a test of $H_0: p=\frac{3}{4}$ against $H_1: p>\frac{3}{4}$ of size not greater than $\alpha=0.05$ because the extreme upper-tail value $r=10$ has a probability exceeding 0.05. Thus, there is no critical region (other than an empty region) and we would never reject H_0. There is no problem for a one-tail test of $H_0: p=\frac{3}{4}$ against $H_1: p<\frac{3}{4}$ at the same level, since $\Pr(S \leq 4) = 0.0162 + 0.0031 + 0.0004 = 0.0197$. The actual size of the appropriate region of nominal size 0.05 is therefore 0.0197. If we adopt the rule that the appropriate level for a two-tail test is twice that for a one tail test, then, if we get a result in the lower tail the actual two-tail test level is $2 \times 0.0197 = 0.0394$ (3.94%). This presents a dilemma, for there is no observable upper tail area corresponding to that in the lower tail. What this means is that if a two-tail test is appropriate we shall in fact only be likely to detect departures from the null hypothesis if they are in one direction; there may well be a departure in the other direction, but if so we will be unlikely to detect it. This is not surprising when the appropriate one-tail test must also fail to detect it, for generally a one-tail test at a given significance level is more powerful at detecting departures in the appropriate direction than is a two-tail test. The conclusion is that we need a larger sample to detect departures of the form $H_1 : p>\frac{3}{4}$ in this example.

There is not complete agreement among statisticians about the desirability of doubling a one-tail probability to get the appropriate two-tail significance level (see, for example, Yates (1984) and the discussion thereon). As an alternative to doubling the one-tail probability, some writers suggest that once the exact size of a one-tail region has been determined, we should, for a two-tail test, add the probability associated with an opposite tail situated equidistant from the mean value of the test statistic to that associated with our observed statistic value. In the symmetric case this is equivalent to doubling the probability, but it seems inappropriate with a non-symmetric distribution. In Example 1.3 the region $r \leq 4$ is appropriate for a lower-tail test. The mean of the test statistic (the binomial mean np) is 7.5. Since $7.5-4=3.5$, the corresponding deviation above the mean is $7.5+3.5=11$. Now $\Pr(r \geq 11)=0$, so the two tail-test based on equidistance from the mean and the one-tail test would have the same actual significance level. We discuss this point further in relation to a specific computer package in Section 1.5.

1.3.2 Permutation tests

Example 1.4

In Section 1.2 we envisaged a situation with nine patients, four being selected at random to receive a new drug. After three weeks all nine patients are examined by a skilled consultant who, on the basis of various tests and clinical observations is able to 'rank' the patients' condition in order from least severe (rank 1) to most severe (rank 9). In this situation, if there really is no beneficial effect of the new drug, what is the probability that the patients who received the new drug are those ranked 1,2,3,4?

Selection of four patients 'at random' means that any four are equally likely to be chosen for the new drug. If there really is no effect one would expect some of those chosen to end up with low ranks, some with moderate or high ranks, in the post-treatment assessment. From a group of nine patients there are 126 ways of selecting a set of four. This is an application of the result in Section A2 for calculating the number of ways of selecting r objects from n. We give all 126 selections in Table 1.1. Ignore for the moment the numbers in brackets after each selection.

If the new drug were ineffective the set of ranks associated with the four patients receiving it are equally likely to be any of the 126 quadruplets listed in Table 1.1. Thus, if there is no treatment effect there is only 1 chance in 126 that the four showing greatest improvement (ranked 1,2,3,4 in order of condition after treatment) happen to be the four patients allocated to the new drug. It seems more plausible that such an outcome reflects a beneficial effect of the drug.

In a hypothesis testing framework we have a group of 4 treated with the new drug and a group of 5 (the remainder) given none or a standard treatment in what is called a *two independent sample* experiment. We discuss such experiments in detail in Chapter 5. The most favourable evidence for the new drug is that those receiving it are ranked 1,2,3,4; the least favourable evidence is that they are ranked 6,7,8,9. Each of these extreme outcomes has a probability of 1/126 of occuring if there is no real effect.

If we consider a test of

$$H_0 : \textit{new drug has no effect}$$

against the two sided alternative

$$H_1 : \textit{new drug has an effect (beneficial or deleterious)}$$

and observe the outcome 1,2,3,4: it is one of two extremes (the other being 6,7,8,9) with a total associated probability of 2/126=0.0158 if H_0 is true. Thus we would reject H_0 at an actual 1.58% significance level if we observed either of these outcomes. Most people would regard this as strong evidence that the new drug is effective. What if the patients receiving the new drug were ranked 1,2,3,5? Intuitively this evidence favours the new drug, but how do we test this?

We seek a statistic (some function of the four ranks) that has a low value if all ranks are low, a high value if all ranks are high and an intermediate value if there is a mix of ranks for those receiving the new drug. An intuitive choice is the sum of the ranks. If we sum the ranks for every quadruplet in Table 1.1, and count how many times each sum occurs we can easily work out the probability of getting any particular sum, and hence the distribution of our test statistic.

In Table 1.1 the number in brackets after each quadruplet is the sum of the ranks for that quadruplet, e.g. for 1,2,7,9 the sum is 1+2+7+9=19. The lowest sum is 10 for 1,2,3,4 and the highest is 30 for 6,7,8,9. Table 1.2 gives the numbers of quadruplets having each given sum.

Clearly since there are 126 different sets of ranks the probability that the rank sum statistic, which we denote by S, takes a particular value is obtained by dividing the number of times that value occurs by 126. Thus $\Pr(S=17)=9/126=0.0714$. If we wish to determine a critical region of nominal size $\alpha=0.05$, because of the discontinuity in the distribution of S, we can only determine a region having a smaller actual size. We select a region in *each* tail (since H_1 implies a two-tail test is appropriate) with the

Table 1.1 Possible selections of four individuals from nine labelled 1 to 9 and the sum of the ranks in brackets

1,2,3,4 (10)	1,2,3,5 (11)	1,2,3,6 (12)	1,2,3,7 (13)	1,2,3,8 (14)	1,2,3,9 (15)	1,2,4,5 (12)
1,2,4,6 (13)	1,2,4,7 (14)	1,2,4,8 (15)	1,2,4,9 (16)	1,2,5,6 (14)	1,2,5,7 (15)	1,2,5,8 (16)
1,2,5,9 (17)	1,2,6,7 (16)	1,2,6,8 (17)	1,2,6,9 (18)	1,2,7,8 (18)	1,2,7,9 (19)	1,2,8,9 (20)
1,3,4,5 (13)	1,3,4,6 (14)	1,3,4,7 (15)	1,3,4,8 (16)	1,3,4,9 (17)	1,3,5,6 (15)	1,3,5,7 (16)
1,3,5,8 (17)	1,3,5,9 (18)	1,3,6,7 (17)	1,3,6,8 (18)	1,3,6,9 (19)	1,3,7,8 (19)	1,3,7,9 (20)
1,3,8,9 (21)	1,4,5,6 (16)	1,4,5,7 (17)	1,4,5,8 (18)	1,4,5,9 (19)	1,4,6,7 (18)	1,4,6,8 (19)
1,4,6,9 (20)	1,4,7,8 (20)	1,4,7,9 (21)	1,4,8,9 (22)	1,5,6,7 (19)	1,5,6,8 (20)	1,5,6,9 (21)
1,5,7,8 (21)	1,5,7,9 (22)	1,5,8,9 (23)	1,6,7,8 (22)	1,6,7,9 (23)	1,6,8,9 (24)	1,7,8,9 (25)
2,3,4,5 (14)	2,3,4,6 (15)	2,3,4,7 (16)	2,3,4,8 (17)	2,3,4,9 (18)	2,3,5,6 (16)	2,3,5,7 (17)
2,3,5,8 (18)	2,3,5,9 (19)	2,3,6,7 (18)	2,3,6,8 (19)	2,3,6,9 (20)	2,3,7,8 (20)	2,3,7,9 (21)
2,3,8,9 (22)	2,4,5,6 (17)	2,4,5,7 (18)	2,4,5,8 (19)	2,4,5,9 (20)	2,4,6,7 (19)	2,4,6,8 (20)
2,4,6,9 (21)	2,4,7,8 (21)	2,4,7,9 (22)	2,4,8,9 (23)	2,5,6,7 (20)	2,5,6,8 (21)	2,5,6,9 (22)
2,5,7,8 (22)	2,5,7,9 (23)	2,5,8,9 (24)	2,6,7,8 (23)	2,6,7,9 (24)	2,6,8,9 (25)	2,7,8,9 (26)
3,4,5,6 (18)	3,4,5,7 (19)	3,4,5,8 (20)	3,4,5,9 (21)	3,4,6,7 (20)	3,4,6,8 (21)	3,4,6,9 (22)
3,4,7,8 (22)	3,4,7,9 (23)	3,4,8,9 (24)	3,5,6,7 (21)	3,5,6,8 (22)	3,5,6,9 (23)	3,5,7,8 (23)
3,5,7,9 (24)	3,5,8,9 (25)	3,6,7,8 (24)	3,6,7,9 (25)	3,6,8,9 (26)	3,7,8,9 (27)	4,5,6,7 (22)
4,5,6,8 (23)	4,5,6,9 (24)	4,5,7,8 (24)	4,5,7,9 (25)	4,5,8,9 (26)	4,6,7,8 (25)	4,6,7,9 (26)
4,6,8,9 (27)	4,7,8,9 (28)	5,6,7,8 (26)	5,6,7,9 (27)	5,6,8,9 (28)	5,7,8,9 (29)	6,7,8,9 (30)

Table 1.2 Sums of ranks of four items from nine

Sum of ranks	10	11	12	13	14	15	16	17	18	19	20
Number of occurrences	1	1	2	3	5	6	8	9	11	11	12
Sum of ranks	21	22	23	24	25	26	27	28	29	30	
Number of occurrences	11	11	9	8	6	5	3	2	1	1	

total associated probability not exceeding 0.025. Clearly if we select in the lower tail $S=10$ and 11, the associated probability is 2/126 and if we add to this $S=12$ the associated total probability, i.e. $\Pr(S\leq 12)=4/126=0.0317$. The latter is greater than 0.025, so our lower-tail critical region is $S\leq 11$ with associated probability $2/126=0.0159$. By symmetry the upper-tail critical region is $S\geq 29$ also with probability 0.0159. Thus for the two-tail test of nominal size 0.05 the critical region is $S=10$, 11, 29, 30 and the actual size is $2\times 0.0159=0.0318$.

Some statisticians suggest choosing a critical region with probability as close as possible to a nominal level rather than the more conservative choice of one no larger. In this example if we add $S=12$ and the symmetric $S=28$ to our critical region the two-tail probability is $8/126=0.0635$. This is closer to 0.05 than the size (0.0318) of the region chosen above. We reaffirm that ideally it is best to quote the actual size of any region used. We reiterate that the practical argument (though there are further theoretical ones) for quoting nominal sizes is that many tables give only these. Computer programs giving actual sizes overcome this difficulty.

In Exercise 1.5 at the end of this chapter we explore appropriate critical regions for a one-tail test in this example. Unless there were firm evidence before the experiment was started that an effect, if any, of the new drug could only be beneficial, a two-tail test is appropriate.

In Section 1.2 we suggested that in preliminary testing of drugs for a rare disease our population may be in a strict sense only the cases we have. However, if we regard our patients as fairly typical of all who might have the disease, it is not unreasonable to assume that any findings from our small experiment may hold for other patients with a similar condition *providing* other treatment factors (nursing attention, supplementary treatments, diagnosis, etc.) are comparable. When our experiment involves what is effectively the whole population and we only have rankings, a permutation test is the only test available. Random allocation of treatments is essential for the test to be valid. In clinical trials this requirement may clash with ethical considerations.

If, for example, there were high hopes that the new drug might greatly relieve suffering in severe cases of the disease and only enough doses were available to treat four patients, ethical considerations suggest the four patients to receive the drug should be those with the most severe symptoms. In a situation like this, the drug may reduce suffering, but such patients may still, after treatment, be ranked 6,7,8,9 because, although their condition may have improved, their symptoms are still more severe than those of patients not given the drug. This problem might be alleviated if we had rankings of ‘degree of improvement’ after treatment rather than rankings only for condition after treatment.

At the other extreme, unethically, an experimenter might manipulate allocation so that the patients with initally the least severe symptoms were the ones to receive the new drug; then, even if it were ineffective, these patients may still be ranked 1,2,3,4 after treatment. We deplore this unethical choice, but some ethical choices must be made on medical grounds.

Even when patients are allocated to treatments at random and we reject the null hypothesis of ‘no treatment effect’, it is possible that the statistically significant outcome may be of no practical importance, or that there may be ethical reasons for ignoring the outcome. A doctor would not be justified in prescribing the new treatment if it merely prolonged by three days the life expectation of terminally-ill patients suffering considerable distress, but may feel bound to prescribe it if it gave better survival prospects with reasonable life quality.

Tests based on permutation of ranks or on certain functions of ranks (including as a special case original ordered measurements) are central to nonparametric methods and are called **permutation** or **randomization** tests. Small scale tests of a drug like that in Example 1.4 are often referred to as *pilot studies*. Efficacy of a drug in wider use may depend on factors such as severity of disease, treatment being administered sufficiently early, the age and sex of patients, etc., all or none of which may be reflected in a small group of 'available patients'. An encouraging result with the small group would suggest further experiments were desirable.

1.3.3 Pitman efficiency

In Section 1.3 we indicated that the power of a test depends upon the sample size, n, the choice of α and the magnitude of departure from H_0. Most 'intuitively' reasonable tests have good power to detect a true alternative that is far removed from the null hypothesis. We sometimes want tests to have as much power as possible for detecting alternatives close to H_0: even when such alternatives are of little practical importance a test which has a better chance of detecting these will also be better at detecting larger departures, a desirable state of affairs. Pitman (1948), in a series of unpublished lecture notes, introduced the concept of asymptotic relative efficiency for comparing two tests. If α is the probability of a Type I error and β is the probability of a Type II error (the power is $1-\beta$), then the efficiency of a test T_2 relative to a test T_1 is the ratio n_1/n_2 of the sample sizes needed to obtain the same power for the two tests with these values of α, β. In practice, we usually fix α; then β depends on the particular alternative *as well as* the sample sizes. Fresh calculations of relative efficiency are required for each particular value of the parameter or parameters of interest in H_1 and for each choice of α, β.

Pitman considered sequences of tests T_1, T_2 in which we fix α, but allow the alternative in H_1 to vary in such a way that β remains constant as the sample size n_1 increases. For each n_1 we determine n_2 such that T_2 has the same β for the particular alternative considered. It can be shown that as we take larger samples we usually increase the power for alternatives closer to H_0, so that for very large samples Pitman studied the behaviour of the efficiency, n_1/n_2, for steadily improving tests for detecting small departures from H_0. He showed under very general conditions that in these sequences of tests n_1/n_2 tended to a limit as $n_1 \to \infty$, and more importantly, that this limit was the same for all choices of α, β. He called this limit the **asymptotic relative efficiency** (ARE) of the two tests. Bahadur (1967) proposed an alternative definition that is less widely used, so for clarity and brevity in this book we refer to Pitman's concept simply as the **Pitman efficiency**. The practical usefulness of Pitman efficiency lies in the fact that when comparing many tests the small sample relative efficiency is often close to, or even higher, than the Pitman efficiency.

The Pitman efficiency of the sign test relative to the t-test when the latter is appropriate is $2/\pi=0.64$. Lehmann (1975, p.173) shows that for samples of size 10 and a wide range of values of the median θ relative to θ_0 with α fixed, the relative efficiency exceeds 0.7, while for samples of 20 it is nearer to, but still above, 0.64. Here Pitman efficiency gives a pessimistic picture of the performance of the sign test at realistic sample sizes.

When the t-test for a mean is relevant and valid it is known to be the most powerful test for any H_0 against any alternative mean specified in H_1. If the t-test is not appropriate, the sign test may have higher efficiency. Indeed, if our sample comes from a distribution called the double exponential, which has longer tails than the normal, the Pitman efficiency of the sign test relative to the t-test is 2. That is, we do as well (at least for large samples) with a sample of size n using a sign test as we do with one of size $2n$ using a t-test.

1.4 ESTIMATION

1.4.1 Confidence intervals

The sample mean is widely used as a **point** estimate of a population mean. Since the sample mean differs from sample to sample we need some measure of the accuracy of our point estimate. A **confidence interval** is one such measure. Although this aspect of the estimation problem is usually formulated somewhat differently, one way of specifying a $100(1-\alpha)\%$ confidence interval for a location parameter θ (see Section A1.2) is to define it as a set of all values θ which would be accepted as supporting H_0 using a test at significance level $100\alpha\%$.

We noted in Section 1.3.1 that for a sign test for a median using a sample of 10 and a two-tail test at the nominal 5% (actual 2.16%) level we accept H_0 if we get between 2 and 8 plus signs. Consider the data in Example 1.2, i.e.

49, 58, 75, 110, 112, 132, 151, 276, 281, 362*

where the asterisk represents a censored observation: clearly we have between 2 and 8 plus signs if the median θ specified in H_0 has any value greater than 58 but less than 281. Thus the interval (58, 281) constitutes a nominal 95% confidence interval for θ, the population median survival time. Since we would accept an H_0 that specified any value for the median geater than 58 but less than 281 there is considerable doubt about the true value. It is an understatement to say our estimate lacks precision.

The more usual interpretation of a $100(1-\alpha)\%$ confidence interval is in terms of its property that if we form such intervals for repeated samples, then in the long run $100(1-\alpha)\%$ of such intervals would contain (or cover) the true but unknown θ. A confidence interval is useful for two main reasons:

1. It tells us something about the precision with which we estimate a parameter.
2. It helps us decide (a) whether a significant result is likely to be

important or (b) whether we need to do more experimental work before we decide whether or not it is.

We elaborate on these points in the next section.

1.4.2 Precision, statistical significance and practical importance

Example 1.5

Doctors treating hypertension are often interested in the decrease in systolic blood pressure after administration of a drug. They may know of several drugs that will reduce blood pressure by between 15 and 20 mm Hg. When testing a more expensive new drug they will be interested if it reduces systolic blood pressure by more than 20 mm Hg. Such differences would be of **practical importance.** Two clinical trials (I and II) are carried out to test the efficacy of a new drug (A) in reducing blood pressure. A third trial (III) is carried out with a second new drug (B). In each trial the hypothesis H_0 : *drug does not reduce blood pressure* is rejected at a nominal 5% significance level. Trial I involved only a small number of patients, but trials II and III involved larger numbers. The 95% confidence intervals for mean blood pressure reduction (mm Hg) after treatment at each trial were:

Drug A	Trial I (3, 35)
Drug A	Trial II (9, 12)
Drug B	Trial III (21, 25)

Trial I is imprecise; we would accept in a significance test at the 5% level any mean reduction between 3 and 35 units. The former is not of practical importance; the latter is. This small trial only answers questions about the 'significant' mean reduction with **low precision.** The larger Trial II, using the same drug, indicates an average reduction between 9 and 12 units, a result of high statistical significance but not of practical importance. Compared to the result of trial I, it has **high precision**; all other things being equal, increasing the size of a trial increases the precision. Trial III using drug B also has high precision and tells us our mean reduction is likely to be between 21 and 25 units. This finding is of practical importance.

From the hypothesis testing viewpoint increasing sample size increases the probability that small departures from H_0 will be found to be significant, whether or not they are of practical importance. The art of designing experiments is to take sufficient observations to ensure a good chance of detecting with reasonable precision departures from H_0 of practical importance, but to avoid wasting resources by taking so many observations that trivial departures from H_0 become highly significant. Practical design of experiments, whether one uses parametric or nonparametric methods of analysis, requires guidance from a trained statistician.

1.4.3 Tolerance intervals

Another useful concept is that of a **tolerance interval.** These are intervals with properties of the form: given p_1, α, there is a probability of at least $1-\alpha$ that the interval includes $p_1\%$ of the population. Given a sample of n observations $x_1, x_2, \ldots, x_n$ from a continuous distribution, the tolerance interval concept is useful for answering such questions as: how large should n be so that the probability is at least 0.95 that 90% of the population lies between the smallest and largest sample values? Or, how large a sample is needed to ensure that with probability 0.8 at least 75% of the population exceeds the second smallest sample value? Problems about tolerance intervals and limits are discussed by Conover (1980, Section 3.3). We mention the topic here only because there is sometimes confusion between tolerance and confidence intervals, confusion compounded further because the term 'tolerance limits' is also used in quality control as limits between which measurements must lie if batches of items are to be acceptable.

1.5 COMPUTERS AND NONPARAMETRIC METHODS

Computer packages to carry out exact permutation tests rapidly for small samples and to give accurate estimates of tail probabilities by simulation for larger samples have revolutionized practical application of nonparametric methods. As in all aspects of computing, the software and hardware situation is changing rapidly. In 1990 two packages for personal computers devoted to nonparametric tests – especially permutation tests – became available. These are STATXACT, distributed by Cytel Software Corporation, Cambridge, Mass., USA, and DISFREE, distributed by Biosoft, Cambridge, UK. These were joined in 1992 by TESTIMATE, a package available earlier on the German market and distributed by IDV Datenanalyse und Versuchsplanung, Munich, Germany; also by STATXACT-Turbo, a faster and less restrictive version of STATXACT. Emphasis in these packages is on hypothesis testing, but STATXACT and TESTIMATE provide confidence intervals for some procedures and with ingenuity other programs in these packages can also be used to obtain confidence intervals. It is anticipated that future releases will provide more facilities for interval estimation. There are some nonparametric programs in standard statistical packages such as MINITAB, SPSS, STATGRAPHICS, BMDP, SAS etc. For example, MINITAB provides programs for confidence intervals for several nonparametric procedures. In some of the general packages only a relevant statistic is calculated for a few nonparametric tests. One must then use asymptotic theory or consult tables to determine if significance is indicated, often forcing the use of nominal rather than actual significance levels. However, some general packages, e.g. SAS, do include exact permutation theory tests for a limited range of nonparametric techniques.

The importance of packages such as STATXACT, TESTIMATE and DISFREE is that they allow us to carry out in a reasonable time exact permutation tests in situations where in the past one often had to resort to asymptotic results in cases where these are of doubtful validity; in particular these packages overcome many of the difficulties associated with ties in ranks or with situations when samples are unbalanced or very skew. The three packages do not cover precisely the same range of tests, but STATXACT and TESTIMATE tend to be quicker for all but very small samples. The output allows flexibility in determining exact significance levels for one- and two-tail tests. STATXACT and DISFREE both provide options for sample-based (called 'Monte Carlo') estimates of exact probabilities when complete calculation is too time consuming, and all three packages give tail probabilities for relevant asymptotic or large-sample tests. It is instructive to consider the differences between exact and asymptotic results for specific data sets. At various places in this book we illustrate situations where these differences are quite marked. The efficiency of STATXACT programs stems from the use of algorithms based on the work of Mehta and coauthors in a series of papers including Mehta and Patel (1983; 1986), Mehta, Patel and Tsiatis (1984), Mehta, Patel and Gray (1985), Mehta, Patel and Senchaudhuri (1988). Similar and other efficient algorithms are used in TESTIMATE. Understanding these algorithms is not essential when using STATXACT or TESTIMATE.

DISFREE is closely associated with a specific text (Krauth, 1988); while the manual recommends that the package should be used in conjunction with Krauth's book many of the standard tests can be performed without reference to it, although one or two non-standard modifications have been made and these may cause minor difficulties.

Example 1.6

We indicate some essential components of output from DISFREE and STATXACT for the test in Example 1.4 when the ranks, after treatment, of the patients receiving the new drug are 1,2,3,6 and the exact permutation test is performed. From that example we know that the rank sum is $S=12$ and from Table 1.2 it is easily verified that $\Pr(S \leq 12)=4/126=0.0317$. DISFREE gives not the value of S, but that of a closely related and often used equivalent statistic derived by subtracting its minimum possible value (here 10) from S, i.e. in this case the statistic, M, say, is $M=2$. It also gives $\Pr(M \leq 2)=0.03175$ and $\Pr(M \geq 2)=0.98413$ (these are referred to in output as pL and pU respectively). These are equal to $\Pr(S \leq 12)$ and $\Pr(S \geq 12)$ respectively. The former indicates an actual critical region of size 0.03175 for a one-tail test. We double this probability for a two-tail test and thus would not reject H_0 at the nominal 0.05 significance level.

For the same data STATXACT gives both $M=2$ and $S=12$ as test statistics and the values $\Pr(S \leq 12)=0.0317$, $\Pr(S=12)=0.0159$, and for a two-tail test that the probability that the magnitude of the difference between S and its mean (here 20) exceeds that observed (here $|12-20|=8$) is 0.0635. Symmetry of the distribution of S means that this is double the single tail probability; this would not be the case if the test statistic has a skew distribution. Thus, in general, the output from STATXACT gives us an actual one- and two-tail test probability and effectively a choice of two-tail tests based on either doubling of a one-tail probability or adding probabilities of results differing in magnitude from the mean by more than that observed (these two options being equivalent only for symmetric distributions of the test statistic). Knowing the probability $\Pr(S=12)=$ 0.0159 is useful for two purposes. It gives an indication of the degree of discontinuity at the observed value (this discontinuity tends to decrease rapidly for most tests as the number of observations increases). It is also useful if one takes the approach to hypothesis testing that splits the probability of the observed value equally between the critical region and the acceptance region. This is sometimes looked upon as a 'continuity correction' or as an allowance for discontinuity. As indicated in Section 1.3.1, if using this approach one should do so regularly and consistently. TESTIMATE output differs in detail to that from STATXACT but provides broadly similar information. All three packages exhibit some differences in asymptotic results, in particular in the handling of ties and in the way they deal with what are called continuity corrections.

Users of any package may find it informative to test programs in it using examples from this book and other sources to ensure that they can interpret relevant output. In some cases the output will inevitably be different, being either more or less extensive than that given in the source of the examples. For instance, output may give actual rather than nominal significance levels, or these and confidence intervals may be adjusted to bring actual and nominal levels closer than the basic theory allows. We mentioned in Example 1.6 that different statistics (S and M) may be used for equivalent tests.

In sophisticated uses of nonparametric methods, like some outlined in Chapter 9 and later chapters, a computer with appropriate software is virtually essential for implementation.

Neave and Worthington (1988) incorporate listings for BASIC programs for a number of nonparametric tests; most of these require reference to tables to determine significance levels. Daniel (1990) gives in most chapters a list of packages and references to individual programs for many procedures. Siegel and Castellan (1988) include computer program listings for a few procedures. In this book we refer most frequently to STATXACT as typical of a package suitable for the serious user of nonparametric methods; users of TESTIMATE, DISFREE or other packages including permutation tests should have little difficulty adapting comments or recommendations to programs that cover the relevant procedure.

1.6 FURTHER READING

This text is a guide to practical applications of basic nonparametric methods, giving simple explanations of the rationale behind procedures, but omitting detailed theory (often involving complicated mathematics). There are a number of texts that cover specific aspects. Conover (1980) gives more background for many of the procedures described here and precise information about when each is applicable, and is recommended for more detailed study. Lehmann (1975) discusses the rationale and theory carefully, using simple illustrative numerical examples without advanced mathematics; his book repays careful reading for those who want to pursue the logic of the subject in more depth. Randles and Wolfe (1979) and Maritz (1981) are excellent books covering the theory at a more advanced mathematical level. Hollander and Wolfe (1973), Gibbons (1985), Krauth (1988), Neave and Worthington (1988) and Daniel (1990) are general books on applied nonparametric methods. A pioneer text that still contains much relevant material is Bradley (1968). There are several good books on applications in the social sciences including Marascuilo and McSweeney (1977), Leach (1979) and more recently the second edition of Siegel and Castellan (1988) – an update of a book with the same title written by Siegel more than 30 years ago. Noether (1991) is a general introductory statistical text that makes use of a nonparametric approach to introduce basic concepts. Although dealing basically with rank correlation methods, a topic we cover in Chapter 7, Kendall and Gibbons (1990) give an insight into the relationship between many nonparametric methods. Agresti (1984; 1990) and Everitt (1992) give detailed accounts of various models, parametric and nonparametric, used in analysis of categorical data. A sophisticated treatment of randomization tests with emphasis on biological applications is given by Manly (1991). With a few exceptions most of the books listed pay more attention to hypothesis testing than to estimation. There are a number of advanced texts and reports of conference proceedings that are mainly for the specialist. In this book we give many references to source papers in journals.

EXERCISES

1.1 As in Example 1.1, suppose that one machine produces rods with diameters normally distributed with mean 27 mm and standard deviation 1.53 mm, so that 2.5% of the rods have diameter 30 mm or more. A second machine is known to produce rods with diameters normally distributed with mean 25 mm and 2.5% of rods produced by it have diameter 30 mm or more. What is the standard deviation of rods produced by the second machine?

1.2 Calculate the probability of each number of successes, r, for a binomial distribution with $n=12$, $p=\frac{1}{2}$.
I have on my shelves 114 books on statistics. I take a random sample of 12 and want to test the hypothesis that the median number of pages, θ, in all 114 books is 225.

In the sample of 12, I note that 3 have less than 225 pages. Should I accept the hypothesis that $\theta=225$? What should I take as the appropriate alternative hypothesis? What is the appropriate critical region for a test of nominal size $\alpha=0.05$? What is its actual size?

1.3 The numbers of pages in the sample of 12 books in Exercise 1.2 were:

126 142 156 228 245 246 370 419 433 454 478 503

calculate a nominal 95% confidence interval for the median θ.

1.4 Use the sum of ranks given in brackets after each group in Table 1.1 to verify the correctness of the entries in Table 1.2.

1.5 Suppose that the new drug under test in Example 1.4 has all the ingredients of a standard drug at present in use and an additional ingredient which has proved to be of use for a related disease, so that it is reasonable to assume that the new drug will do at least as well as the standard one, but may do better. Formulate an appropriate one-tail test. If the post-treatment ranking of the patients receiving the new drug is 1,2,4,6 should one reject the relevant H_0?

1.6 An archaeologist numbers some articles 1 to 11 in the order he discovers them. He selects at random a sample of 3 of them. What is the probability that the sum of the numbers on the items he selects is less than or equal to 8? (Note that you do not need to list all combinations of 3 items from 11 to answer this question.) If the archaeologist believed that items belonging to the more recent of two civilizations were more likely to be found earlier in his dig and of his 11 items 3 are identified as belonging to that more recent civilization (but the remaining 8 come from an earlier civilization) does a rank sum of 8 for the 3 matching the more recent civilization provide reasonable support for his theory?

1.7 In Section 1.4.1 we associated a confidence interval with a two-tail test. As well as such two-sided confidence intervals, one may define a one-sided confidence interval composed of all parameter values that would not be rejected in a one-tail test. Follow through such an argument to obtain a nominal 95% confidence interval based on the sign test criteria for the 12 book sample values given in Exercise 1.3 relevant to a one-sided alternative $H_1 : \theta>225$.

1.8 We wish to compare a new treatment with a standard treatment and only 6 patients are available. We allocate 3 to each treatment at random and after an appropriate interval rank the patients in order of their condition. What is the situation (i) for testing H_0 : *treatments do not differ* against H_1 : *the new treatment is better* and (ii) for testing the same H_0 against H_1 : *the two treatments differ in effect*?

2

Location tests for single samples

2.1 COUNTS AND RANKS

In Chapters 2 and 3 we consider data derived from single samples, usually from a continuous distribution. These may be actual measurements but most of the hypothesis tests require only ranks specifying size order or preference. Full numerical data are usually needed to establish confidence intervals for measurements. While most practical problems involve comparison of, or studying relations between, two or more samples the logic of many basic nonparametric procedures is easily explained for single samples.

In the rest of this book we illustrate the logic using specific examples in all but the simplest of which we discuss points under the headings:

The problem
Formulation and assumptions
Procedure
Conclusion
Comments
Computational aspects

The summary and exercises at the end of each chapter are preceded by an indicative, rather than exhaustive, list of fields of application.

Nonparametric tests are commonly based on counts, ranks or transformations of ranks. In the sign test in Example 1.2 we counted the number of observations above the hypothesized median. The permutation test in Example 1.4 was a two-sample test based on ranks. This chapter covers single-sample tests based on counts and ranks. Transformations of ranks are dealt with in Chapter 3.

2.1.1 The sign test and the effect of sample size

We look more closely at assumptions and possible complications and extend the concepts developed in Example 1.2. For the sign test – like nearly all methods – the larger the sample the better the power and the shorter the confidence interval for a given confidence level.

Example 2.1

The problem. Research and review papers in scientific journals list references to related work. For the journal *Biometrics* in the years 1956–57 in a random sample of 12 from 70 published papers the numbers of references, arranged in ascending order, were

2 2 4 5 5 5 6 6 9 16 20 26

In a second independent sample of 24 the numbers were

0 1 2 2 2 3 3 3 3 3 4 6 6 7 7 8 9 10 12 15 20 22 68 72

For each sample use the sign test at a nominal 5% significance level to test the hypothesis that the population median number of references is 11.

Formulation and assumptions. If the population median is 11, then any paper in a random sample is equally likely to contain either more or less than 11 references (ignoring temporarily the real possibility that it contains exactly 11 references). If we associate a plus sign with *more than 11 references*, and the hypothesis $H_0 : \theta=11$ holds, the number of plus signs has a binomial distribution with $p=\frac{1}{2}$, $n=12$ for the first sample and $p=\frac{1}{2}$, $n=24$ for the second sample. If the alternative hypothesis is $H_1 : \theta\neq 11$, a two-tail test is appropriate.

Procedure. For the sample of 12, there are 3 values greater than 11 (giving 3 plus signs). We need the probability of 3 or fewer successes (plus signs) for a binomial distribution with $n=12$ and $p=\frac{1}{2}$ (see Exercise 1.2). Table 2.1 gives relevant probabilities.

Table 2.1 Binomial probabilities, P, for r successes, when $n=12$, $p=\frac{1}{2}$

r	0	1	2	3	4	5	6
P	0.000	0.003	0.016	0.054	0.121	0.193	0.226

The binomial distribution is symmetric when $p=\frac{1}{2}$, so if X represents the number of plus signs when $n=12$, then $\Pr(X=12-r)= \Pr(X=r)$: e.g. $\Pr(X=9)=\Pr(X=3)=0.054$. From Table 2.1 we see that the probability of 3 or less plus signs is $0.000+0.003+0.016+0.054=0.073$. By symmetry the corresponding upper tail probability is also 0.073: thus under H_0 the probability of getting a number of plus signs differing from the expected number (here 6) by as great or greater a number than that observed is $2\times 0.073=0.146$.

Table 2.2 gives probabilities of r successes for a binomial distribution with $n=24$, $p=\frac{1}{2}$.

Table 2.2 Binomial probabilities P for r successes, when $n=24$, $p=\frac{1}{2}$

r	0	1	2	3	4	5	6	7	8	9	10	11	12
P	0.000	0.000	0.000	0.000	0.001	0.003	0.008	0.021	0.044	0.078	0.117	0.149	0.161

If the median is 11, the sample of 24 gives 6 plus signs corresponding to the values 12, 15, 20, 22, 68, 72. From Table 2.2 we easily calculate the probability of 6 or fewer plus signs to be 0.012. The relevant exact probability for a two-tail test is $2\times0.012=0.024$.

Conclusion. With the sample of 12, since $\alpha=0.146$ we do not reject H_0 since this clearly exceeds the conventional $\alpha=0.05$. With a sample of 24 we reject H_0 since $\alpha=0.024$, corresponding to a 2.4% significance level.

Comments. 1. We used a two-tail test since there was no *a priori* reason to confine alternatives to only greater than, or only less than, the value specified in H_0.

2. Our results are consistent with our remark in Section 1.3 that increasing sample size generally increases the power of a test.

3. Especially if we reject H_0, our interest switches to what the true population median might be. This we consider in Example 2.2.

4. We were given numbers of references in this example, but for a sign test we only need the sample size and the number of plus or minus signs.

5. By rounding to 3 decimal places in Table 2.2 we lose some accuracy in determing the exact tail proability. A more accurate computation would give the relevant one-tail probability as 0.011 to 3 decimal places. In this book there are many instances where different rounding conventions would alter the final significant digit without affecting the broad conclusions. Anderson (1992) discusses the effect of rounding on precision and accuracy.

Computational aspects. Given adequate tables of binomial probabilities for $p=\frac{1}{2}$ and various n (e.g. Table I at the end of this book) computer programs are hardly necessary. STATXACT, TESTIMATE and DISFREE include programs for this or equivalent procedures. STATXACT (Version 2.0) includes a program for a test called McNemar's test which may be adapted for a sign test. STATXACT also provides an alternative computational approach discussed in Section 2.1.5.

If one or more sample values coincide with the value of θ specified in H_0 we cannot logically assign either a plus or minus to these observations. They may be ignored and the sample size reduced by 1 for each, e.g. if in the sample of 12 in Example 2.1 we specify $H_0: \theta=6$ we treat our problem (see Exercise 2.1) as one with 10 observations giving 4 plus signs. Doing this rejects evidence which strongly supports the null hypothesis, yet is uninformative about the direction of possible alternatives. Another approach is to toss a coin and allocate a plus to a value equal to θ if the coin falls heads and a minus if it falls tails: yet another is to assign a plus or minus to such a value in a way that rejection of H_0 is less likely. The former approach usually makes only a small difference and has little to commend it; the second approach is ultraconservative. Lehmann (1975, p. 144) discusses pros and cons of these choices in more detail.

If a relevant computer program is not available, tables give exact probabilities for various n. Table I at the end of this book may be used for sample sizes between 6 and 20. It differs from Tables 2.1 and 2.2 by giving not the individual probabilities for each r, but the **cumulative** binomial probabilities, i.e. the probabilities that the number of plus signs takes a value less than or equal to a given r. These are obtained by summing the probabilities for 0, 1, 2, . . . , r plus signs; i.e. if we observe $P=4$ plus signs when $n=12$ we see from Table 2.1 that $\Pr(P \leq 4)=0.000+0.003+0.016+0.054+0.121=0.194$. This is the entry in the column headed $r=4$ and the row labelled $n=12$ in Table I. A simpler table in Neave (1981, p. 29) gives critical values for a wide range of n at *nominal* 5% and 1% levels for both one- and two-tail tests. Neave also gives tables analogous to our Table I if exact probabilities are required for $n \leq 20$. In Section 2.1.3 we give an approximation that works well when $n > 20$.

2.1.2 A confidence interval

Example 2.2

The problem. Obtain nominal 95% confidence intervals for the median number of references per paper using the sign test with the samples in Example 2.1. Determine the exact confidence level in each case.

Formulation and assumptions. We seek all values for the population median θ that would not be rejected in a two-tail test at a nominal 5% significance level.

Procedure. We use the argument developed in Section 1.4. Consider first the sample of 12. From Table 2.1 we immediately see that a critical region consisting of 0, 1, 2, 10, 11, 12 plus signs has size $2\times(0.000+0.003+0.016)=0.038)$ and that this is the largest exact size less than 0.05. Thus we reject H_0 if it leads to 2 or fewer or 10 or more plus signs. We accept H_0 if we get between 3 and 9 plus signs. It is clear from the sample values 2, 2, 4, 5, 5, 5, 6, 6, 9, 16, 20, 26 given in Example 2.1 that we have between 3 and 9 plus signs if θ lies in the open interval from 4 to 16, i.e. $\theta > 4$ and $\theta < 16$. What happens if $\theta=4$ or 16? We effectively reduce the sample size by 1 and consider a situation when $n=11$ and the number of minus or plus signs respectively is 2. Table I gives the relevant one-tail probability in each case as 0.033 (giving a two-tail critical region of size $\alpha=2\times0.033=0.066$). However, if θ has a value just below 4 or just above 16 we again have a sample of 12 with now 2 minus or plus signs and so $\alpha=0.038$ for a two-tail test, implying a confidence level of $100(1-0.038)=96.2\%$. Formally, we may describe the confidence interval as consisting of all θ such that $4\leq\theta\leq16$, i.e. the closed interval (4, 16), as a nominal 95% or actual 96.2% confidence interval. The implication is that for any θ **outside** this interval we reject the hypothesis that it is the population median at the $100-96.2=3.8\%$ significance level. An interval (a, b) is closed if it includes the end points a, b and open if these end points are not included.

We leave it as an exercise for the reader, using Table 2.2, to show that for the sample of 24 in Example 2.1 the closed interval (3, 10) is an actual 97.6% confidence interval (or 97.8% with the greater accuracy mentioned in comment 5 on that example).

Conclusion. For the sample of 12 an actual 96.2% confidence interval is (4, 16) and for the sample of 24 an actual 97.8% interval is (3, 10).

Comments. 1. The sample median is an appropriate point estimator of the population median. For our samples these are respectively (see Section A1.2) 5.5 and 6. The rationale for this choice is that it gives equal numbers of plus and minus signs, the strongest supporting evidence for H_0.

2 Increasing the sample size shortens the confidence interval.

3. Both samples contain some values greatly in excess of the median, suggesting a skew distribution of numbers of references with a long upper tail. This is confirmed in the full set of population values (not quoted here). This is not surprising since a few papers (e.g. those reviewing a particular topic) are likely to include many references, whereas the majority of papers on new theory or applications generally contain fewer references.

4. Skewness makes normal theory (t-distribution) tests or confidence intervals inappropriate. The mean and median do not coincide for a skew distribution. In passing we note that normal theory estimation for the sample of 12 in this example give a mean of 8.83 with a 95% confidence interval (4.0, 13.7) and for the sample of 24 a mean of 12 and a 95% confidence interval (4.1, 19.9) with the somewhat bizarre property that the longer interval is associated with the larger sample. This is attributable mainly to the two very high values 68, 72 in that sample.

5. In this example we discussed at length whether the end points were included in the interval; in practice there is usually little need to worry about this unless the test statistic distribution has a marked discontinuity at these end points. In general this only occurs for very small samples.

Computational aspects. Given suitable tables there is little call for a computer program to calculate sign-test confidence intervals. Many general statistical packages will generate the relevant binomial probabilities if these are not available in tables. In comment 4 above we considered t-distribution based 95% confidence intervals. For direct comparison with the nonparametric intervals, many statistical packages provide parametric (normal theory, t-distribution) confidence intervals at any specified level such as 96.2% or 97.8%. If no appropriate computer package is available 95% normal theory confidence intervals can be obtained using tables as described in Section A4.1. In Section 2.1.5 we give an alternative approach to 'sign test' confidence intervals.

2.1.3 A large-sample approximation

For any binomial distribution, if X is the number of successes, $E(X)=np$ and $\text{Var}(X)=npq$, where $q=1-p$. If n is reasonably large and p not too small (so that $np>10$ is a useful guidline)

$$Z=\frac{X-np}{\sqrt{(npq)}} \tag{2.1}$$

has approximately a normal distribution with mean 0 and standard deviation 1 (the standard normal distribution). Widely available tables give probabilities that Z takes values less than or equal to any specified value. For the sign test $p=q=\frac{1}{2}$ and (2.1) becomes

$$Z=\frac{X-\frac{1}{2}n}{\frac{1}{2}\sqrt{n}} \tag{2.2}$$

The approximation is improved by a continuity correction which adjusts for approximating to a discrete distribution taking the values $X = 0, 1, 2, \ldots, n$ by a continuous one which may take any real value. If r is the greater of the number of plus or minus signs we subtract $\frac{1}{2}$ from r, i.e. put $X=r-\frac{1}{2}$ in (2.2), while if r is the lesser of the number of plus or minus signs we add $\frac{1}{2}$ to r, i.e. put $X=r+\frac{1}{2}$.

When testing at the conventional 5% significance level, we reject H_0 in a two-tail test if $Z\geq 1.96$ or $Z\leq -1.96$, i.e. if Z takes values outside the open interval $(-1.96, 1.96)$. We refer to $|Z|\geq 1.96$ as the rejection criterion, where the vertical bars indicate we take the magnitude (ignoring the sign) of Z. For a one-tail test at the 5% level we reject H_0 if $Z\geq 1.64$ (upper tail) or $Z\leq -1.64$ (lower tail) as appropriate. Corresponding critical values for rejection at the 1% level are 2.58 (two-tail) and 2.32 or -2.32 (one-tail).

Example 2.3

The problem. Using the sample of 24 observations in Example 2.1, test the hypothesis that the population median is 11 using the normal approximation to the two-tail sign test.

Formulation and assumptions. As in Example 2.1, but we now use (2.2) with a continuity correction.

Procedure. We have 18 minus and 6 plus signs, thus $r=18$; with the appropriate continuity correction, $X=18-\frac{1}{2}=17.5$. Since $n=24$, (2.2) gives $Z=(17.5-12)/(0.5\times\sqrt{24})=2.25$, indicating significance at the 5% level. Tables of the normal distribution give $\Pr(|Z|\geq 2.25)=0.024$, equal to the two-tail probability found in Example 2.1, but see comment 5 on that example.

Conclusion. We reject $H_0: \theta=11$ at the 2.2% significance level.

Comment. We arrive at the same conclusion if we work with the number of plus signs, taking $X=6+\frac{1}{2}$.

2.1.4. Large-sample confidence intervals

We may rearrange equation (2.2) to obtain an approximation to the critical value r_0, say, of the lesser of the number of plus or minus signs that would just give significance at a specific level, e.g. 5%. This corresponds to $Z=-1.96$. For a given n we may calculate r_0. Using a continuity correction we put $X=r_0+\frac{1}{2}$. Simple algebraic manipulation of (2.2) with these values of X, Z gives

$$r_0=\tfrac{1}{2}n-0.98\sqrt{n}-\tfrac{1}{2} \tag{2.3}$$

Substituting $n=24$ in (2.3) gives $r_0=6.70$. Since r can take only integer values we reject H_0 if $r\leq 6$, in agreement with the exact criterion. Similar calculations can be used for other significance levels for one- or two-tail tests. Once the critical value r_0 is determined, confidence intervals may be obtained as for the exact test.

2.1.5 Some modifications to the sign test

Example 2.4 illustrates an alternative approach to the sign test and associated confidence intervals using a computer program that calculates confidence limits for p, the population probability of success, when we have a sample of n observations of which r ($\leq n$) are successes. In the case of the sign test if we associate a plus sign with success, then we accept the hypothesis that the median θ has the value specified under H_0 at significance level α if the $100(1-\alpha)$% confidence interval for p includes $p=\frac{1}{2}$.

Example 2.4

The problem. Using the first data set in Example 2.1, i.e. 2, 2, 4, 5, 5, 5, 6, 6, 9, 16, 20, 26, test $H_0: \theta=11$ against $H_1: \theta\neq 11$ at the 5% significance level. Also obtain a nominal 95% confidence interval for the median θ.

Formulation and assumptions. We base our solution on the sign test using the approach outlined at the start of this section.

Procedure. Assume we have a program that produces a 95% confidence interval for p, the proportion of successes in the population (corresponding to plus signs in this context) when $n=12$ and we observe (as is the case when $\theta=11$) $r=3$ successes (plus signs). Such a program is included in several packages including STATXACT, version 2.0. Using that program the 95% confidence interval turns out to be (0.055, 0.57). Since this includes the value 0.5 we accept H_0. To obtain a nominal 95% confidence interval for θ we repeat the calculation decreasing r by 1 each time until we first get a confidence interval that does not include the value 0.5. Proceeding in this way we find that for $n=12$, when $r=2$ the interval is (0.02. 0.48). Since this does not contain 0.5 we conclude that $r=3$, implying at least three plus signs, is necessary if we are to accept H_0. The symmetry in this case implies that we also require 3 or more negative signs (i.e. not more than $r=12-3=9$ plus signs) if we are to accept H_0. This is consistent with our findings in Examples 2.1 and 2.2 and as in the latter example we conclude that the appropriate *nominal* 95% confidence interval is (4, 16).

Conclusion. We accept $H_0 : \theta=11$ testing at a nominal 5% significance level and the nominal 95% confidence interval for θ is the interval (4,16).

Comments. This procedure leads only to nominal signficance levels and confidence intervals. With some programs it is possible to refine these to actual levels but this is tedious.

Computational aspects. Programs that give confidence intervals for a binomial probability may be used for testing whether any specified value of p is acceptable, not just $p=0.5$ as in this example.

We may modify the sign test to test hypotheses about any given p. For example, the kth quantile of a continuous distribution is the value q_k of the random variable X such that $\Pr(X<q_k)\leq k$ and $\Pr(X>q_k)\leq 1-k$. Clearly $q_{½}$ is the median. For continuous distributions q_k is unique; for discrete distributions special conventions are needed to define unique quantiles; see Section A1.2.

In particular if $k=0.1r$, where r is any integer between 1 and 9, q_k is the rth decile (tenth) and if $k=0.25r$, where $r=1$, 2 or 3, q_k is the rth quartile (quarter). The fifth decile or the second quartile both equal the median. We illustrate a sign test specifically for the first quartile. Extension to other quantiles follows similar lines.

Example 2.5

The problem. A central examining body awards a mark in each subject and publishes the information that 'three quarters of the candidates achieved a mark of 40 or more'. One school entered 32 candidates of whom 13 scored less than 40. The president of the Parents' Association argues that the school's performance is below national standards. The headmaster counters by claiming that in a random sample of 32 candidates it is quite likely that 13 would score less than the lower quartile mark even though 8 out of 32 is the national proportion. Is his assertion justified?

Formulation and assumptions. The headmaster's assertion may be formulated as test $H_0 : q_{1/4}=40$ against $H_1 : q_{1/4}\neq 40$.

Procedure. We associate a minus with a mark below 40; thus for our 'sample' of 32 we have 13 minuses (and 19 pluses). If the first quartile is 40 then the probability is $p=0.25$ that each candidate in a random sample has a mark below 40 (and thus is scored as a minus). The distribution of minuses is therefore binomial with $n=32$ and $p=0.25$. Using any program that gives a 95% confidence interval for p given $r=13$ 'successes' when $n=32$ we find the relevant confidence interval is (0.24, 0.59). This includes $p=0.25$.

Conclusion. Having observed 13 minus signs we do not reject the hypothesis that the population first quartile is 40. In this sense the headmaster's assertion is justified.

Comments. 1. Recall that non-rejection of a hypothesis does not prove it true. It is only a statement that the evidence to date is not sufficient to reject it. We may make an error of the second kind by non-rejection. Indeed, since the above confidence interval includes $p=0.5$ we would not reject the hypothesis that the median is 40.

2. We used a two-tail test. A one-tail test would not be justified unless we had information indicating the school performance could not be better than the national norm. For example, if most schools devoted three periods per week to the subject but the school in question only devoted two, we might argue that lack of tuition could only depress performance. Indeed, a one-tail test at the 5% level rejects H_0 if there are more than 12 minuses. If we observe 13 minuses and feel a one-tail test at the 5% significance level is appropriate and want to use the method described in this example we might determine a 90% confidence interval for p and reject H_0 if $p=0.25$ is below the lower limit for this interval. Think carefully about this to be sure you see why and whether you have any reservations about the accuracy of this approach.

3. Although we do not recommend the normal approximation (2.1) for

values of $np<10$ it can be used if one is prepared to sacrifice a little accuracy. Here $n=32, p=0.25$, whence $np=8$, $npq=32\times\frac{1}{4}\times\frac{3}{4}=6$. Since the number of minus signs, 13, is above the expected number $np=8$ we subtract the continuity correction of ½. Thus

$$Z=(12.5-8)/(\sqrt{6})=1.84$$

and since $Z<1.96$ we accept H_0 in a two-tail test. As in the more exact test we reject H_0 in a one-tail test at the 5% level since $Z=1.64$ is then the critical value (p.33).

4. The headmaster's claim said 'if one took a random sample'. Pupils from a single school are in no sense a random sample from all candidates. Our test only establishes that results for this particular school are not out of line with national results in the sense that they might well arise if one took a random sample of 32 candidates from all entrants.

5. With no computer program to calculate the confidence limits used here, one may adopt an approach like that in Examples 2.1 and 2.2, requiring a table of binomial probabilities for $p=0.25$ and $n=32$ (not common, although Neave (1981) and others give tables for $n\leq 20$) or a willingness to compute the relevant tail probabilities for this distribution (Exercise 2.2).

6. Tests for the third quartile are symmetric with those for the first quartile if we interchange plus and minus signs.

2.1.6 A sign test for trend

Cox and Stuart (1955) proposed an ingenious way to test for a monotonic trend, i.e. an increasing or decreasing trend. A straight line is the simplest monotonic trend and with well-known assumptions about departures from linearity least squares regression is appropriate for making inferences. A monotonic trend need not be linear; it may simply express a tendency for observations to increase or decrease subject to certain local or random irregularities. Consider a set of **independent** observations $x_1, x_2, \ldots, x_n$ ordered in time. If we have an even number of observations, $n=2m$, say, we take the differences $x_{m+1}-x_1, x_{m+2}-x_2, \ldots, x_{2m}-x_m$. For an odd number of observations, $2m+1$, we may proceed as above omitting the middle value x_{m+1} and calculating $x_{m+2}-x_1$, etc. If there is an increasing trend we would expect most of these differences to be positive, whereas if there were no trend and observations differed only by random fluctuations about some median these differences (in view of the independence assumption) are equally likely to be positive or negative. A preponderance of negative differences suggests a decreasing trend.

This implies that under the null hypothesis of no trend, the plus (or minus) signs have a binomial distribution with parameters m and $p=\frac{1}{2}$.

Example 2.6

The problem. The US Department of Commerce publishes estimates obtained from independent samples each year of the mean annual mileages covered by various classes of vehicles in the United States. The figures for cars and trucks (in thousands of miles) are given below for each of the years 1970–83. In either case is there evidence of a monotonic trend?

Cars	9.8	9.9	10.0	9.8	9.2	9.4	9.5	9.6	9.8	9.3	8.9	8.7	9.2	9.3
Trks	11.5	11.5	12.2	11.5	10.9	10.6	11.1	11.1	11.0	10.8	11.4	12.3	11.2	11.2

Formulation and assumptions. As the figures for each year are based on independent samples we may use the Cox–Stuart test for trend. Without further assumptions a two-tail test is appropriate as a trend may be increasing or decreasing.

Procedure. For cars relevant differences are 9.6−9.8, 9.8−9.9, 9.3−10.0, 8.9−9.8, 8.7−9.2, 9.2−9.4, 9.3−9.5 and all are negative. From Table 1 for $n=7$ we find a critical region of size 0.016 corresponding to zero or 7 plus signs.

Conclusion. There is evidence of a downward monotonic trend at an actual 1.6% significance level.

Comments. 1. For trucks the corresponding differences have the signs −, −, −, −, +, +, +. A moment's reflection shows that 3 plus and 4 minus (or 3 minus and 4 plus) signs provides the strongest possible evidence in support of the null hypothesis of no monotonic trend. The fact that the first four differences are all negative and the last three all positive suggests the possibility of a decreasing trend followed by an increasing trend (i.e. a non-monotonic trend) rather than random fluctuations. The sample here is too small to establish this, but in Section 3.6 we describe a 'runs test' appropriate to testing whether fluctuations are random in circumstances like these.

2. Periodic trends are common. For example, at many places in the northern hemisphere mean weekly temperature tends to increase from February to July and decrease from August to January. A Cox–Stuart test applied to data of this type might either miss such a trend (because it is not monotonic) or indicate a monotonic trend for records over a limited period (e.g. from February to June). Conover (1980, Example 5, p.137) shows how in certain circumstances the Cox–Stuart test may be adapted to detect a periodic trend by re-ordering the data.

3. If the same samples of cars and trucks had been used each year the independence assumption would not hold; inference would then only be valid for vehicles in that sample, for anything atypical about the sample

would influence observations in all years. With independent samples for each year, anything atypical about the sample in any one year will be incorporated in the random deviation from trend in that year. If the samples are not random there is a possibility of bias and the samples may not properly reflect population characteristics of all cars or trucks in the USA.

2.2 INFERENCES ABOUT MEDIANS BASED ON RANKS

The sign test uses little of the information in a data set like that for numbers of references in Example 2.1. In this section we assume that the population distribution is symmetric. The population mean and median then coincide at the point of symmetry and we can make inferences in ways that make greater use of the measurements.

Ideally, we assume our population distribution is continuous. In theory this means there is zero probability of two sample values coinciding. This is often unrealistic, but it is a complication we can cope with by modifications of the test.

2.2.1 The Wilcoxon signed rank test

Given a sample of n independent measurements we may determine the magnitude of departures from a mean μ or median θ specified in H_0. The mean and median coincide in a symmetric distribution, so for simplicity we speak of the mean in what follows. If μ is the true population mean of a **symmetric** distribution, departures of any given magnitude are equally likely to be positive or negative; e.g. a given value between 4 and 5 units above μ is as likely as the corresponding value between 4 and 5 units below μ. Thus, if we arrange all sample deviations from a mean μ_0 specified in H_0 in order of **magnitude** and replace them by their ascending ranks (1 for the smallest, n for the largest) and attach a negative sign to ranks corresponding to values below μ_0 we expect a good scatter of positive and negative ranks if μ_0 is the true mean. This implies that the sum of all positive ranks and the sum of all negative ranks should not differ greatly. A high sum of the positive (negative) ranks relative to that of the negative (positive) ranks implies μ_0 is unlikely to be the population mean. These ideas are incorporated in a **signed rank test** proposed by Wilcoxon (1945).

Example 2.7

The problem. Data are available (see e.g. Anon, 1991) for the percentage of the population aged 60 or over for more than 200 countries. For a random sample of 12 of these the percentages are

4.9 6.0 6.9 17.6 4.5 12.3 5.7 5.3 9.6 13.5 15.7 7.7

Assuming population symmetry, use the Wilcoxon signed rank test to test $H_0 : \mu = 12$ against the alternative $H_1 : \mu \neq 12$.

Formulation and assumptions. We arrange the deviations from 12 in ascending order of magnitude and rank these, associating with each rank the sign of the corresponding deviation. We calculate the lesser of the sum of positive and negative ranks and either compare this with tabulated values or use appropriate computer software to assess significance.

Procedure. Subtracting 12 from each sample value gives the deviations −7.1, −6.0, −5.1, 5.6, −7.5, 0.3, −6.3, −6.7, −2.4, 1.5, 3.7, −4.3. We rearrange these in increasing order of magnitude while retaining signs, i.e. 0.3, 1.5, −2.4, 3.7, −4.3, −5.1, 5.6, −6.0, −6.3, −6.7, −7.1, −7.5, whence the signed ranks are 1, 2, −3, 4, −5, −6, 7, −8, −9, −10, −11, −12. The sum of the positive ranks, $S_+ = 14$, is less than the sum, S_-, of the negative ranks. We may use Table II (or tables in Neave (1981, p.29) to see if this is significant. Entries in these tables have been calculated by a method we outline in Section 2.2.2. Table II shows that if $n = 12$ we reject H_0 in a two-tail test at the nominal 5% significance level only if the lesser of S_-, S_+ does not exceed 13. We found $S_+ = 14$, so we do not reject H_0. Since 14 is only just above the critical value 13 for significance at a nominal 5% level one expects intuitively that $\Pr(S_+ \leq 14)$ will not greatly exceed 0.05. We discuss this further under *computational aspects* below.

Conclusion. We do not reject H_0 at the nominal 5% significance level.

Comments. 1. It is only approximately true that the percentage of people aged over 60 is symmetrically distributed in more than 200 countries. There are several tests for symmetry when we are given only the sample values. One proposed by Randles *et al.* (1980) is described by Siegel and Castellan (1988, Section 4.4) who also give a BASIC program listing to carry it out; some caution is needed in assessing significance if $n < 20$ when the asymptotic result is on the borderline of significance. However, for these data the test statistic has a value well below any critical value. The data for all countries do, however, show a marked departure from normality. The mean for all countries is close to 10% and had the distribution been near normal a fairly high proportion of countries would have about 10% of the population aged over 60. In fact, relatively few countries are near the 10% figure. Many are either in the range from 5% to 8% or from 12% to 15%, something that is broadly reflected in our sample values. Because of these two 'peaks' statisticians refer to the distribution as **bimodal**. Despite a slight lack of symmetry (there are more countries with percentages below the mean than there are with percentages above) the Wilcoxon test should not be seriously misleading.

2. The two peaks in the distribution of population percentage aged over 60 reflect national differences often related to climate, the economy and health services. Countries with a temperate climate, sound economies and good health services (the three often go together) tend to have a relatively high proportion of the population aged over 60, whereas countries with poor economies and health services, factors that again often go together and which may be associated with adverse climatic conditions, usually have a relatively smaller proportion of their population aged over 60.

3. The data provide a useful illustration of the test, but it is hard to see why anyone should be interested in the mean percentage aged over 60. To estimate the proportion of the population in all countries aged over 60 one would base an estimate on a *weighted sample mean*, weighting the proportion for each country in the sample by the population of that country. The data here might be of interest in a comparison with similar data 10 years earlier or later. This type of problem is considered in Chapter 4.

Computational aspects. Neave and Worthington (1988, pp. 77–8) give a program listing in BASIC to calculate S_-, S_+ for any data set; unless n is large, when they calculate a relevant probability asymptotically, tables must be used to establish significance at a nominal level. MINITAB also has a program to calculate S_+ or S_- which gives relevant probabilities based on asymptotic theory (see Section 2.2.5); tables should be used to verify significance for small samples. A comprehensive test program is given in STATXACT, version 2.0, Chapter 7. For these data it gives the probability for $S_+ \leq 14$ appropriate to a two-tail test to be 0.0522. The program also gives the asymptotic test probability analogous to that given by MINITAB. TESTIMATE and SAS also provide an exact test and several other packages (e.g. STATGRAPHICS) compute S_+ or S_- but leave the user to consult tables for significance levels.

2.2.2 Theory of the Wilcoxon signed rank test

We describe the exact permutation test for a sample of 7 observations. If H_0 is true, the symmetry assumption implies the sums of ranks for positive and negative deviation should be nearly equal; one sum should be much higher than the other if H_0 is not true. In the extreme case when all sample values are above the mean hypothesized in H_0 all ranks will be positive so $S_-=0$ and $S_+=1+2+3+4+5+6+7=28$. A moment's reflection shows that for any mixture of positive and negative signed ranks $S_+=28-S_-$ when $n=7$. More generally, for n observations with no ties and no observations equal to the hypothesized mean, $S_+=\frac{1}{2}n(n+1)-S_-$, since the sum of the ranks 1, 2, 3, . . . , n is $\frac{1}{2}n(n+1)$. This implies symmetry between S_+ and S_- in the sense that $\Pr(S_-=s) = \Pr[S_+=\frac{1}{2}n(n+1)-s]$.

Since any rank may have a positive or negative sign, signs may be attached to the 7 ranks in $2^7=128$ different ways. If signs are allocated entirely at random the probability that all will be positive (i.e. $S_-=0$) will be 1/128. If only rank 1 is negative, $S_-=1$, again with probability 1/128. If only rank 2 is negative, $S_-=2$, again with probability 1/128. If ranks 1 and 2 are the only negative ranks, or if rank 3 is the only negative rank, in either case $S_-=3$; since $S_-=3$ can occur in these two mutually exclusive ways the associated probability is 2/128. The same probabilities clearly apply to $S_+=0$, 1, 2 or 3 and in view of symmetry the probabilities that S_- or S_+ take the values 25, 26, 27 or 28 are again respectively 2/128, 1/128, 1/128 or 1/128. These extreme values are relevant to a critical region for a two-tail test; they are improbable if the population mean has the value in H_0. A nominal critical region of size 0.05 (actual size 6/128=0.047) is found by taking the rank sum values 0, 1, 2 from the lower tail and 26, 27, 28 from the upper tail. In view of symmetry this implies rejection if the lesser of S_+, S_- does not exceed 2; this is the value indicated in Table II. For a one-tail test at the nominal 5% (actual 3.9%) significance level we reject H_0 if the sum of the positive or negative ranks (whichever is smaller) does not exceed 3.

Manual computation of the probabilities associated with all possible values of S_+ or S_- for all but small n is time consuming and error prone. STATXACT obtains these values rapidly for sample sizes of practical importance. Table 2.3 gives the probabilities for each sum from 0 to 28 for sample size $n=7$.

Using Table 2.3 we evaluate $\Pr(S_-\leq k)$ by adding the probabilities for S_- equal to 0, 1, 2, . . . k. For example, $\Pr(S_-\leq 5)=(1+1+1+2+2+3)/128=$ $10/128=0.0781$. Symmetry implies that this is equal to $\Pr(S_+\leq 5)$ and also to $\Pr(S_-\geq 23)$ or $\Pr(S_+\geq 23)$.

As the sample size n increases so does the number of possible values of the rank sums and the discrete probabilities associated with each sum become smaller. Also the distribution of S_+, S_- approaches that of a normal distribution as n increases. When $n=12$ there are $2^{12}=4096$ possible associations of signs with ranks and the possible signed rank sums range from 0 to 78. Symmetry means we only need to know the probabilities for sums between 0 and 39 since e.g. $\Pr(S_-=k)=\Pr(S_-=78-k)$ $=\Pr(S_+=k)=\Pr(S_+=78-k)$. Table 2.4 is adapted from STATXACT output and gives sufficient information to calculate the complete permutation test distribution of S_+ or S_-. Using Table 2.4 we easily find that $\Pr(S_-\leq 13)=87/4096=0.02124$ while $\Pr(S_-\leq 14)=107/4096=0.02612$. For two tail tests we double these probabilities giving 0.04248 and 0.05224. Values of the smaller signed rank sum less than or equal to 13 indicate significance at a nominal 5% or actual 4.25% significance level. Again from Table 2.4 we see that $\Pr(S_-\leq 9)=33/4096=0.00806$, while $\Pr(S_-\leq 10)$ $=43/4096=0.0105$; it follows that for a one tail test at a nominal 1% level

Table 2.3 Exact probabilities that S_+, S_- take a given value k in a Wilcoxon signed rank test for a sample of size $n=7$ when H_0 is true

k	*Probability*	k	*Probability*
0	1/128	15	8/128
1	1/128	16	8/128
2	1/128	17	7/128
3	2/128	18	7/128
4	2/128	19	6/128
5	3/128	20	5/128
6	4/128	21	5/128
7	5/128	22	4/128
8	5/128	23	3/128
9	6/128	24	2/128
10	7/128	25	2/128
11	7/128	26	1/128
12	8/128	27	1/128
13	8/128	28	1/128
14	8/128		

Table 2.4 Exact probabilities that S_+, S_- equal a given value k in a Wilcoxon signed rank test for a sample of size $n=12$ when H_0 is true. Probabilities for $k>39$ follow from the symmetry property $\Pr(S_-=k)=\Pr(S_-=78-k)$

k	*Probability*	k	*Probability*	k	*Probability*
0	1/4096	13	17/4096	26	78/4096
1	1/4096	14	20/4096	27	84/4096
2	1/4096	15	24/4096	28	89/4096
3	2/4096	16	27/4096	29	94/4096
4	2/4096	17	31/4096	30	100/4096
5	3/4096	18	36/4096	31	104/4096
6	4/4096	19	40/4096	32	108/4096
7	5/4096	20	45/4096	33	113/4096
8	6/4096	21	51/4096	34	115/4096
9	8/4096	22	56/4096	35	118/4096
10	10/4096	23	61/4096	36	121/4096
11	12/4096	24	67/4096	37	122/4096
12	15/4096	25	72/4096	38	123/4096
				39	124/4096

(actual level 0.806%) our critical region for rejection is $S_-\leq 9$.

Figure 2.1, based on Table 2.4, shows the probability mass function of S_+ or S_- for $n=12$ as a bar chart. The shape resembles that of the familiar probability density function for the normal distribution. In Exercise 2.3 we ask for a bar chart based on Table 2.3 for the case $n=7$.

2.2.3 The Wilcoxon test with ties

In theory, when we sample from a continuous distribution both the probabilities of tied observations and of getting a sample value equal to the population mean are zero. In practice observations never have a strictly continuous distribution even if only due to rounding or limited precision of measurement. We measure lengths to the nearest centimetre or millimetre; weights to the nearest kilogram, gram or milligram; the number of pages in a book in complete pages, although chapter layout often results in part-pages of text. These realities may produce rank ties or zero departures from a hypothesized mean. If they do, the exact distribution of S, the lower of the positive or negative rank sums, requires fresh computation for different numbers of ties and ties in different positions in the rank order. Prior to the advent of software that allows computation of exact probabilities for permutation distributions with ties, a common procedure was, after adjusting the scoring method for ranks, to use a modification of the normal approximation given in Section 2.2.5. This works well for large n and only a few ties, but may have bizarre consequences for small n or large numbers of ties. We describe the modification to scoring appropriate for ties and then demonstrate the effect of rank ties on the exact permutation distribution. When two or more ranks are equal in magnitude, but not necessarily of the same sign, we replace them by their mid-rank, an idea best explained by an example. If 7 observations are 1, 1, 5, 5, 8, 8, 8 and we want to perform a Wilcoxon signed rank test with $H_0 : \theta=3$ the relevant signed differences from 3 are -2, -2, 2, 2, 5, 5, 5. The magnitude of the smallest difference is 2 and there are four differences of that magnitude. The mid-rank rule is to assign to these ties the mean of the four smallest ranks 1, 2, 3, 4; i.e we allocate the signed ranks -2.5, -2.5, 2.5, 2.5 to these observations. The remaining three differences are all 5 and we give each the mean of the remaining ranks 5, 6 and 7; i.e. each is ranked 6. The exact permutation test is based on the signed mid-ranks -2.5, -2.5, 2.5, 2.5, 6, 6, 6. Our test statistic is $S_-=5$, the sum of the negative mid-ranks.

If we base our test on the permutation distribution for the 'no-tie' case given in Table 2.3 we find $\Pr(S_- \leq 5)=10/128=0.078$ and we would not reject H_0 at a nominal 5% level even in a one-tail test. However, with ties, the distribution of S_- under H_0 is **not** that in Table 2.3. The exact distribution with ties depends both on the numbers and positions of ties in the rank sequence. Prior to modern computer developments it was not practical to work out exact significance levels with ties in any but small samples. For any tie pattern STATXACT, TESTIMATE and DISFREE give exact significance levels for one- and two-tail tests and STATXACT will compute the complete distribution of S_+, S_- for samples that are not too large. For the tie pattern above for a sample of 7, STATXACT gives the exact distribution in Table 2.5.

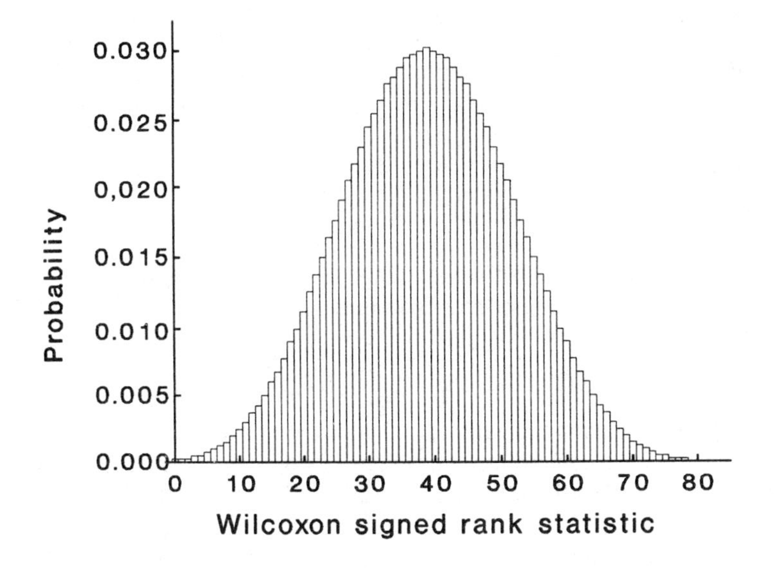

Figure 2.1 A bar chart illustrating the distribution of the Wilcoxon signed rank statistic when $n=12$.

Table 2.5 Exact probabilities that S_+, S_- equal a given value k in a Wilcoxon signed rank test for a sample of size $n=7$ when H_0 is true and mid-ranks are 2.5 (four times) and 6 (three times)

k	*Probability*	*k*	*Probability*
0	1/128	14.5	12/128
2.5	4/128	16	3/128
5	6/128	17	18/128
6	3/128	18	1/128
7.5	4/128	19.5	12/128
8.5	12/128	20.5	4/128
10	1/128	22	3/128
11	18/128	23	6/128
12	3/128	25.5	4/128
13.5	12/128	28	1/128

Comparing Tables 2.3 and 2.5 we see that in each case the distribution of the test statistic is symmetric (implying that we may base our test on either S_+ or S_-). However, whereas the distribution with no-ties is unimodal with values of the statistic confined to integral values increasing in unit steps, that in Table 2.5 is heavily multimodal with S taking unevenly spaced and not necessarily integral values. The discontinuities are also more marked. From Table 2.5 we easily deduce that $\Pr(S_- \leq 5) = 11/128$, compared to the no-tie probability 10/128 given by Table 2.3. Despite the heavy incidence of ties in this example and the marked difference in the distribution of the test statistic in the two cases, it may appear that ties have little effect on our conclusion. However, the position is not quite as comforting as this illustration seems to imply.

Example 2.8

Ties often arise due to rounding. Data 1, 1, 5, 5, 8, 8, 8 could result from rounding to the nearest integer in either of the sets:

0.9, 1.1, 5.2, 5.3, 7.9, 8.0, 8.1
0.9, 1.0, 4.8, 4.9, 7.9, 8.0, 8.1

Performing the Wilcoxon test with $H_0 : \theta = 3$ we now have no tied ranks and for the first set $S_- = 3$, while for the second set $S_- = 7$. From Table 2.3 we find $\Pr(S_- \leq 3) = 5/128 = 0.0391$ and $\Pr(S_- \leq 7) = 19/128 = 0.1484$. Thus a one-tail test at a nominal 5% level would indicate significance in the first case but not in the second. Rounding to the nearest integer thus loses a significant result in one case and in the other it brings a clearly non-significant result closer to significance.

This illustration highlights difficulties of interpretation with ties arising from rounding, a process that makes small perturbations to most observations. Many statistical analyses are sensitive to such data changes, an aspect we consider further in Section 11.1 where we describe what are called 'robust' methods. These are in general less sensitive to such perturbations and other data peculiarities. The degree of tying in this example is fairly extreme, giving only two different magnitudes for ranks. Generally speaking, for larger samples and not too many ties, the effect on significance is less marked.

Example 2.9

The problem. Consider the data in Example 2.7 for the percentage of the population aged 60 or over in a random sample of 12 countries, i.e.

4.9 6.0 6.9 17.6 4.5 12.3 5.7 5.3 9.6 13.5 15.7 7.7

Use the Wilcoxon signed rank test to test $H_0 : \mu = 12.9$ against $H_1 : \mu \neq 12.9$.

Formulation and assumptions. To justify the Wilcoxon test we assume the population is symmetric and determine the lesser of S_+, S_- after ranking, with due regard to sign, deviations of observations from 12.9 using mid-ranks for ties.

Procedure. The actual signed deviations from 12.9 arranged in ascending order of magnitude are −0.6, 0.6, 2.8, −3.3, 4.7, −5.2, −6.0, −6.9, −7.2, −7.6, −8.0, −8.4 and the corresponding signed ranks using mid-ranks for the tie at 0.6 are −1.5, 1.5, 3, −4, 5, −6, −7, −8, −9, −10, −11, −12. This gives $S_+=1.5+3+5=9.5$. A two-tail test is appropriate. If we have no program to determine the permutation distribution of S with ties we may use tables of critical values for $n=12$ strictly relevant to the 'no ties' case. Table II indicates significance at the 5% level if $S\leq 13$ and at the 1% level if $S\leq 7$. If we accept that a single tie is likely to have little effect on the probability that S is at or below any specified value we can be reasonably confident of significance at the 5% level. The complete distribution of S for the 'no tie' case given in Table 2.4 involves only integral values of S, and if the distribution there were valid we would have $\Pr(S\leq 9.5)=\Pr(S\leq 9)$, and that table gives the relevant probability as $33/4096=0.00806$. This probability is relevant to a one-tail test and we double it for a two-tail test, giving 0.01612. STATXACT gives the exact probability of observing $S\leq 9.5$ with a tie for the lowest rank when $n=12$ to be 0.0085 whence the actual significance level for a two-tail test is $2\times 0.0085=0.017$ (1.7%).

Conclusion. We reject H_0 at a nominal 5% signficance level using tables; the exact size of the critical region is $\alpha=0.017$ corresponding to an actual significance level of 1.7%.

Comments. 1. Our conclusion would not be seriously affected by the single tie if we assumed our test statistic based on mid-ranks has the same distribution as that relevant for no ties. However, our discussion of the extreme tie situation in Example 2.8 indicates the need for caution in interpreting results when there are ties. With many ties the exact probabilities associated with a critical region may be of less interest than a study of how these probabilities vary if ties are broken in different ways as in Example 2.8.

2. Before the advent of computer programs to calculate the exact distribution of the test statistic when there are ties, the advice often given was to resort to asymptotic (large sample) results of a type we describe in Section 2.2.5. There is little justification for this with small samples.

Computational aspects. If a package such as STATXACT is available it is worth obtaining exact distributions for a number of tied situations for small sample sizes in order to develop a feeling for the effect of ties on standard results. An extreme case arises with the observations 4, 4, 8, 8, 8, 8, 8 using the Wilcoxon test for $H_0 : \theta=6$. We then have all ranks of deviations tied apart from signs and it is not difficult to see (Exercise 2.4) that the Wilcoxon test statistic has a distribution equivalent to that for the sign test. In this sense a sign test is a special case of the Wilcoxon test.

If one or more sample values equal the mean hypothesized under H_0, 'deviations' associated with such observations are zero. Statisticians are divided on the most appropriate action in this case. We may proceed, as proposed for the sign test, by omitting such observations as being uninformative about any alternative hypothesis, and apply the test to a reduced sample. An alternative recommendation is to rank all observations temporarily giving any zero the rank 1 because it is the smallest difference (or the appropriate tied rank if there is more than one zero). After all signed ranks are allocated we then change the rank(s) associated with zero difference to 0, leaving the other ranks as allocated. To obtain exact tests in the latter situation a suitable computer program is needed for the exact distribution of the test statistic, else we must resort to asymptotic results in the way indicated in Section 2.2.5. Deleting zero differences is the simplest procedure if no suitable program for exact probabilities of S for the modified rank statistic is available, but the second procedure has theoretical advantages in that it provides a more powerful test.

2.2.4 Confidence intervals based on signed ranks

For many nonparametric methods calculation of confidence intervals lacks the simplicity of hypothesis testing, but the procedures, once mastered, are straightforward and easily programmed if no package program is available. Surprisingly, at the time of writing there are few packages that deal in a completely satisfactory way with confidence intervals based on Wilcoxon signed rank test theory, but the position is likely to improve rapidly during the 1990s. Meantime, with ingenuity, existing packages can be used to obtain intervals.

'Trial and error' approaches may be avoided by using an alternative way of calculating the statistic S_- (or S_+). We outline the theory, but the reader prepared to take it on trust may move directly to Example 2.10 where we discuss its application both to hypothesis testing and determining confidence intervals.

For convenience we assume a sample of n observations with no ties has been arranged in **ascending** order $x_1, x_2, \ldots, x_n$. We specify $H_0: \mu=\mu_0$ as the null hypothesis. In practice μ_0 will usually lie between x_1 and x_n, but this is not a theoretical requirement. A deviation $x_i-\mu_0$ will have a negative signed rank for any x_i less than μ_0; since we have arranged the x_i in ascending order, among all **negative** ranks, if there are any, that of greatest magnitude will be associated with x_1, that of next greatest magnitude with x_2, and so on. Similarly, all x_j greater than μ_0 have positive signed ranks and of these x_n will have the highest rank, x_{n-1} the next highest and so on.

Consider now the paired averages of x_1 with each of $x_1, x_2, \ldots, x_n$. Suppose x_1 has associated signed rank $-p$. A moment's reflection shows that each of the averages $\frac{1}{2}(x_1+x_q)$ will be less than μ_0 providing the deviation associated with x_q has either a negative rank or a positive rank less than p. Since there are no rank ties no observation has positive rank p (since the rank p is already associated with x_1). If x_q is the smallest observation associated with a positive rank greater than p, this implies $|x_q - \mu_0| > |x_1 - \mu_0|$ and it follows that the average of x_1 and x_q is greater than μ_0. Then, because of the ascending order, so is the average of x_1 with any $x_r > x_q$. Thus the number of averages involving x_1 which are less than μ_0 is equal to the negative rank associated with x_1. Similarly, if we form all the paired averages of any x_i less than μ_0 with each of $x_i, x_{i+1}, \ldots, x_n$ the number of these averages less than μ_0 will equal the (negative) rank associated with x_i. Clearly, when x_i has a positive signed rank none of the averages with itself or greater sample values will be less than μ_0. The complete set of averages $\frac{1}{2}(x_i + x_j)$ for $i = 1, 2, \ldots, n$ and all $j \geq i$ are often called the **Walsh averages.** Clearly the number of Walsh averages less than μ_0 equals S_- and the number greater than μ_0 equals S_+. We may use Walsh averages to calculate the test statistic S for a given H_0 and to obtain confidence limits for the population mean.

Example 2.10

The problem. Given the percentages of population aged over 60 for 12 countries in Example 2.7, viz. 4.5, 4.9, 5.3, 5.7. 6.0, 6.9, 7.7, 9.6, 12.3, 13.5, 15.7, 17.6, obtain an estimate of the population mean with nominal 95% and 99% confidence intervals, assuming a symmetric distribution.

Formulation and assumptions. We use Walsh averages and the principle enunciated in Section 1.4.1 that the $100(1-\alpha)$% confidence interval contains those values of the parameter that would be accepted in a significance test at the 100α% significance level.

Procedure. The Walsh averages may conveniently be set out in a tableau like Table 2.6. Both the top row and the first column of that table consist of the ordered sample values written in the same ascending order. The triangular matrix forming the body of the table consists of the Walsh averages. In comment 2 below we discuss computation of these if a computer program is not available.

An intuitively reasonable point estimator of the population mean is either the mean of the sample values or the median of the Walsh averages since it follows from the arguments given before this example that for the latter the sums of the negative and positive rank deviations are equal, both being equal to the number of Walsh averages above (or below) this median. The median of the Walsh averages may be determined without difficulty by inspecting Table 2.6 if we note that entries in any row increase from left to right and entries in any column increase as we move down, resulting in the smallest entries being at the top left of the table and the largest at the

Table 2.6 Walsh averages for population percentages aged over 60

	4.5	4.9	5.3	5.7	6.0	6.9	7.7	9.6	12.3	13.5	15.7	17.6
4.5	4.5	4.7	4.9	5.1	5.25	5.7	6.1	7.05	8.4	9.0	10.1	11.05
4.9		4.9	5.1	5.3	5.45	5.9	6.3	7.25	8.6	9.2	10.3	11.25
5.3			5.3	5.5	5.65	6.1	6.5	7.45	8.8	9.4	10.5	11.45
5.7				5.7	5.85	6.3	6.7	7.65	9.0	9.6	10.7	11.65
6.0					6.0	6.45	6.85	7.8	9.15	9.75	10.85	11.80
6.9						6.9	7.3	8.25	9.6	10.2	11.3	12.25
7.7							7.7	8.65	10.0	10.6	11.7	12.65
9.6								9.6	10.95	11.55	12.65	13.6
12.3									12.3	12.9	14.0	14.95
13.5										13.5	14.6	15.55
15.7											15.7	16.65
17.6												17.6

bottom right. For a sample of size n there are $\frac{1}{2}n(n+1)$ Walsh averages, i.e. 78 in this example. Thus the median of these lies between the 39th and 40th ordered Walsh average and inspection of the table establishes that both these Walsh averages equal 9.0; this is an appropriate estimate of the population mean. To establish a 95% confidence interval we select as end points values of μ that will just be acceptable. We found in Example 2.7, using Table 2.4, that we reject H_0 if the sum of the lesser of positive or negative ranks does not exceed 13. Thus if we choose a value less than the 14th smallest Walsh average we would reject H_0. Similarly if we choose a value greater than the 14th largest Walsh average we also reject H_0 since in either case our test statistic S would be such that $S \leq 13$. Thus for $n=12$ a nominal 95% confidence interval is the interval (14th smallest Walsh average, 14th largest Walsh average). From Table 2.6 we easily see that this is the interval (5.7, 12.25). We noted in Example 2.7 that using $S \leq 13$ as a criterion for significance corresponded to an actual significance level of 4.25% (two tails). It would seem intuitively reasonable therefore to regard the interval (5.7, 12.25) as an actual 95.75% confidence interval. Owing to discontinuities in the distribution of the test statistic and the fact that the end points of the confidence interval often coincide with observed values we must, in an analogous way to that described in Example 2.2 for the sign test, regard values outside this interval as ones indicating rejection at a 4.25% significance level. We must also, if our exact level is critical, test the end values to see if they would be accepted or rejected at the 5% siginficance level. Exact tests using STATXACT show that we would reject at that level both 5.7 and 12.25 as hypothesized means. Thus in this case all points in the open interval $5.7 < \mu < 12.25$ would be acceptable, but the closed interval (5.7. 12.25) is a a conservative 95% interval, having a confidence level exceeding 95%. In practice it is usual to cite a confidence interval simply as the interval (5.7, 12.25) without being specific as to whether it is open or closed, bearing in mind that discontinuities inherent in the Wilcoxon test statistic distribution may imply (and it will vary from case to case) that the end points may or may not be included as **acceptable** values at the nominal level. One should not get too excited about subtleties due to discontinuity at the end-point of intervals. If relevant programs are available it may be worthwhile checking the actual confidence level associated with any quoted interval. It is left as an exercise to establish that a nominal 99% confidence interval is (5.3, 13.6).

Conclusion. An appropriate point estimator of the population mean (median) is 9.0 and actual 95.75% and 99.08% confidence intervals are respectively (5.7, 12.25) and (5.3, 13.6), where it is understood that for points outside these intervals we reject H_0 in a hypothesis test at the corresponding relevant significance level.

Comments. 1. An acceptable alternative point estimator to the median of the Walsh averages is the mean of all sample values. It is unbiased in the sense that if we sampled repeatedly and took the mean of the sample values as our estimate in all cases, the mean of these estimates would converge to the population mean. However, the median of the Walsh averages is in a certain technical sense, which we do not discuss here, generally closer to the population mean; this estimator is often referred to as the **Hodges–Lehmann estimator.**

2. If confidence limits only are required we need write down only a few of the Walsh averages in the top left and bottom right of the triangular matrix in Table 2.6. If a computer program is available for forming this matrix there is no problem in computing all the averages. If the matrix has to be formed manually note that once we form the Walsh averages in the first row, the required averages in the second row may be obtained by adding the same constant, namely half the difference x_2-x_1, to the average immediately above it. To verify this denote by x_{ij} the average of x_i and x_j. This is the entry in row i and column j of the matrix of Walsh averages. Now for any $p>2$

$$x_{2p}=\tfrac{1}{2}(x_2+x_p)=\tfrac{1}{2}(x_2-x_1+x_1+x_p)=\tfrac{1}{2}(x_2-x_1)+x_{1p}.$$

The idea generalizes for all subsequent rows, each entry in row $r+1$, say, being obtained by adding the same constant $\tfrac{1}{2}(x_{r+1}-x_r)$ to the entry immediately above it.

3. In Example 2.7 we found when testing $H_0 : \mu=12$ that $S_+=14$. In Table 2.6 there are 14 Walsh averages exceeding 12.0.

Computational aspects. MINITAB includes a program for computing Walsh averages, but does not print these out. It gives asymptotic confidence intervals (see Section 2.2.5). These may not be satisfactory for small n, especially if there are many ties. If one has the matrix of Walsh averages generated by some available program one may easily establish nominal 95% or 99% confidence limits using tables of critical values of the statistic S. SAS includes a program for exact confidence intervals based on Walsh averages.

In the case of ties in signed ranks the methods for confidence intervals in this section need modification, just as they did for hypothesis testing. Pending the availability of comprehesive computer software to cover confidence intervals in these circumstances we suggest tentatively using the Walsh average method described above, then using an available computer program to study appropriate endpoint adjustments to obtain a suitable interval. Such adjustments are likely to be relatively small except in cases of heavy tying. We give an example in Exercise 2.5. For the data sets in Example 2.8 nominal 90% confidence intervals are respectively (3.05, 8) and (2.85, 8). The intervals are very similar but the marked differences in one-tail significance levels when testing the hypothesis $\mu=3$ are clearly attributable to the fact that this value falls just outside the relevant confidence interval in the first case and just inside that interval in the second case.

2.2.5 A large-sample approximation

Figure 2.1 suggests that for moderate or large n we might approximate to the distribution of S_+ or S_- by a normal distribution. In practice the relevant approximation is very good for $n>20$. Denoting the **magnitude** of the lesser of positive and negative rank sums by S, it can be shown that, if there are no ties, then under H_0, S has mean $\frac{1}{4}n(n+1)$ and variance $n(n+1)(2n+1)/24$, whence

$$Z=\frac{S+\frac{1}{2}-\frac{1}{4}n(n+1)}{\sqrt{[n(n+1)(2n+1)/24]}} \tag{2.4}$$

has approximately a standard normal distribution; the term ½ in the numerator is a continuity correction. Critical values of Z in a two-tail test are $Z\leq-1.96$ at the 5% level and $Z\leq-2.58$ at the 1% level; the corresponding one-tail values are $Z\leq-1.64$ and $Z\leq-2.33$.

In Example 2.7 we only had a sample of 12, but even here a normal approximation is reasonable. Substituting $n=12$ and $S=14$ in (2.4) gives $Z=-1.92$. This is in good agreement with the permutation test probability of 0.0522 obtained in Example 2.7 since, from tables of the standard normal distribution, we may verify that $\Pr(|Z|>1.92)=0.0548$. For technical reasons relating to greater generality STATXACT performs the normal approximation test (there called the asymptotic test) **without** the continuity correction. This gives $Z=1.961$, just significant at the 5% level. As we stressed in Section 1.3, not too much emphasis should be put on rigid adherence to critical values such as 5%, unless there is a firm requirement for rejection if significance is shown at some prespecified level. Such legalistic requirements are increasingly being incorporated in health, safety or consumer protection legislation.

The asymptotic test needs modification for ties. The continuity correction of ½ is removed since the statistic S is no longer confined to integral values. No adjustment to the mean is needed for tied ranks unless we replace an observed value or values equal to the mean hypothesized under H_0 by a zero rank. Generally, the mean is $\frac{1}{2}\Sigma|s_i|$, where summation is over all ranks or mid-ranks, s_i, after any zero allocation of ranks if applicable. The standard deviation in the denominator needs modification for ties. Adjustment formulae that take account of the position of ties and the numbers of observations at each tie are available but modern computational methods make it easy to compute the standard deviation as $\sqrt{(\Sigma s_i^2/4)}$ where the summation is over all s_i. In the 'no tie' case it is easily verified that this reduces to the denominator in (2.4).

Example 2.11

The problem. Use the appropriate asymptotic formula to perform the test requested in Example 2.9.

Formulation and assumptions. As in Example 2.9. Note the one tie involving 2 observations.

Procedure. As before, we find $S=9.5$. When $n=12$ the mean is 39. We modify the denominator in (2.4), calculating $\Sigma(s_i^2/4)=(1.5^2+1.5^2+3^2+\ldots+12^2)/4=162.375$, whence $Z=(9.5-39)/\sqrt{162.375}=-2.315$. From tables of the standard normal distribution we verify that for a two-tail test this indicates significance at the 2.06% level.

Conclusion. We reject the hypothesis that the population mean is 12.9 at a 2.06% significance level.

Comments. 1. In Example 2.9 we found the exact permutation test significance level was 1.7%, indicating that even for this relatively small sample size the asymptotic approximation is quite good. It will tend to improve further as the sample size increases.

2. The difference in the denominator with only one pair of ties from that for the no-tie situation given in (2.4) is here trivial. Using (2.4) we find the variance to be 162.5 rather than 162.375. For only a few ties use of Table II, or for larger samples equation (2.4) without modification, is unlikely to be seriously misleading but actual, or even nominal, significance levels may

no longer hold. For numerous ties the effect on the denominator of (2.4) may be appreciable.

3. If we apply the sign test in this example (see Exercise 2.7) we would confirm significance only at an actual 14.6% level. Using more of the available information sharpens our indication of significance.

Computational aspects. STATXACT output always includes the asymptotic 'tail' probabilities (without a continuity correction).

We may use (2.4) to determine asymptotic values for the statistic S required for significance at the 5% level for a given n by putting $Z=-1.96$ in (2.4). When $n=12$ we get

$$-1.96=\frac{S+\frac{1}{2}-39}{\sqrt{(12\times 13\times 25/24)}}$$

whence $S=-1.96\times 12.7475+38.5=13.51$. Rounding down to the integer below gives $S=13$. This is effectively the way MINITAB establishes confidence limits. Note that in this case the asymtotic value of S is identical with that given in tables, sugggesting that for samples of 12 or more nominal 95% limits determined using asymptotic results with no ties may well be close to those based on permutation theory. In Exercise 2.6 we ask you to check whether the situation also obtains for a 99% confidence interval.

2.2.6 Robustness

In Example 2.10 we found nominal 95% and 99% confidence intervals for the population mean using the Wilcoxon signed-rank procedure to be respectively (5.7, 12.25) and (5.3, 13.6). Not surprisingly, intervals based on the sign test are longer, namely (5.3, 13.5) and (4.9, 15.7). We pointed out in Example 2.7 that our sample could not be regarded as one from a normal distribution, yet it is of some interest to compute confidence intervals based on normality. These turn out to be (6.26, 12.03) and (5.07, 13.21); each is slightly shorter than the Wilcoxon intervals. Thus, despite the departure from normality, normal theory does better. Although the Wilcoxon approach does not seem to have an advantage for this particular departure from normality, both the Wilcoxon and the sign-test procedures exhibit a property called **robustness** which we discuss more fully in Chapter 11. We regard test and estimation procedures as robust if they are little influenced by fairly blatant departures from assumptions. In Example 2.12 below we indicate that the Wilcoxon test may be more robust than the t-test if there is a marked departure from symmetry. We also see that in such circumstances the sign test may, not unexpectedly, do even better since it requires no assumption of symmetry.

Example 2.12

Suppose that the data in Example 2.7 are amended by omitting the observation 9.6 and replacing it (i) by 26.6 and (ii) by 46.6. The data are now obviously skew with indications of a long upper tail. Table 2.7 gives the nominal 95% and 99% confidence intervals based on the normal theory t, the Wilcoxon signed rank statistic and the sign test in each case.

Table 2.7 95% and 99% confidence intervals for data sets (i) and (ii) in Example 2.12 based on normal theory (t), Wilcoxon signed ranks (W) and the binomial sign test (B). For each data set the shortest interval at each given confidence level is indicated by an asterisk

		95%	99%
	t	(6.24, 14.87)*	(4.47, 16.65)
Data set (i)	W	(5.7, 15.7)	(5.3, 16.75)*
	B	(5.3, 15.7)	(4.9, 17.6)
	t	(4.76, 19.69)	(1.69, 22.75)
Data set (ii)	W	(5.7, 16.65)	(5.3, 26.75)
	B	(5.3, 15.7)*	(4.9, 17.6)*

The first data set shows moderate skewness and here we have a mixed picture, the t-test doing best at the 95% level and Wilcoxon marginally better at the 99% level. For data set (ii) which is very skew, with an observation at 46.6, the t limits are highly unsatisfactory, and those based on Wilcoxon are moderately satisfactory at the 95% level but not satisfactory at the 99% level. However, the binomial limits are the same as those for set (i) and are not influenced by the extreme observation of 46.6. This is not surprising since no assumption of symmetry is needed for validity of the sign test.

We may query whether such extreme values as that in data sets (i) and (ii) are realistic for proportions of a population aged over 60. The entry of 26.6 in set (i) is indeed a true value for one country that might occur in a sample. That of 46.6 in set (ii) is a data error. While it is good statistical practice to check data carefully to eliminate error, one is often required to analyse data where there is no indication that an extreme value (often referred to as an outlier) is not a correct observation. In such circumstances one may do better with the simple sign test than with a test requiring more stringent assumptions.

We have assumed for the Wilcoxon signed rank test that our sample of n observations came from a distribution that is specified as being symmetric, but the test is still valid in the less common but more general situation where each x_i may come from a different symmetric distribution, providing only that each such distribution has the same mean. Similarly, for the sign test it suffices to assume each x_i comes from a different distribution, asymmetric or not, providing only that each has the same median. See also Section 3.5.

2.2.7 An alternative test statistic

We based tests on the lesser of S_+, S_-. We may use an alternative statistic W given by $W = |S_+ - S_-|$, i.e. the magnitude of the difference between the sum of the positive and negative ranks. Since $S_+ + S_- = \frac{1}{2}n(n+1)$ it is easily shown that $W = \frac{1}{2}n(n+1) - 2S_-$ in the case where $S_- < S_+$, so that W is a linear function of S_-, i.e. there is a straight-line relationship between S_- and W. Different tables, which can be derived from Table II, are needed if W is used for testing. The normal approximation corresponding to W has the simple form

$$Z = W/[\sqrt{\Sigma(s_i^2)}] \tag{2.5}$$

since W has expectation (mean) zero under the null hypothesis. If there are no ties the denominator reduces to $\sqrt{[n(n+1)(2n+1)/6]}$.

Throughout this book we draw attention to cases where alternative but equivalent test statistics may be used. Which form is used often depends on available tables or computer software.

2.3 OTHER LOCATION ESTIMATORS

There exist other location tests and estimators with intuitive appeal that are closely related to the Wilcoxon signed rank estimator. They are often more difficult to calculate and may give little gain in efficiency. Some are described in Chapter 3.

2.4 FIELDS OF APPLICATION

Insurance

The median of all motor policy claims paid by a company in 1991 is £670. Early in 1992 the management thinks claims are higher. To test this, and estimate the likely rise in mean or median, a random sample of 25 claims is taken. The distribution of claims is skew, so a sign test is appropriate.

Medicine

The median pulse rate of a group of boys prior to physical training is known. If the pulse rate is taken for a sample after exercise the sign test could be used to test for a shift in median. Would you consider a one- or a two-tail test appropriate? If it appears reasonable to assume a symmetric distribution a Wilcoxon test or even a normal theory *t*-test would be more appropriate. Even if an assumption of symmetry in pulse rates before exercise is reasonable, this may not be so after exercise. For instance, the rate increase after exercise might be relatively higher for those with above median rest rates and this could give rise to skewness. A medic asking questions about changes should know if this is likely.

Engineering

Median noise level under the flight path to an airport might be known for aircraft with a certain engine type (the actual level will vary from plane to plane and from day to day depending on weather factors, the precise height each plane flies over the measuring point, etc.). If the engine design is modified a sample of measurements under similar conditions to that for the old engine may indicate a noise reduction. A one-tail test would be appropriate if it were clear the modification could not increase noise. We are unlikely to be able to use a true random sample here, but if taken over a wide range of weather conditions, the first 40 approaches using the new engine may broadly reflect characteristics of a random sample.

Biology

Heartbeat rates for female monkeys of one species in locality A may have a symmetric distribution with known mean. Given heartbeat rates for a sample of similar females from locality B, the Wilcoxon test could be used to detect a shift or to obtain confidence limits for the true mean.

Physics

Specimens of a metal are heated and the hardness measured on independent samples at 5° intervals over a 90°C temperature range. The Cox–Stuart test could be used for indications of a monotone relation between hardness and temperature.

Education

A widely used test of numerical skills for 12-year-old boys gives a median mark of 83. A new method of teaching such skills is used for a class of 42. The asymptotic approximation to the Wilcoxon test could be used to test for a shift in median if symmetry could be assumed. If not, a sign test would be preferred. See comment 4 on Example 2.5 for reservations about regarding the test group as a random sample.

Management

Records give the mean and median number of days absent from work for all employees in a large factory for 1991. The number of days absent for a random sample of 20 is noted in 1992. Do you think such data are likely to be symmetric? In the light of the answer to this question one may select the appropriate test for indications of changes in the absentee pattern.

Geography and environment

Estimates of the amount of cloud cover at the site of a proposed airport are taken at a fixed time each day over a period. The site might be rated unsuitable if the median cover were too high. A confidence interval for the median would be useful.

Psychology

A psychologist is told that the 'national average IQ of drug abusers aged between 16 and 18 is 103'. He assesses the IQ of a sample of abusers in that age group from an area where drug abuse is rife. He might use a Wilcoxon test to assess whether it is reasonable to assume the mean is 103 for that area.

Industry

The median time people stay in jobs in a large motor assembly plant in Germany is known to be 5.2 years. Times for all 30 employees who have left a UK plant in the last 3 months are available. Although this may not be a random sample it may be reasonable to test whether the median time for UK workers is also 5.2 years. If the distribution appears skew a sign test would be appropriate.

2.5 SUMMARY

Sign test. The test statistic is the number of observations above or below a median specified in H_0 (see e.g. Example 2.1). Significance is determined from Table I or a suitable computer program. Table I may also be used to obtain confidence intervals (Section 2.1.2). Normal theory approximations (Section 2.1.3) may be used for sample sizes greater than 20. Modifications are required to deal with values equal to the median specified in H_0 (p.30) No assumption of symmetry is required. Related methods include the Cox–Stuart test for trend (Section 2.1.6) and tests for quantiles other than the median (Section 2.1.5).

Wilcoxon signed rank test. The test statistic S is the lesser of the number of positive and negative ranked differences from the mean or median specified in H_0 (Section 2.2.1). Table II or appropriate computer software may be used for significance tests. Ties require special treatment (Section 2.2.3) and suitable computer software is then needed to determine exact permutation distributions. Walsh averages (Section 2.2.4) provide an alternative estimation and hypothesis testing procedure; they are particularly useful for calculating confidence intervals. An assumption of symmetry is required for validity of the test and estimation procedures. Asymptotic normal approximations (Section 2.2.5) work well for sample sizes $n=12$ or more if there are no (or very few) ties; modification of the denominator in the test statistic is required for numerous ties even for relatively large n.

EXERCISES

2.1 For the sample of 12 in Example 2.1 if θ is the population median test the hypothesis $H_0 : \theta=6$ against the alternative $H_1 : \theta\neq 6$.

2.2 Compute $\Pr(X \leq 13)$ when X is a binomial variable with $n=32, p=0.25$ and use your result to confirm the conclusions reached in Example 2.5

2.3 Using the data in Table 2.3 for the distribution of the Wilcoxon S when $n=7$, construct a bar chart analogous to that in Figure 2.1 showing the probability function for S. Discuss the similarity, or lack of similarity, to a normal distribution probability density function.

2.4 Establish that the permutation distribution of the Wilcoxon signed rank statistic for testing the hypothesis $H_0 : \theta=6$, given the observations 4, 4, 8, 8, 8, 8, 8 has a distribution equivalent to that for the sign test of the same hypothesis.

2.5 Establish nominal 95% confidence intervals based on the Wilcoxon signed rank test for the following data sets. If an appropriate computer program is available use it to comment on the discontinuities at the end points of your estimated intervals based on Walsh averages.

Set I 1, 1, 1, 1, 1, 3, 3, 5, 5, 7, 7
Set II 1, 2, 2, 4, 4, 4, 4, 5, 5, 5, 7

2.6 Determine an $\alpha=0.01$ two-tail test critical region for the Wilcoxon statistic S when $n=12$ based on the asymptotic result using equation (2.4). Does it equal the value in Table II?

2.7 Apply the sign test to the data and hypothesis test in Example 2.9.

2.8 The first application listed in Section 2.4 involved insurance claims. The 1991 median was £670. A random sample of 14 claims from a large batch received in 1992 showed the following amounts (in £):

275 283 427 681 692 724 877 1024 1583 1742 1813 2519 4450 6125

What test do you consider appropriate for a shift in median? Would a one-tail test be appropriate? Obtain a 95% confidence interval for the median based upon these data.

2.9 The weight losses in kilograms for 16 overweight women who have been on a diet for two months are as follows:

4 6 3 1 2 5 4 0 3 6 3 1 7 2 5 6

The firm sponsoring the diet advertises 'Lose 5 kg in two months'. In a consumer watchdog radio programme they claim this is an 'average' weight loss. You may be unclear about what is meant by 'average', but assuming the sample is effectively random does it indicate a median weight loss of 5 kg in the population of dieters? Test this without an assumption of symmetry. What would be a more appropriate test with an assumption of symmetry? Carry out this latter test.

2.10 I counted the numbers of pages in a random sample of 12 of my statistics books. The numbers were

126 142 156 228 245 246 370 419 433 454 478 503

Use the Wilcoxon signed rank test to test the hypothesis that the mean number of pages in my collection of statistics books is 400. Obtain a 95% confidence interval for the mean number of pages based on the Wilcoxon test and compare it with the interval obtained on an assumption of normality.

2.11 A pathologist counts the numbers of diseased plants in randomly selected areas each 1 metre square on a large field. For 40 such areas the numbers of diseased plants are:

21	18	42	29	81	12	94	117	88	210
44	39	11	83	42	94	2	11	33	91
141	48	12	50	61	35	111	73	5	44
62	11	35	91	147	83	91	48	22	17

Use an appropriate nonparametric test to find whether it is reasonable to assume the median number of diseased plants per square metre might be 50 (i) without assuming population symmetry, (ii) assuming population symmetry.

2.12 Before treatment with a new drug a number of insomniacs have a median sleeping time of 2 hours per night. A drug is administered and it is known that if it has an effect it will increase sleeping time, but some doctors doubt if it will have any effect. Are their doubts justified if the hours per night slept by insomniacs after taking the drug are:

3.1 1.8 2.7 2.4 2.9 0.2 3.7 5.1 8.3 2.1 2.4

2.13 The UK Meteorological Office monthly weather summaries published by HMSO give the following annual rainfalls in mm for 15 stations in the UK during 1978. The stations are listed below in order of increasing latitude. Is there evidence of a monotonic trend in rainfall as one moves from South to North?
Margate, 443; Kew, 598; Cheltenham, 738; Cambridge, 556; Birmingham, 729; Cromer, 646; York, 654; Carlisle, 739; Newcastle, 742; Edinburgh, 699; Callander, 1596; Dundee, 867; Aberdeen, 877; Nairn, 642; Baltasound, 1142.

2.14 In a small pilot opinion poll 18 voters in one electorate were selected at random and asked if they thought the British Prime Minister was doing a good job. Six (one-third) said 'yes' and twelve (two-thirds) said 'no'. Is this sufficient evidence to reject the hypothesis that 50% of the electorate think the Prime Minister is doing a good job? The pilot results were checked by taking a larger sample of 225 voters. By a remarkable coincidence 75 (one-third) answered 'yes' and 150 (two-thirds) answered 'no'. Do we draw the same conclusion about the hypothesis that 50% of the electorate think the Prime Minister is doing a good job? If not, why not?

2.15 A traffic warden is required to note the time a car has been illegally parked after its metered time has expired. For 16 offending cars he records the time in minutes as:

10 42 29 11 63 145 11 8 23 17 5 20 15 36 32 15

Obtain an appropriate 95% confidence interval for the median overstay time of offenders prior to detection. What assumptions were you making to justify using the method you did? To what population do you think the confidence interval you obtained might apply?

2.16 Kimura and Chikuni (1987) give data for lengths of Greenland turbot of various ages sampled from commercial catches in the Bering Sea as aged and measured by the Northwest and Alaska Fisheries Center. For 12-year-old turbot the numbers of each length were:

Length (cm)	64	65	66	67	68	69	70	71	72	73	75	77	78	83
No. of fish	1	2	1	1	4	3	4	5	3	3	1	6	1	1

Would you agree with someone who asserted that, on this evidence, the median length of 12-year-old Greenland turbot was almost certainly between 69 and 72 cm?

2.17 The journal *Biometrics* 1985, **41**, p. 830, gives data on numbers of medical papers published in that journal for the period 1971–81. These data are extended below to cover the period 1969–85. Is there evidence of a monotonic trend in numbers of medical papers published?

11 6 14 13 18 14 11 22 19 19 25 24 38 19 25 31 19

2.18 Knapp (1982) gives the percentage of births on each day of the year averaged over 28 years for Monroe County, New York State. Ignoring leap years (which make little difference), the median percentage of births per day is 0.2746. Not surprisingly, this is close to the expected percentage on the assumption that births are equally likely to be on any day, that is, 100/365=0.274. We give below the average percentage for each day in the month of September. If births are equally likely to be on any day of the year this should resemble a random sample from a population with median 0.2746. Do these data confirm this?

0.277 0.277 0.295 0.286 0.271 0.265 0.274 0.274 0.278 0.290 0.295 0.276
0.273 0.289 0.308 0.301 0.302 0.293 0.296 0.288 0.305 0.283 0.309 0.299
0.287 0.309 0.294 0.288 0.298 0.289

2.19 A commentator on the 1987 Open Golf Championship asserted that on a good day 10% of top-class players could be expected to complete a round with a score of 68 or less. On the fourth day of the championship weather conditions were poor and the commentator remarked before play started that one might expect the weather to increase scores by four strokes per round. This would suggest 10% of players might be expected to return scores of 72 or less. In the event 26 of the 77 players competing returned scores of 72 or less. Regarding the players as a sample of 77 top class players and assuming the commentator's assertion about scores on a good day is correct, do these fourth-day results suggest the commentator's assertion about scores in the poor weather conditions prevailing was (i) perhaps correct, (ii) almost certainly optimistic or (iii) almost certainly pessimistic?

2.20 Rogerson (1987) gave the following annual mobility rates (percentage of population living in different houses at the end of the year than at the beginning) for people of all ages in the USA for 28 consecutive post-war years. Is there evidence of a monotonic trend?

18.8 18.7 21.0 19.8 20.1 18.6 19.9 20.5 19.4 19.8 19.9
19.4 20.0 19.1 19.4 19.6 20.1 19.3 18.3 18.8 18.3 18.4
17.9 17.1 16.6 16.6 16.1 16.8

3

Rank transformations and other tests for single samples

3.1 USING FULL INFORMATION

Given a sample of n observations from a continuous distribution, the Wilcoxon signed rank test does not use all information in the original data. The signed ranks are a unique transformation that preserves the order of the actual deviations. Also these ordered deviations are a non-unique transformation of the ranks; many different data sets give the same ranks.

We may carry out a permutation test analogous to the Wilcoxon signed rank test using the signed deviations. This is sometimes called a **raw scores** test or a **Pitman** test. We illustrate the process by a simple example. The method has a practical limitation that is removed only if appropriate computer software is available (see comment 2 on Example 3.1). Even if we overcome this difficulty the raw score test is less robust than the Wilcoxon test against even moderate departures from symmetry. Using the extra information is of limited value in that we nearly always get results similar to that from the Wilcoxon test when both are valid, i.e. when symmetry for the population distribution is assumed. The approach is of historical interest because it was developed before the Wilcoxon test in an attempt to justify normal distribution theory even when normality could not be assumed. Pioneers were Fisher (1935) and Pitman (1937a; 1938) who first developed 'raw data' permutation tests, calling them **randomization tests**.

Example 3.1

The problem. Given the observations 98, 107, 112, 93, 149, 85, 122 from a distribution assumed symmetric with unknown mean μ, test the hypothesis $H_0 : \mu=95$ against $H_1 : \mu>95$, basing the test on the deviations of observations from 95 (the raw scores).

Formulation and assumptions. Symmetry implies that under H_0 we expect sample values to have a nearly symmetric scatter above and below $\mu=95$, whereas samples with a preponderance of deviations above 95 are more likely under H_1. We discuss how to assess the evidence under *procedure*. Since H_1 is a one-sided alternative a one-tail test is appropriate.

Procedure. Under H_0 the deviations from the hypothesized mean, 95, are 3, 12, 17, −2, 54, −10, 27. Two are negative, five positive; four of the positive deviations are larger in magnitude than either negative deviation. Intuitively this makes us suspicious of H_0. We ask formally: given the magnitudes of these deviations and considering all possible positive and negative signs attached to them, which allocations fall in a critical region for which we reject H_0? Our test statistic is the sum of the negative deviations, an analogue of the Wilcoxon signed rank statistic. This is appropriate for the one-tail test above. For a two-tail test we use the lesser of the sum of positive or the sum of negative deviations.

Clearly the lowest negative sum occurs when all deviations are positive, then $S_-=0$. It is easy to see groups with next lowest sums are 2 only (sum 2), 3 only (sum 3), 2 and 3 (sum 5), 10 only (sum 10) then 12 or 2 and 10 (both with sum 12). Thus there are seven groups with sums less than or equal to 12. We observed $S_-=12$. As in the Wilcoxon signed rank test (p. 42) there are $2^7=128$ equally likely allocations of signs to the 7 deviations, whence $\Pr(S_- \leq 12)=7/128=0.055$. Similarly, we see that $\Pr(S_- \leq 10)=5/128=0.039$. This implies that for significance at the nominal 5% level we require $S_- \leq 10$ in the relevant one-tail test.

Conclusion. Since, in our sample, $S_-=12$ we do not reject H_0, but $\alpha=0.055$ brings us close to rejection at the 5% level.

Comments. 1. Modifications for a two-tail test are straightforward: see Exercises 3.1 and 3.2.

2. Even for a small sample it is tedious to list all cases in the critical region, let alone the complete distribution of S_- or S_+. Fortunately computer software does this for small or moderate samples, but fresh computations are required for each data set even for fixed sample size n.

3. The approach has intellectual appeal as it uses all relevant information in the data, but it lacks robustness against departures from symmetry.

4. Because the distribution of the test statistic depends upon the *actual* deviations from μ specified in H_0 the test is said to be a **conditional** test and the distribution of the test statistic is different for every sample despite the fact that many of these give rise to the same ranks.

Computational aspects. STATXACT (directly) and DISFREE (indirectly) both allow computation of exact tail probabilities for permutation tests using raw scores for a specifed μ for moderate sample sizes. Asymptotic tail probabilities based on (3.1) below may be used for large samples.

Using a general result given by Lehmann, (1975, pp. 330−1) we may show that **any** sum of scores statistic S for testing location parameters (e.g. the statistics S_+ and S_- used for the Wilcoxon signed rank test, with or without ties, the raw data scores and scores we propose in Section 3.2 using rank transformations) are distributed under H_0 with mean $\frac{1}{2}\Sigma|s_i|$ and variance $\frac{1}{4}\Sigma s_i^2$, where the s_i are the scores allocated and the summations are over all n scores. For large n the distribution of

$$Z=\frac{S-\frac{1}{2}\Sigma|s_i|}{\sqrt{[\frac{1}{4}\Sigma s_i^2]}} \tag{3.1}$$

approaches the standard normal distribution; this is the basis for asymptotic tests. For ranks (3.1) reduces to (2.4). Also (2.2) is a special case of (3.1) with all scores s_i being either $+1$ or -1.

In pioneer work, Fisher and Pitman showed that under a wide range of conditions the results of permutation tests using raw scores and parametric tests assuming normality (e.g. *t*-tests) were very similar and their arguments in that sense justified normal theory tests and estimation when normal theory was not justified in its own right.

Modern studies have indicated that there is generally little loss in efficiency if we replace deviations by their ranks and use the Wilcoxon test. More importantly, there may be gains in robustness by not using the full information if the assumption of symmetry is unjustified. This arises because the full data permutation test then behaves more like the *t*-test. We indicated in Section 2.2.6 that the *t*-test may be unsatisfactory in the presence of outliers. For the data sets in Example 2.12 the raw scores permutation test behaves almost identically to the *t*-test for hypothesis testing and confidence intervals. Thus, despite the feasibility of carrying out such tests with modern computer programs, the lack of robustness against departures from symmetry, plus the fact that when there is symmetry, there is usually little gain in efficiency over the Wilcoxon test, weakens the case for using **raw scores** in preference to ranks.

3.2 TRANSFORMATION OF RANKS

3.2.1. Normal scores

Transformation of continuous ordered data to ranks in essence replaces sample values by something like those for a sample from a uniform distribution over $(0, n)$ – the rank r corresponding to the $r/(n+1)$th quantile of such a distribution. Might further transformations either increase the Pitman efficiency or enable us to use asymptotic results for smaller samples without having to resort to exact permutation distributions of our test statistics? We saw in Section 2.2.5 that the distribution of the Wilcoxon signed rank statistic under H_0 rapidly approached normality when $n \geq 12$ in the no-tie situation.

Intutively, it seems that if we transform ranks to give a sample something like one from a normal distribution, then the asymptotic results may work well for smaller n. A possible transformation is suggested by our remark above that ranks correspond to quantiles of a uniform distribution. Why not transform ranks to values (often called scores) corresponding to quantile values for a normal distribution? More specifically, why not choose the standard normal distribution, for then it is easy to make the transformation using tables of the standard normal cumulative distribution function (in this chapter we abbreviate **cumulative distribution function** to **cdf**). For example, if we have the very non-normal sample 2, 3, 7, 21, 132 we replace the values by ranks 1, 2, 3, 4, 5 and then transform these ranks to 'normal' scores which are the 1/6th, 2/6th, 3/6th, 4/6th and 5/6th quantiles of the standard normal distribution. We read these from standard normal cdf tables or an appropriate computer program. The normal score corresponding to the 1/6=0.1667th quantile is the x value such that the standard normal cdf, commonly denoted by $\Phi(x)$, is $\Phi(x)=0.1667$. From tables (e.g. Neave, 1981, p.18) we find $x=-0.97$. For the 2/6=0.3333th quantile we find $x=-0.43$ and for the 3/6th (the mean or median) we find $x=0$, since, for the standard normal distribution, $\Phi(0)=0.5$. By symmetry, the 4/6th and 5/6th quantiles are $x=0.43$ and $x=0.97$. These quantile scores are often referred to as **van der Waerden scores**, having been proposed by van der Waerden (1952; 1953).

Alternatives to the above normal scores are discussed by Conover (1980, pp. 316–27) and others. These include **expected normal scores** where the ith ordered sample value is replaced by the expectation of the ith order statistic for a standard normal distribution; the ith order statistic is the ith smallest value in a sample of n observations. When observations are arranged in ascending order we often denote the ith order statistic by $x_{(i)}$. Fisher and Yates (1957, Table XX) give expected normal scores corresponding to ranks for $n \leq 50$. In practice, van der Waerden or expected normal scores usually lead to similar conclusions and we illustrate the use of the former.

A complication precludes direct application of standard normal scores in an analogue to the Wilcoxon signed rank test since the rationale of the test demands that we allocate signs to **magnitudes** of ranks but van der Waerden scores are of equal magnitude but opposite signs. We overcome this difficulty by adding a constant, k, to all scores so that they are non-negative. An appropriate choice is $k=3$. Then, for any sample of size $n \leq 700$, the amended scores will be positive. Adding 3 gives normal scores for a distribution of mean 3 and standard deviation 1. We use these modified van der Waerden scores, after allocating appropriate signs, in a test with basic similarities to the Wilcoxon signed rank test (or the raw score equivalent in Section 3.1). Programs such as STATXACT generate exact tail probabilities, but because our data are so like a sample from a normal distribution asymptotic results are reasonable in hypothesis testing even for fairly small samples. If we denote the modified scores with $k=3$

by s_i, $i=1,2,\ldots,n$, and the test statistic by S, where this is the lesser of the sums of the scores corresponding to positive or negative deviations, then if there are no ties, it is easily established that S has mean $3n/2$ and variance $\Sigma(s_i^2)/4$, whence (3.1) becomes

$$Z=\frac{S-3n/2}{\sqrt{[\Sigma(s_i^2/4)]}} \tag{3.2}$$

An equivalent test statistic is an analogue of (2.5) in Section 2.2.7, where W now refers to our modified van der Waerden scores.

In the earlier edition of this book we gave an alternative scoring procedure to avoid the problem of negative deviates. This alternative is described by Marascuilo and McSweeney (1977, pp. 342–4).

Example 3.2

The problem. Using modified van der Waerden scores, test $H_0: \mu=12$ against $H_1: \mu\neq 12$ for the data on percentage of populations aged over 60 given in Example 2.7, namely

4.9 6.0 6.9 17.6 4.5 12.3 5.7 5.3 9.6 13.5 15.7 7.7

Formulation and assumptions. We replace the signed ranks in Example 2.7 by signed scores corresponding to the $r/(n+1)$th quantiles, $n=12$, $r=1, 2, \ldots, 12$, of a normal distribution with mean 3, standard deviation 1. Our test is based on the lesser of the sums of positive or negative scores.

Procedure. The signed rank deviations (see Example 2.7) are 1, 2, −3, 4, −5, −6, 7, −8, −9, −10, −11, −12. Using tables of the standard normal cdf we easily find the $1/13=0.0769$th quantile is -1.43 since $\Pr(Z<-1.43)=0.0764$ and it suffices to express scores to 2 decimal places. Similarly, the remaining van der Waerden scores are −1.02, −0.74, −0.50, −0.27, −0.10, 0.10, 0.27, 0.50, 0.74, 1.02, 1.43. Adding 3 to each to make all scores positive and then assigning a negative sign to a score corresponding to any negative signed rank we get as the relevant scores for our test 1.57, 1.98, −2.26, 2.50, −2.73, −2.90, 3.10, −3.27, −3.50, −3.74, −4.02, −4.43. The test statistic is $S_+=1.57+1.98+2.50+3.10=9.15$. With a computer program such as that in STATXACT, version 2.0 or later, one may compute the exact permutation probability, $\Pr(S_+\leq 9.15)$ for these scores. It is 0.051. We double this probability for a two tail test, giving $\alpha=0.102$.

Tables are not readily available for this scores test so we resort to the asymptotic result (3.2) if the exact probability is not obtainable. Here (3.2) gives $Z=(9.15-18)/(\sqrt{28.9829})=-1.644$. From tables we easily verify that $\Pr(Z<-1.644)=0.050$, in close agreement with the 'exact' value 0.051.

Conclusion. The two tail probability of 0.102 (exact) and 0.100 (asymptotic) indicates non-significance at the 5% level, so we accept H_0.

Comments. 1. The result is similar to that for the Wilcoxon test in Example 2.7 although the tail probability is slightly higher. Not unexpectedly, the asymptotic and exact probabilities are in close agreement. For the Wilcoxon test the asymptotic result may work quite well when $n=12$ (Section 2.2.5). For normal scores it is a reasonable approximation for even smaller n.

2. Discontinuities in the normal scores test statistic for a given sample size are generally less marked than they are for the Wilcoxon test. In this example for our observed value, 9.150, $\Pr(S_+=9.150)=0.0002$ for normal scores, compared to $\Pr(S_+=14)=0.0049$ for the observed value 14 in the Wilcoxon test. The former probability was obtained from STATXACT output. The latter may be obtained from Table 2.4. The reason for the smaller 'normal scores' discontinuity is that the test statistic may take many more values. The Wilcoxon statistic only takes integral values in the range 0 to 78 when $n=12$, but the normal score statistic takes 1,413 distinct values (not, however, equally spaced) in the range 0 to 36. In each case the distribution is symmetric.

3. Normal score tests in this and many other situations have a Pitman efficiency of 1 when the alternative normal theory t-test is appropriate, and in any other case the Pitman efficiency is at least 1. Any euphoria so generated must be tempered by the realization that there are other tests with Pitman efficiency greater than 1 relative to normal theory tests for certain appropriate distributional assumptions, and also that an asymptotic result does not guarantee equivalent small-sample efficiency, although Pitman efficiency is usually a good guide.

3.2.2 Other scores

In certain circumstances scores other than normal scores may be appropriate. We meet one or two of these in more sophisticated problems. The need for them arises with data that are basically non-normally distributed or where there are complications due to censoring of data. An example is given in Section 5.1.8. In dealing with survival data the use of what are called Savage or exponential scores may be appropriate. These are discussed by Lehmann (1975, p. 103).

3.2.3 Confidence intervals using raw or other scores

Pending development of computer programs giving confidence intervals based on **any** scores, if a hypothesis testing program that gives tail probabilities in tests for a location parameter θ specified under H_0 is available, a crude approach is one of trial and error. For instance, for the data in Example 3.2, if we required a 95% confidence interval for the mean based on raw scores we might take as a first approximation the Wilcoxon limits or the t-test limits. In Example 2.10 we found the Wilcoxon nominal 95% confidence interval to be (5.7, 12.5). If we now use raw scores to test the hypothesis $\theta=5.7$ we find using, say, STATXACT, that the one-tail critical region is of size 0.0127. For an appropriate 95% confidence interval we require a one-tail region of size not exceeding 0.025. This suggests we should set θ in H_0 at a higher value. It is easily verified using the program in STATXACT that $\theta=6.7$ gives $\alpha=0.041$. Setting $\theta=6.2$, 6.0 and 6.1 respectively gives $\alpha=0.0273$. 0.0210, 0.0244; indicating a lower limit for the 95% confidence interval at 6.1 is appropriate. Similar trial and error methods for the upper limit starting with $\theta=12.5$ lead to $\theta=12.1$ as an upper limit. The appropriate interval is thus (6.1, 12.1). In Section 2.2.6 we noted that the normal theory t-based interval was (6.26, 12.03), in fairly close agreement. This is in line with our remarks in Section 3.1 on the tendency for raw score permutation tests to approximate closely to parametric tests based on the normal distribution (even when these may not be appropriate); therefore normal theory confidence limits may prove the most appropriate starting point for a trial and error method to determine corresponding limits based on raw scores.

3.3 MATCHING SAMPLES TO DISTRIBUTIONS

We have considered tests and estimation procedures for location – the median when we do not assume symmetry and either the mean or median if we do. We now consider the more general problem of matching data to particular distributions. Populations, whether or not they have identical locations, often differ widely in other characteristics.

We may ask whether observations are consistent with their being a sample from some specified continuous distribution. Kolmogorov (1933; 1941) devised a test for this purpose and Smirnov (1939; 1948) extended it to test whether two independent samples may reasonably be supposed to come from the same distribution. We consider here the Kolmogorov test.

Given observations $x_1, x_2, \ldots, x_n$, we ask if the values are consistent with our sample being from some **completely specified** distribution, e.g. a uniform distribution over (0, 1) or over (20, 30), or a normal distribution with mean 20 and standard deviation 2.7. Kolmogorov's test is distribution-free in that the same procedure is used for any distribution completely specified under H_0. The test requires modification to answer questions like 'Can we suppose these data are from a normal distribution, mean and variance unspecified?' We consider one such modification in Section 3.3.3.

3.3.1 Kolmogorov's test

The **continuous uniform distribution** (sometimes called the **rectangular distribution**) is a simple continuous distribution. It arises when a random variable is equally likely to take a value within any given fixed length segment within a specified finite interval. For example, suppose pieces of thread each 6 cm in length are clamped at each end and a force is applied until they break; if each thread breaks at the weakest point and this is equally likely to be anywhere along its length, then the breaking points will be uniformly distributed over the interval (0, 6), where the distance to the break is measured in centimetres from the left-hand clamp.

The probability density function (see Section A1.1) has the form

$$f(x)=0,\ x\leq 0;\quad f(x)=1/6,\ 0<x\leq 6;\quad f(x)=0,\ x>6.$$

The rectangular form of $f(x)$ gives rise to the name 'rectangular distribution'. If the thread always breaks, the probability is 1 that it breaks somewhere in the interval (0, 6). Thus the total area under the density curve must be 1; clearly only the density function above satisfies both this condition and that of equal probability of a break in segments of the same length no matter where they lie within the interval. The function is graphed in Figure 3.1 and is essentially a line running from 0 to 6 at height 1/6 above the x-axis. The lightly shaded area represents the total probability of 1 associated with the complete distribution. The heavily shaded rectangle *PQRS* between $x=3.1$ and $x=3.8$ represents the probability that the random variable X, the distance to the break, takes a value between 3.1 and 3.8. Since $PQ=3.8-3.1=0.7$ and $PS=1/6$, clearly this area is $0.7/6=0.1167$. This is also the probability of a break occurring in any segment of length 0.7 lying entirely in the interval (0, 6). It is conventional to use X, Y as names for random variables, and the corresponding lower-case letters x, y (with or without subscripts) for specific values of these variables.

The Kolmogorov test uses not the probability density function, but the cumulative distribution function (cdf), a function giving $\Pr(X\leq x)$ for all x in (0, 6). In our example the probability that the break occurs in the first two centimetres is $\Pr(X\leq 2)$. Clearly this is 1/3. The cdf is written $F(x)$ and for any x between 0 and 6, $F(x)=x/6$. It has the value 0 at $x=0$ and 1 at $x=6$. It is graphed in Figure 3.2. These notions generalize to a uniform distribution over any interval (a, b) and the cdf is $F(x)=(x-a)/(b-a)$, $a<x\leq b$, specifying a straight line rising from zero at $x=a$ to 1 at $x=b$. Clearly $F(a)=0$ and $F(b)=1$. This is illustrated in Figure 3.3.

For most distributions the cdf is less simple. For any continuous distribution it is a curve starting at zero for some x and increasing as we move from left to right until it attains the value 1. It never decreases as x increases and is said to be **monotonic non-decreasing** (or **monotonic increasing**). Figure 3.4, shows the cdf for the standard normal distribution.

In essence the Kolmogorov test compares a population cdf with a related curve $S(x)$ based on sample values and called the **sample** (or **empirical**) **cumulative distribution function.** For a sample of n observations

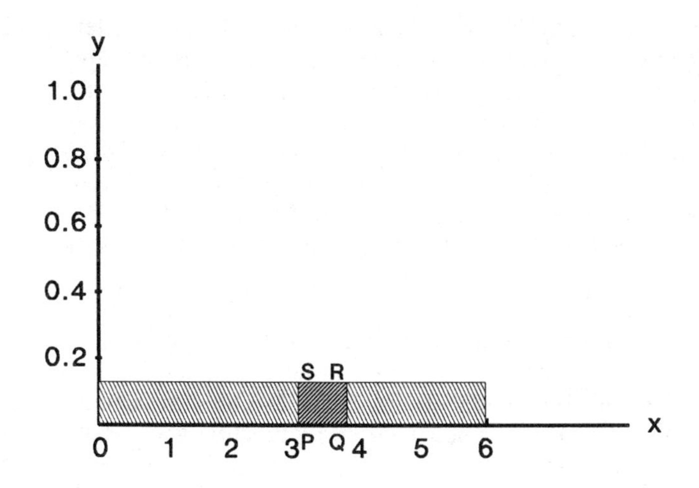

Figure 3.1 Probability density function for a continuous uniform distribution over (0, 6).

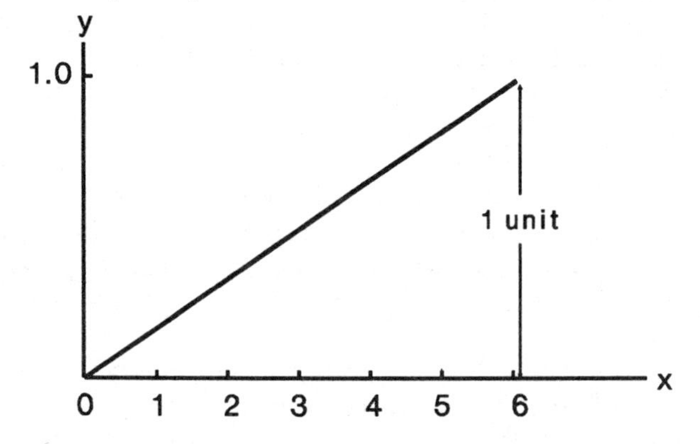

Figure 3.2 Cumulative distribution function for a uniform distribution over (0, 6).

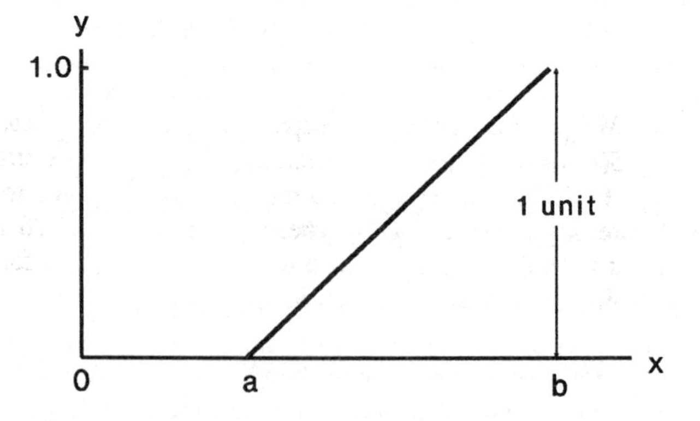

Figure 3.3 Cumultive distribution function for a uniform distribution over (a, b).

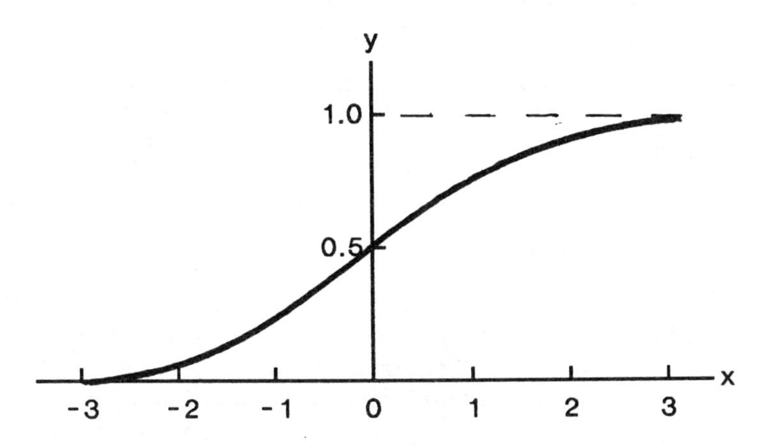

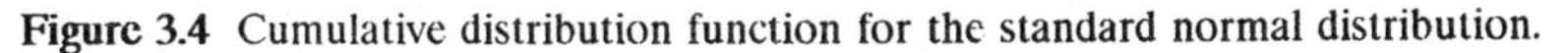
Figure 3.4 Cumulative distribution function for the standard normal distribution.

$$S(x) = \frac{\text{number of sample values less than or equal to } x}{n} \tag{3.3}$$

Example 3.3

The problem. Below are the distances from one end at which each of 20 threads 6 cm long break when subjected to strain. Form and graph $S(x)$. For convenience the distances are ordered.

0.6	0.8	1.1	1.2	1.4	1.7	1.8	1.9	2.2	2.4
2.5	2.9	3.1	3.4	3.4	3.9	4.4	4.9	5.2	5.9

Formulation and assumptions. From (3.3) it is clear that $S(x)$ increases by 1/20 at each unique x value corresponding to a break, or by $r/20$ if r break distances coincide.

Procedure. When $x=0$, $S(x)=0$; it keeps this value until $x=0.6$, the first break. Then $S(0.6)=1/20$, and $S(x)$ maintains the value 1/20 until $x=0.8$ when it jumps to 2/20, a value retained until $x=1.1$. It jumps in steps of 1/20 at each break value until $x=3.4$, where it increases by 2/20 since two breaks occur at $x=3.4$. $S(x)$ is referred to as a step function for obvious reasons. Its value at each step is given in Table 3.1.

Conclusion. For these data $S(x)$ takes the form in Figure 3.5.

Table 3.1 Values of $S(x)$ at step points

x	0.6	0.8	1.1	1.2	1.4	1.7	1.8	1.9	2.2	2.4	2.5	2.9
$S(x)$	1/20	2/20	3/20	4/20	5/20	6/20	7/20	8/20	9/20	10/20	11/20	12/20
x	3.1	3.4*	3.9	4.4	4.9	5.2	5.9					
$S(x)$	13/20	15/20	16/20	17/20	18/20	19/20	1					

*Repeated sample value

Comments. $S(x)$ is a sample estimator of the population cdf $F(x)$. If a sample comes from a specified distribution, the step function $S(x)$ should not depart too markedly from the population cdf $F(x)$. In this problem if breaks are uniformly distributed over (0, 6) one expects, for example, about half of these to be in the interval (0, 3), so that $S(3)$ should not have a value very different from 0.5 and so on. In Figure 3.5 we show also the cdf for a uniform distribution over (0, 6) reproduced from Figure 3.2. Intuitively one feels that this is not a good fit.

Recognizing that $S(x)$ should not depart violently from $F(x)$ for a sample from a distribution with this cdf is fundamental to Kolmogorov's test. The test statistic is the maximum difference in magnitude between $F(x)$ and $S(x)$.

Example 3.4

The problem. Given the 20 breaking points and the corresponding $S(x)$ values in Table 3.1, is it reasonable to suppose breaking points are uniformly distributed over (0, 6)?

Formulation and assumptions. We obtain the maximum difference in magnitude between $F(x)$ and $S(x)$; this is compared with a tabulated value to see if significance is indicated.

Procedure. Table 3.2 shows values of $F(x)$ and $S(x)$ at each break point given in Table 3.1. It also gives at each of these points the difference $F(x_i)-S(x_i)$. In view of the stepwise form of $S(x)$ we cannot guarantee that the maximum difference will be included in this set. Inspection of Figure 3.5 makes it clear that a greater difference may occur immediately before such a step. There $S(x)$ has the value attained at the previous step, so a maximum may occur among the $F(x_i)- S(x_{i-1})$. These differences are also recorded in Table 3.2.

Figure 3.5 indicates that for much of its course $S(x)$ lies above $F(x)$. The entry of greatest magnitude in the last two columns of Table 3.2 is -0.18 when $x_i=3.4$. From Table III, we see that when $n=20$ the largest difference

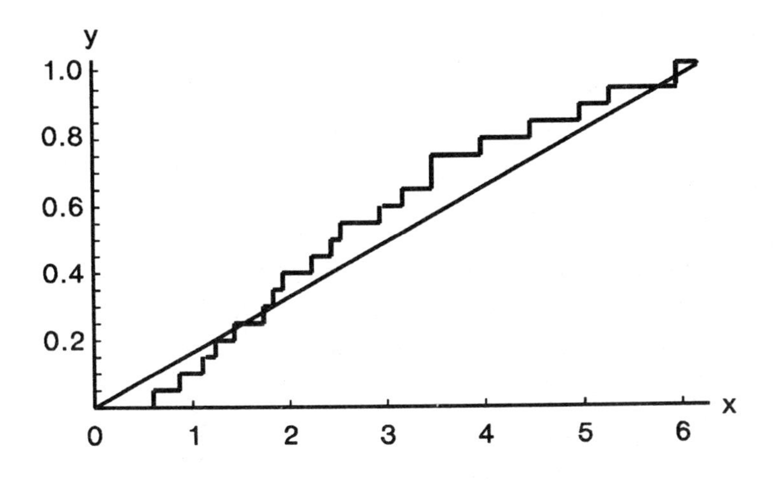

Figure 3.5 Sample cumulative distribution function for thread breaks together with the cdf for a uniform distribution over (0,6).

(whether positive or negative) must be at least 0.294 for significance at the 5% level in a two-tail test.

Conclusion. We have insufficient evidence to reject H_0 : *breaks are uniformly distributed over (0, 6).*

Comments. A two-tail test is appropriate if the alternative hypothesis is that the observations may come from any other unspecified distribution. Practical situations may arise where the only alternative is that the cdf must lie either wholly above or wholly below that specified in the null hypotheis. For example, in the thread problem we may know that the test equipment places strains on the thread that, if not uniformly distributed, will be greater at the left and decrease as we move to the right. This might increase the tendency for breaks to occur close to the left end of the string. Then the cdf would everywhere lie above the straight line representing $F(x)$ for the uniform distribution. Then a maximum difference with $S(x)$ greater than $F(x)$ for the uniform distribution would, if significant, favour the acceptable alternative. A one-tail test is then appropriate and we reject the null hypothesis if the largest difference is significant and of appropriate sign.

If the alternative hypothesis were that the population cumulative distribution function were always at or below that for the uniform distribution, a positive difference would be appropriate in Table 3.2. Table III indicates that when $n=20$ the magnitude must then exceed 0.265 for significance at the 5% level in a one-tail test.

Table 3.2 Comparison of $F(x)$ and $S(x)$ for thread breaks

x_i	$F(x_i)$	$S(x_i)$	$F(x_i)-S(x_i)$	$F(x_i)-S(x_{i-1})$
0.6	0.10	0.05	0.05	0.10
0.8	0.13	0.10	0.03	0.08
1.1	0.18	0.15	0.03	0.08
1.2	0.20	0.20	0.00	0.05
1.4	0.23	0.25	−0.02	0.03
1.7	0.28	0.30	−0.02	0.03
1.8	0.30	0.35	−0.05	0.00
1.9	0.32	0.40	−0.08	−0.03
2.2	0.37	0.45	−0.08	−0.03
2.4	0.40	0.50	−0.10	−0.05
2.5	0.42	0.55	−0.13	−0.08
2.9	0.48	0.60	−0.12	−0.07
3.1	0.52	0.65	−0.13	−0.08
3.4	0.57	0.75	−0.18	−0.08
3.9	0.65	0.80	−0.15	−0.10
4.4	0.73	0.85	−0.12	−0.07
4.9	0.82	0.90	−0.08	−0.03
5.2	0.87	0.95	−0.08	−0.03
5.9	0.98	1.00	−0.02	0.03

Computational aspects. Some general statistical packages, e.g. STATGRAPHICS, include programs for the Kolmogorov test. These rely on asymptotic tests for significance which are not very satisfactory for sample sizes less than $n=20$.

The Kolmogorov test looks to waste information by using only the difference of greatest magnitude. Tests that take account of all differences exist but tend to have few, if any, advantages. This is not as contrary as intuition suggests, for the value of $S(x)$ at any stage depends on how many observations are less than the current x and therefore at each stage we make the comparison on the basis of accumulated evidence.

The theory behind the test is beyond the scope of this book. For a discrete rather than a continuous distribution the test tends to give too few significant results and in this sense is conservative. We consider preferred tests for discrete distributions in Section 9.4.

3.3.2 Comparison of distributions: confidence regions

If, in Figure 3.5 we drew two further step functions everywhere at distances 0.294 units above or below that representing $S(x)$, we would, since 0.294 is the critical distance for significance at the 5% level, have a 95% confidence region for $F(x)$ in the sense that any $F(x)$ lying entirely between these two new step functions would be an accepted $F(x)$ when testing at a 5% significance level in a two-tail test. This interval is not very useful in practice. A more common practical situation is one where it is reasonable to assume that our data may have come from one of a (usually small) number of completely specified distributions. We might then use a Kolmogorov test for each. In each test the relevant $S(x_i)$ will be the same for a given data set. As a result of these tests we may reject some of the hypotheses but still find more than one acceptable. It is then of some interest to get at least an overall eye comparison of how well the match of $S(x)$ is to each acceptable $F(x)$. Ross (1990, p. 85) gives a useful procedure for such eye comparisions. Basically it consists of plots of the maximum of the two deviations given in the last two columns of each row in a table like Table 3.2 against the $S(x_i)$ for that row. Plots may easily be done for several different hypothesized population distributions on the same graph.

Example 3.5

The problem. Compare the fit of the thread break data in Example 3.3 to (i) a uniform distribution over (0,6) and (ii) the distribution with cumulative distribution function over (0,6) given by

$$F(x)=x/5,\ 0<x\leq 3;\quad F(x)=0.2+4x/30,\ 3<x\leq 6. \tag{3.4}$$

Formulation and assumptions. All information needed for the Kolmogorov test procedure for the distribution (i) were obtained in Example 3.4. We require similar information for the alternative distribution in (ii). We make a graphical comparision using the format suggested by Ross.

Procedure. For (i) the values of $S(x)$ of interest are those in column 3 of Table 3.2. In each case the corresponding deviation is the value of greatest magnitude in columns 4 and 5. For (ii) the values of $S(x)$ are again those in Table 3.2. However, fresh values of $F(x)$ for each x_i must be calculated using (3.4). For example, when $x_i=1.2$, (3.4) gives $F(1.2)=1.2/5=0.24$. Using this and the values of $S(1.1)$ and $S(1.2)$ in Table 3.2 it follows that $F(1.2)-S(1.2)=0.24-0.20=0.04$ and $F(1.2)-S(1.1)=0.24-0.15=0.09$. Thus corresponding to $x=1.2$, the maximum deviation is 0.09 just before the step in $S(x)$ at $x=1.2$. In Table 3.3 we set out for each x_i the corresponding $S(x_i)$ and maximum deviations for the population distributions specified in (i) and (ii). These deviations are plotted against S(x) in Figure 3.6. Although it is not essential we have joined consecutive points in each case by straight line segments. The parallel lines at $y=\pm 0.294$ represents the critical value for significance at the 5% level. We reject H_0 at the 5% significance level only if we observe a deviation outside these lines.

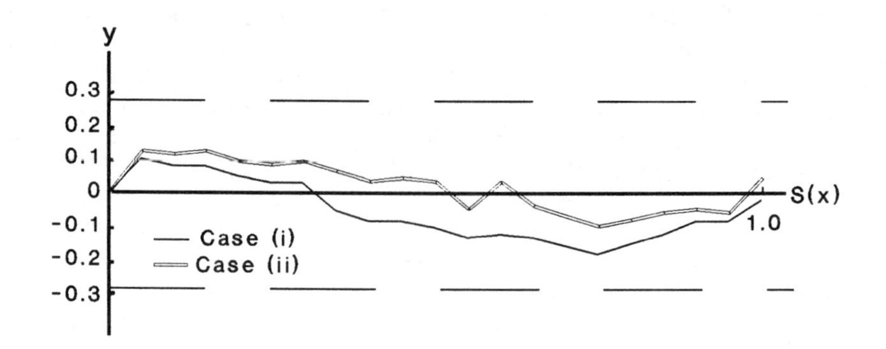

Figure 3.6 Plot of deviations of $F(x)$ from $S(x)$ against $S(x)$ for the two $F(x)$ functions specified in Example 3.5.

Table 3.3 Values of x_i, $S(x_i)$, $d_i = \max[F(x_i) - S(x_i), F(x_i) - S(x_{i-1})]$ for population distributions specified in (i) and (ii) of Example 3.5

x_i	$S(x_i)$	d_i(i)	d_i(ii)
0.6	0.05	0.10	0.12
0.8	0.10	0.08	0.11
1.1	0.15	0.08	0.12
1.2	0.20	0.05	0.09
1.4	0.25	0.03	0.08
1.7	0.30	0.03	0.09
1.8	0.35	−0.05	0.06
1.9	0.40	−0.08	0.03
2.2	0.45	−0.08	0.04
2.4	0.50	−0.10	0.03
2.5	0.55	−0.13	−0.05
2.9	0.60	−0.12	0.03
3.1	0.65	−0.13	−0.04
3.4	0.75	−0.18	−0.10
3.9	0.80	−0.15	−0.08
4.4	0.85	−0.12	−0.06
4.9	0.90	−0.08	−0.05
5.2	0.95	−0.08	−0.06
5.9	1.00	0.0	0.04

Conclusion. We do not reject either of the population hypotheses specified in (i) and (ii).

Comment. The deviations, except for the early observations, are rather smaller for (ii), suggesting that this may be a slightly better fit. The hypothesis (ii) implies a uniform but higher probability of breakages between 0 and 3 relative to a uniform but lower rate between 3 and 6 cm. If other hypotheses are of interest relevant deviations for these may be entered in Figure 3.6.

3.3.3 A test for normality

The Kolmogorov test is applicable for any completely specified continuous distribution, including the normal with pre-specified mean and standard deviation. If we have to estimate these parameters from the data the same test statistic, the maximum deviation of sample cdf from the population cdf, may be used, but the critical values in Table III no longer apply.

For this important practical situation a test proposed by Lilliefors (1967) is useful. We assume observations are a random sample from some unspecified continuous distribution, and test whether it is reasonable to suppose this is a member of the normal family. In this sense our test is not distribution-free.

The basic idea can be extended to test compatibility with other families of distributions such as a gamma distribution with unspecified parameters, but separate tables of critical values are required for each family of distributions. Conover (1980, Section 6.2) describes some difficulties in obtaining critical values and also discusses the Shapiro–Wilk test for normality which again requires special tables. This test has good power against a range of alternatives but the rationale is less easy to describe by intuitive arguments. Bowman and Shenton (1975) describe tests for normality based on estimates of skewness and kurtosis. This approach has intuitive appeal and one of their tests is shown by Jarque and Bera (1987) to be powerful against a wide range of alternatives. However, departures from normality do not always reflect themselves in marked changes in skewness and kurtosis, especially when there is bimodality.

While tests for normality are not nonparametric, they may be important in deciding whether or not to apply nonparametric methods in a specific problem. We describe only Lilliefors' test and demonstrate its similarity to the Kolmogorov test. Approximate critical values for Lilliefors' test for normality for rejection at the 5% and 1% level are given in Table IV.

Example 3.6

The problem. In the Badenscallie burial ground in Wester Ross, Scotland, the ages at death of males were noted on all tombstones for four clans in the district (see Section A7). From all 117 ages recorded a random sample of 30 was taken and the ages for that sample were, in order:

11 13 14 22 29 30 41 41 52 55 56 59 65 65 66
74 74 75 77 81 82 82 82 82 83 85 85 87 87 88

Is it reasonable to suppose the death ages are normally distributed?

Formulation and assumptions. In Lilliefors' test the Kolmogorov statistic is used to compare $\Phi(z)$ and $S(z)$, the standardized cumulative distribution function based on the transformation

$$z_i = \frac{x_i - m}{s}$$

where m is the sample mean and s the usual estimate of population standard deviation, i.e.

$$s^2 = [\Sigma(x_i^2) - (\Sigma x_i)^2/n]/(n-1)$$

Procedure. We find $m=61.43$, $s=25.04$. Successive z_i are calculated: e.g. for $x_1=11$, we find $z_1=(11-61.43)/25.04=-2.014$. We treat this as a sample value from a standard normal distribution and tables (e.g. Neave, 1981, pp. 18–19) show $\Phi(-2.014)=0.022$. Table 3.4 is set up in analogous manner to Table 3.2. Here $\Phi(z)$, $S(z)$ represent the standard normal and sample cdfs respectively. Although not essential, we have included the x values, an asterisk against a value implying it occurs more than once. The largest difference is 0.192, occuring in the final column when $x=74$. From Table IV we find the 1% critical value when $n=30$ is 0.187.

Conclusion. The maximum difference between the step function and the normal cdf is 0.192 (exceeding the critical 0.187), so we reject at a nominal 1% level the hypothesis that the sample is from a normal distribution.

Comments. A glance at the sample data suggests that a few males died young and that the distribution is skew with a large number of deaths occurring after age 80. Figure 3.7 shows these data on a histogram with class interval 10. If you are familar with histograms for samples from a normal distribution you will not be surprised that we reject the hypothesis of normality. Indeed 'life-span' data, whether for man, animal or the time a machine functions without a breakdown, will often characteristically have a non-normal distribution.

Computational aspects. Neave and Worthington (1988) give a BASIC program listing for calculating the statistic for Lilliefors' test.

3.4 PRACTICAL IMPLICATIONS OF EFFICIENCY

It is intuitively reasonable that the more **relevant** information we use the better a test or estimation procedure should be in terms of power and

Table 3.4 The Lilliefors normality test. Badenscallie ages at death

x	z	$\Phi(z)$	$S(z)$	$\Phi(z_i)-S(z_i)$	$\Phi(z_i)-S(z_{i-1})$
11	−2.014	0.022	0.033	−0.011	0.022
13	−1.934	0.026	0.067	−0.044	−0.007
14	−1.894	0.029	0.100	−0.071	−0.038
22	−1.575	0.058	0.133	−0.075	−0.042
29	−1.295	0.098	0.167	−0.069	−0.035
30	−1.255	0.105	0.200	−0.095	−0.062
41*	−0.816	0.207	0.267	−0.060	−0.007
52	−0.377	0.353	0.300	0.053	0.086
55	−0.257	0.399	0.333	0.066	0.099
56	−0.217	0.414	0.367	0.047	0.081
59	−0.097	0.461	0.400	0.061	0.094
65*	0.142	0.556	0.467	0.089	0.156
66	0.183	0.572	0.500	0.072	0.105
74*	0.502	0.692	0.567	0.125	0.192
75	0.542	0.706	0.600	0.106	0.139
77	0.622	0.733	0.633	0.100	0.133
81	0.781	0.782	0.667	0.115	0.149
82*	0.821	0.794	0.800	−0.006	0.127
83	0.861	0.805	0.833	−0.028	−0.005
85*	0.942	0.827	0.900	−0.073	−0.006
87*	1.021	0.846	0.967	−0.121	−0.054
88	1.061	0.856	1.000	−0.144	−0.111

*repeated value

efficiency. It is now well established that in cases where the Pitman efficiency is not too low in the worst possible circumstances, yet high in more favourable ones, then, if there is any doubt about distributional assumptions, we seldom lose much efficiency, if any, by using the most appropriate nonparametric test. For the Wilcoxon signed rank test compared with the t-test the Pitman efficiency is $3/\pi=0.955$, when the t-test is optimal. This rises to 1 for a sample from a uniform distribution and to 1.5 if the data are from the sharply peaked and long-tailed double exponential distribution. More importantly, under no circumstances is the Pitman efficiency of the Wilcoxon test relative to the t-test less than 0.864.

We now compare confidence intervals obtained for a population median by several methods using a large body of data.

Example 3.7

The problem. For one particular clan – it would be invidious to give it the real name, so we shall call it the McAlpha clan (but the data are real) – ages at death were recorded for all males of that clan buried in the Badenscallie burial ground. For all 59 burials the ages (arranged for convenience in ascending order) were:

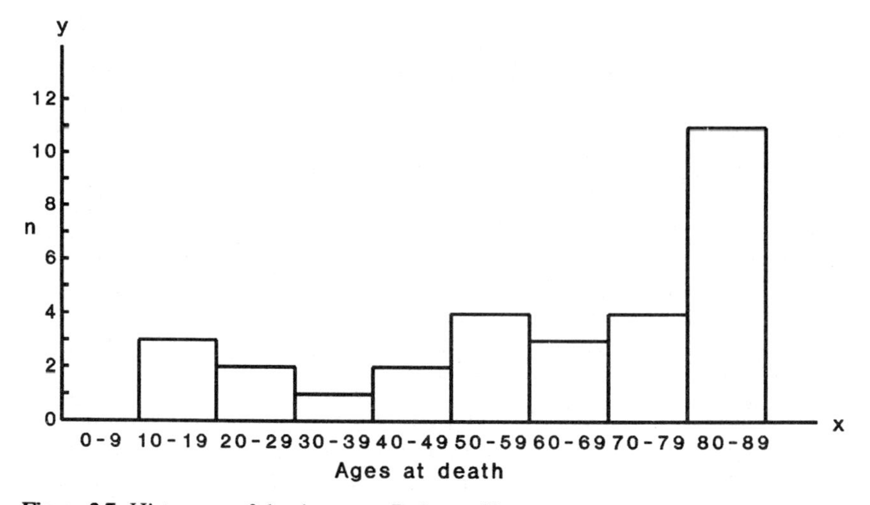

Figure 3.7 Histogram of death ages at Badenscallie.

0 0 1 2 3 9 14 22 23 29 33 41 41 42 44 52 56 57 58 58 60
62 63 64 65 69 72 72 73 74 74 75 75 75 77 77 78 78 79 79 80 81
81 81 81 82 82 83 84 84 85 86 87 87 88 90 92 93 95

Obtain and compare nominal 95% confidence limits for the population median using (i) the sign test procedure, (ii), the Wilcoxon signed rank procedure, and (iii) normal distribution theory.

Formulation and assumptions. We apply the standard methods already given, using large sample approximations as appropriate. In the comments below we consider the validity of assumptions for each.

Procedure. Using the normal approximation to the binomial distribution with $n=59, p=\frac{1}{2}$ and the method for sign test confidence intervals given in Section 2.1.2 a 95% confidence interval is (62, 78). The Wilcoxon procedure in Section 2.2.4 is time consuming without a computer as one must eliminate the 624 largest and 624 smallest Walsh averages, however the pattern of the Walsh averages tableau described in comment 2 on Example 2.10 means not all these need be evaluated. The resulting confidence interval is (56.75, 73.5). The normal theory confidence interval is (54.7, 68.9).

Conclusion. The intervals are somewhat differently positioned; that based on normal assumptions is a little narrower than the other two, which are of similar width. Care is needed in interpreting these contradictory results.

Comments. 1. The sign-test interval is essentially an interval for a median; the other two assume symmetry for validity and apply to a mean or median. What about a normality assumption? Lilliefors' test, tedious without a computer program for a sample this size, indicates an assumption of normality is not justified. The test statistic for these data is 0.203, compared with a critical value 0.134 for significance at the 1% level. There is strong evidence for rejecting the normality assumption. This is clearly evident also from a histogram for the data with class interval 10 shown in Figure 3.8. The histogram indicates skewness and bimodality. There is a small peak of infant deaths, few deaths in early adulthood, then a build-up to many deaths in old age. This skewness invalidates the symmetry assumption for the Wilcoxon signed rank test. The sign test tells us something valid about the median of a skew distribution. The fact that the other tests push the interval lower and bring it nearer to the sample mean than the median (the mean is 61.8, the median 74) indicates the sort of compromise our tests are being forced to make to accommodate an incompatibility between estimates and assumptions. The normal theory confidence interval tells us something about the mean; the confidence interval is centred at the sample mean, but it is not a relevant interval because of the failure of the normality assumption. The confidence level may not be near 95% and certainly has no relevance to the median.

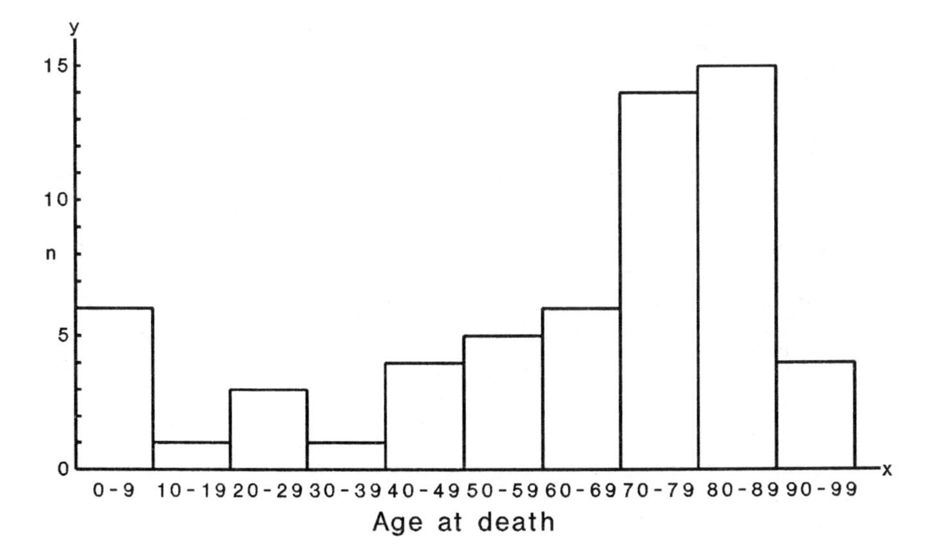

Figure 3.8 Histogram of death ages for clan McAlpha.

2. What is the relevance to reality of any estimation procedure in a case like this? From what population is this a random sample? It is clearly not a random sample of all males buried in the Badenscallie burial ground, for clans and families have differing lifespan characteristics. Perhaps the pattern of death ages is a reasonable approximation to that for McAlphas in Wester Ross, Sutherland and the Western Isles (where the clan McAlpha is well established). The main use of these data is for comparison with patterns for other clans, something we consider in Section 5.1.7.

3. Difficulties like those described in this example arise with data collection generally and are not confined to nonparametric tests. Analyses of the data with various assumptions draw our attention to such problems.

Computational aspects. Modern software makes it relatively easy to apply several tests to the same data. Indeed STATXACT allows one to do not only Wilcoxon and normal scores test, but indeed to base tests on any scores one chooses. Care must be taken not to abuse this freedom; a scoring system should be chosen on logical grounds. If different choices are logical under differing assumptions and these lead to inconsistent results (as in this example) further analysis (e.g. applying Lilliefors' test) or more simply drawing a histogram as in Figure 3.8, or just looking at the data, may suggest the most appropriate choice. In this example clearly a sign test type confidence interval for the median is appropriate.

3.5 MODIFIED ASSUMPTIONS

We have usually assumed our n observations are randomly chosen from one population. Appropriate location tests in this and the previous chapter may be applied to n samples each of one observation from n different populations each having the same unknown mean or median, but differing in other distributional characteristics (see also Section 2.2.6). This situation does not often arise in practice but it may occur with biochemical or mechanical measurements: e.g. daily oil consumption by different machines may have under certain conditions symmetric distributions with a common mean, but variation about this mean may differ from machine to machine. A question that may then arise is whether this mean consumption alters if we change the operating temperature. Testing this would be appropriate if there were physical reasons for believing temperature changes would alter the mean by the same amount for all machines.

3.6 A RUNS TEST FOR RANDOMNESS

Much statistical inference assumes a sample of independent (random) observations. Methods for obtaining a random sample – especially in simulation studies – often depend upon sophisticated mechanisms that purport to be equivalent to repeatedly tossing a coin or to repeatedly selecting one of the 10 digits between 0 and 9, each having a probability 0.1 of selection. One characteristic of the coin tossing situation is that in the long run there should be approximately equal numbers of heads and tails and a characteristic of random digit selection is that in a long string of such digits each digit should occur approximately equally often. In the coin tossing situation a sign test would be appropriate to see whether this frequency requirement is being met. We give in Section 9.4 a method applicable to testing whether there is evidence that digits are not occuring with equal probability. However, in an ordered sequence randomness implies more than compliance with frequency criteria. For example, if the outcomes, in order, of a computer process that purports to simulate 20 tosses of a coin were

H H H H H T T T T T T T T T T H H H H H (3.5)

we would suspect the process did not achieve its aim. We might be equally surprised if the ordered outcomes were

H T H T H T H T H T H T H T H T H T H T (3.6)

but reasonably happy if they were

H H T H T T T H T H H T H T H H H T T H (3.7)

A characteristic that reflects our reservations about (3.5) and (3.6) is the **number of runs**, where a run is a sequence of one or more heads or tails. In (3.5) there are three runs – runs of 5 heads, then 10 tails, then 5 heads. In (3.6) there are 20 runs, each consisting of a single head or a single tail. Intuitively we feel that (3.5) and (3.6) have respectively too few and too many runs for a truly random sequence. The sequence (3.7) has an intermediate number of runs, namely 13. Both numbers of runs and lengths of runs are relevant to tests for randomness. A detailed treatment of runs tests is given by Bradley (1968, Chapters 11, 12). We consider here only a simple test based on the number of runs, r, in a sequence of N ordered observations of which m are of one kind (e.g. H) and $n=N-m$ are of another kind (e.g. T). We reject the hypothesis that the outcomes are independent or random if we observe too few or too many runs. Computation of the probability of observing any given number of runs under the hypothesis of randomness is a subtle application of combinatorial mathematics and we content ourselves with quoting the outcome. The random variable R specifies the number of runs. We consider separately the cases r odd and r even. If r is odd we write $r=2s+1$ and

$$\Pr(R=2s+1)=({}^{m-1}C_{s-1}.{}^{n-1}C_s + {}^{m-1}C_s.{}^{n-1}C_{s-1})/({}^NC_m)$$

If r is even we write $r=2s$ and

$$\Pr(R=2s)=2\times({}^{m-1}C_{s-1}.{}^{n-1}C_{s-1})/({}^NC_m)$$

The test for randomness is based on the relevant tail probabilities associated with small and large numbers of runs. These have been tabulated for $n, m \leq 20$ by Swed and Eisenhart (1943) and tables based on theirs are given by Siegel and Castellan (1988, Table G), Daniel (1990, Tables A5 and A6).

For the sequences in (3.5) and (3.6) we have $N=20$ and $m=n=10$ and in each case tables indicate we reject at a nominal 5% significance level the hypothesis of randomness if $r \leq 6$ or $r \geq 16$. We observed $r=3$ and $r=20$ for these sequences. For the sequence in (3.7) we have $N=20$, $m=11$ and $n=9$ and $r=13$. We accept the hypothesis of randomness, since tables show the critical region for rejection in this case is again defined by $r \leq 6$ or $r \geq 16$.

The above test is not so restricted as might at first appear. For example, in sequences of supposedly random digits between 0 and 9 we may count the numbers of runs above and below the median value of 4.5 and apply the above test. It might also be applied in this case to runs of odd and even digits.

For large samples an asymptotic test may be based on the standard normal distribution using the fact that R has mean $\mathrm{E}(R)=1+2mn/(m+n)$ and $\mathrm{Var}(R)=[2mn(2mn-m-n)]/[(m+n)^2(m+n-1)]$.

3.7 FIELDS OF APPLICATION

Tests for location given in this chapter are relevant to many of the examples given in Section 2.4. For some of these examples more general tests about distributions of the Kolmogorov type, or the Lilliefors test for normality, may also be relevant. We give examples where tests about distributions may be relevant.

Biology

Heart weights are observed for a number of rats used in a drug-testing experiment. The Kolmogorov test could be used to see whether weights are consistent with a normal distribution of mean 11 g and standard deviation 3 g, where these are established values for a large batch of untreated rats from the same source. This approach is appropriate if it were uncertain how the drug might affect heart weight, especially if it were felt that it might affect characteristics such as dispersion and symmetry.

Forestry

The volume of usable timber per tree is obtained for 50 randomly selected trees from a mature forest. If we want to know if it is reasonable to assume volumes are normally distributed with unspecified mean and variance a Lilliefors test or another of those mentioned in Section 3.3.3 would be appropriate.

Testing of economic models

Forecasting models are often used to predict patterns of factors of economic importance, e.g. the distribution of numbers of days off work due to illness in a certain industry. We might compare the observed distribution with that predicted by the model using a Kolmogorov test.

Time-dependent responses

Sometimes tests or observations on different subjects have to be performed one at a time and those tested or observed may have the opportunity of discussing the test with other subjects due to undergo the test later. This might influence the test performance of later candidates. For example, if the test is one of a skill, individuals tested earlier may pass hints to those due for later testing that improve their performance. If we note at the end of the test whether the scores for candidates (taken in the order in which they were tested) are below the median (−) or above the median (+) a runs test may detect any such trend from independent responses (see Exercise 3.10).

3.8 SUMMARY

Raw score permutation tests. These use all information in the data (Section 3.1). Hypothesis testing is analogous to that for Wilcoxon signed ranks with deviations from the mean specified in H_0 replacing ranks. Appropriate computer programs are needed in practice (Example 3.1) as the test statistic distribution is data-dependent. Lack of robustness against departures from symmetry offsets the attraction of using full information.

Normal scores. Scores where a rank r is replaced by the $r/(n+1)$th quantile of a standard normal distribution are called van der Waerden scores. For an analogue to the signed rank test these scores should all be increased by 3 to eliminate negative scores. An asymptotic test statistic when this procedure is used is given by (3.2) on p. 66, and is reasonable for all but very small samples. Normal scores have a Pitman efficiency that is never less than 1 (p.67) but for certain specific distributions there may be tests with higher Pitman efficiency.

Distribution tests. The Kolmogorov test (Section 3.3.1) tests whether data are an acceptable fit to a completely specified continuous distribution. Often several distributions may give an adequate fit. In this case graphical methods (Section 3.3.2) are useful to get an eye comparison of the relative goodness of fit. Lilliefors' test for 'normality' with unspecified mean and variance (Section 3.3.3) uses the Kolmogorov statistic, but separate tables are needed to test for significance.

Runs test. The simplest runs test applies to numbers of runs for dichotomous outcomes (e.g. H or T). In a sequence of N such ordered outcomes the hypothesis of randomness or independence is rejected if there are too few or too many runs.

EXERCISES

3.1 Using the data in Example 3.1, carry out a test of $H_0 : \mu=94$ against $H_1 : \mu \neq 94$, using a randomization test based on deviations from 94.

3.2 Using the data in Example 3.1, carry out a test of $H_0 : \mu=92$ against $H_1 : \mu \neq 92$, using a randomization test based on deviations from 92.

3.3 Are the insurance data in Exercise 2.8 likely to have come from a normal distribution? Test using Lilliefors' test.

3.4 Test whether the data in Exercise 2.8 may have come from a uniform distribution over the interval (250, 6250).

3.5 Assuming symmetry, carry out a test relevant to the situation and data in Exercise 2.12 using a normal scores procedure. (Note: appropriate van der Waerden scores for tied ranks are determined by using mid-rank values of r in the expression $r/(n+1)$ for the relevant quantile).

3.6 For the data in Exercise 2.10 use a van der Waerden normal scores procedure to test the hypothesis $H_0 : \mu=240$ against $H_0 : \mu \neq 240$.

3.7 Use (i) the Wilcoxon signed rank test and (ii) modified van der Waerden normal scores, to test the hypothesis that the median length of 12-year-old turbots is 73.5, using the data in Exercise 2.16.

3.8 The negative exponential distribution with mean 20 has the cumulative distribution function $F(x)=1-e^{-x/20}$, $0 \leq x \leq \infty$. Use a Kolmogorov test to determine if it is reasonable to assume the excess parking times in Exercise 2.15 are a sample from this distribution.

3.9 In Section A7 one set of ages is for 21 members of the McGamma clan. Perform an appropriate test to determine whether it is reasonable to assume age at death is normally distributed for that clan.

3.10 A psychologist is testing 16 applicants for a job one at a time. Each has to perform a series of tests and the psychologist awards an overall point score to each. As each applicant may discuss the tests with later applicants before the latter are tested it is suggested that applicants tested later may have an unfair advantage. Do the test scores below support this assertion?

62 69 55 71 64 68 72 75 49 74 81 83 77 79 89 42

Use an appropriate runs test. Do you consider the Cox–Stuart test (Section 2.1.6) may also be appropriate? If not, why not?

4

Methods for paired samples

4.1 COMPARISONS IN PAIRS

Single-sample problems illustrate basic ideas, but have limited application. Practical problems usually involve two or more samples which may or may not be independent. This chapter considers paired samples, a dependent-sample situation where many problems reduce to a single-sample equivalent.

4.1.1 Why pair?

Responses to stimuli may vary greatly between individuals, so to compare two stimuli there is a case for applying both, if practicable, to each of, say, n individuals. If this is not possible we might use n pairs of individuals so chosen that the members of each pair are as like as possible; in each pair one individual receives the first stimulus and the other the second. In the respective cases we look at response differences within individuals or between members of each pair.

Care is needed if subjecting the same individual to two stimuli, since one stimulus may affect an individual's response to the other. For example, when comparing two drugs for alleviation of pain, if the second is administered before effects of the first have completely worn off the response is unlikely to be the same as it would be if the second were given quite independently. Response to either drug may be affected by the prior or subsequent administration of the other (interaction is the technical statistical term). In an extreme case administration of one drug after the other might make a patient's condition worse, whereas if either were given separately the effect may be beneficial. Such interactions are one reason patients prescribed certain drugs are warned not to consume alcohol.

If the effect of an analgesic drug were known to last only a few hours and the drug to be completely eliminated from the body within 24 hours, it might be reasonable to test each drug on the same patients at intervals 3 days apart and note which gave the greater pain relief. The measure of relief might be a physical measurement, but for pain it is often only patient preference (e.g. first drug gave most relief, second gave most relief, both equally good or both ineffective). One may score '*drug A gave more relief*' as a 'plus', '*drug B gave more relief*' as a 'minus', '*no difference*' as a zero.

Individual patient scores – plus, minus, or zero – provide the basis of a sign test of H_0 : *drugs are equally effective* against an appropriate one- or two-tail alternative.

To avoid patient or physician prejudices influencing responses such clinical trials are usually carried out using the **double blind** procedure. Neither the patient nor the person administering the drug knows the order of presentation (which should be random). Drugs are identified by codes available only to research workers or administrators who have no direct contact with the patient.

4.1.2 Some further examples

Geffen, Bradshaw and Nettleton (1973) wanted to know whether certain numbers presented in random order to individuals were perceived more rapidly in the right (RVF) or left visual fields (LVF), or whether there was no consistent difference, it being a matter of chance whether an individual responded more quickly in one field or the other. For each of 12 subjects the mean response times to digital information in the LVF and RVF were measured. Response times in either field varied much more between individuals than they did between fields for any one individual. The data and the difference LVF−RVF for each individual are given in Table 4.1.

Table 4.1 Mean response time (ms) to digital information

Subject	1	2	3	4	5	6	7	8	9	10	11	12
LVF (1)	564	521	495	564	560	481	545	478	580	484	539	467
RVF (2)	557	505	465	562	544	448	531	458	560	485	520	445
(1)−(2)	7	16	30	2	16	33	14	20	20	−1	19	22

Table 4.1 shows that the response time for all but one individual is quicker in the RVF. No difference exceeds 33 ms, whereas in either field differences between some individuals exceed 100 ms. Without matched pairs these latter differences might swamp the relatively consistent differences between fields for individuals. We explore this in Exercise 5.14.

Educationists often compare two examination or teaching methods by pairing students so that each member of a pair is as similar as possible in age, intelligence and previous knowledge of the subject. Member of a pair are then exposed, one to each technique, and the results of tests interpreted in terms of pairwise differences.

Consistency between two examiners may be compared by asking each to mark the same series of essays without disclosing their assessments to one another. The differences between the marks awarded by each examiner for each essay are then compared to see whether one examiner consistently awards a higher mark or whether differences appear to be purely random or have some other pattern.

Two animal diets might be compared using sets of twin lambs; one diet is fed to each twin, and their growth measured over a period: attention is focused on growth differences between twins in each pair. Because of genetic similarity each of a pair of twin lambs fed on identical diets tends to grow at much the same rate: when fed different diets reasonably consistent differences in growth may be attributed to the effects of diet.

In summary, the aim of pairing is to make conditions, other than the treatment or factor under investigation, as like as possible within each pair; the mean or median of differences for pairs then provides a measure of treatment effect, being indicative of any 'shift' in the distribution.

4.1.3 Single-sample analysis of matched pairs

Differences between paired observations provide a single sample that can be analysed by the methods developed in Chapters 2 and 3. We must, however, consider the assumptions about observations on each member of the pair and precautions in experimental procedure needed to validate our analyses. These points are best brought out by examples.

Example 4.1

The problem. Using the LVF, RVF data in Table 4.1, assess the strength of the evidence for a consistent response difference between the two fields for individuals. Obtain appropriate 95% confidence intervals for that difference.

Formulation and assumptions. We denote the observations on subject i by x_i in the RVF and by y_i in the LVF and analyse the differences $d_i = y_i - x_i$, $i = 1, 2, \ldots, n$. The d_i are independent since each refers to a different individual. Under H_0 : *the median of the differences is zero* the d_i are equally likely to be positive or negative and a sign test is justified. If we assume a symmetric distribution of the d_i under H_0 we may use a Wilcoxon signed rank (or a normal scores) test. There are several different response patterns in the two visual fields that could result in a symmetric distribution of the differences under H_0. In particular, if we assume response times are identically and independently (but not necessarily symmetrically) distributed in the two fields for any individual (but this distribution need not be the same for every individual) the difference for each individual will be symmetric about zero (since if X and Y have the same independent distributions, then $X-Y$ and $Y-X$ will each have the same distribution and therefore must be symmetrically distributed about zero). We also get a symmetric distribution of differences if the LVF and RVF for any individual have different symmetric distributions providing each has the same mean.

It is well to be aware of such subtleties, but by far the most usual situation where a Wilcoxon test is justified is when we assume identical distributions for LVF and RVF under the null hypothesis; the alternative hypothesis of interest is usually that there is a shift in location only, as indicated by a shift in one field or the other in the median of otherwise identical distributions. Often the d_i values themselves indicate whether there is serious asymmetry.

If we assume also that the differences are approximately normally distributed, a t-test is appropriate. We may test whether normality is a reasonable assumption by Lillifors' test or one of its alternatives.

We consider relevant methods in turn.

Procedure. Working with the 'one sample' d_i we proceed as in Chapter 2, so we only sketch details. Denoting our location parameter by θ, the null hypothesis is $H_0 : \theta=0$, so the ordered signed deviations are simply the ordered differences obtainable from Table 4.1, i.e:

$$-1 \quad 2 \quad 7 \quad 14 \quad 16 \quad 16 \quad 19 \quad 20 \quad 20 \quad 22 \quad 30 \quad 33$$

and the corresponding signed ranks are

$$-1 \quad 2 \quad 3 \quad 4 \quad 5.5 \quad 5.5 \quad 7 \quad 8.5 \quad 8.5 \quad 10 \quad 11 \quad 12$$

Using the sign test for zero median difference we have 1 minus and 11 plus signs. Table I indicates significance in a two-tail test at an exact 0.6% level. The nominal 95% confidence interval based on the sign test is (7, 22) since we reject H_0 at the nominal 5% level only for 2 or fewer or 10 or more minus signs. For the Wilcoxon test $S_-=1$ and despite ties one should have no qualms about adjudging this significant in a two-tail test at the 1% level since the critical value (Table II) is $S=7$. STATXACT indicates significance at the 0.1% level in an exact two-tail test that allows for ties.

For a 95% confidence interval we require the 14 greatest and least Walsh differences. Table 4.2 gives all differences in a format similar to that in Table 2.6. Ignoring for the moment the influence of ties, Table II tells us we must eliminate the 13 largest and smallest Walsh averages to establish a nominal 95% confidence interval.

This procedure leads to the interval (9.5, 23.5). We can verify or adjust by fine tuning to ensure a correct 95% level if a suitable computer program is available to test the null hypotheses $H_0 : \theta=9.5$ and $H_0 : \theta=23.5$ (see *Computational aspects* below). The median of the Walsh averages, 17.5, is the Hodges–Lehmann point estimator.

Assuming normality, the t-test indicates significance at the 0.1% level in a two-tail test and the relevant 95% confidence interval is (10.05, 22.95) and the point estimate of the mean is 16.5. Normal theory here gives the shortest confidence interval and the fact that confidence intervals are not seriously displaced for the sign test indicates reasonable symmetry. Lillifors' test statistic, 0.151, is well below the value required for significance.

Table 4.2 Walsh averages for Wilcoxon confidence limits for visual field differences

	−1	2	7	14	16	16	19	20	20	22	30	33
−1	−1	0.5	3	6.5	7.5	7.5	9	9.5	9.5	10.5	14.5	16
2		2	4.5	8	9	9	10.5	11	11	12	16	17.5
7			7	10.5	11.5	11.5	13	13.5	13.5	14.5	18.5	20
14				14	15	15	16.5	17	17	18	22	23.5
16					16	16	17.5	18	18	19	23	24.5
16						16	17.5	18	18	19	23	24.5
19							19	19.5	19.5	20.5	24.5	26
20								20	20	21	25	26.5
20									20	21	25	26.5
22										22	26	27.5
30											30	31.5
33												33

Conclusion. All three tests reject the null hypothesis at a nominal 1% level and we conclude that response times in the RVF are faster. Consistency of the confidence intervals given by the three approaches suggests the mean difference is between about 10 and 22 ms.

Comments. 1. The set of differences we compare are independent (a necessary condition for validity of our location tests) because each difference is calculated for a different individual.

2. If the response rate in the LVF had been measured before that in the RVF for all subjects a difficulty in interpretation would arise. Our result might imply a learning process, people responding more quickly in the RVF because they were learning to react more rapidly; or there could be a mixture of a learning effect and an inherent faster response in the RVF. We avoid this difficulty if we decide at random which field – left or right – is to be tested first for each individual; that was done in this experiment. This should balance out and largely annul any learning factor. Another approach is to achieve balance by selecting six subjects (preferably at random) to be tested first in the LVF. The remaining six are tested first in the RVF. Such balanced designs provide a basis for separating a learning effect from an inherent difference between field responses, although a somewhat larger experiment would be needed to do this by appropriate parametric or nonparametric methods.

Computational aspects. Rejection of the 13 smallest and largest Walsh averages in Table 4.2 to determine a nominal 95% confidence interval ignores ties. Our findings in Chapter 2 indicate that a few ties are unlikely to have a drastic effect on confidence intervals, being confined probably to an effect of discontinuities at the end points. Using STATXACT to test the hypothesis $H_0 : \theta=9.5$ (the lower end point) we find $S_-=14$ and the exact permutation test allowing for ties indicates $\Pr(S_- \leq 14)=0.0247$. Similarly, testing $H_0 : \theta=23.5$ (the upper end point) gives $\Pr(S_+ \leq 13.5)=0.0222$. Each indicates rejection of H_0, implying that if we consider the open interval (9.5, 23.5), i.e. $9.5<\theta<23.5$, we have an exact $100(1-0.0247-0.0222)=95.31\%$ confidence interval. The effect of ties is here trivial.

The modification for testing a hypothesis that a location difference has some specified value D is obvious. As in Chapters 2 and 3, we simply consider deviations from that value. This may be looked upon effectively as shifting the origin to the hypothetical median, so that for our revised data (the deviations) we are testing the hypothesis that the location parameter for that population has a value zero.

Example 4.2

The problem. Eleven children are given an arithmetic test; after three weeks' special tuition they are given a further test of equal difficulty (something we say more about in the *comments*). Their marks in each test and the individual differences are given in Table 4.3. Do these support a claim that the average improvement due to extra tuition is 10 marks?

Table 4.3 Marks in two arithmetic tests

Pupil	A	B	C	D	E	F	G	H	I	J	K
First test	45	61	33	29	21	47	53	32	37	25	81
Second test	53	67	47	34	31	49	62	51	48	29	86
Second−first	8	6	14	5	10	2	9	19	11	4	5

Formulation and assumptions. The question essentially asks whether, if we assume these children are a random sample from some hypothetical population (perhaps children of the same age trained in the same educational system or studying the same syllabus), it is reasonable to suppose the mean mark difference is 10.

Procedure. We consider deviations from 10 for the differences given in the last line of Table 4.3. These deviations are:

$$-2 \quad -4 \quad 4 \quad -5 \quad 0 \quad -8 \quad -1 \quad 9 \quad 1 \quad -6 \quad -5$$

The differences arranged in order of magnitude with appropriate signs are

$$0 \quad -1 \quad 1 \quad -2 \quad -4 \quad 4 \quad -5 \quad -5 \quad -6 \quad -8 \quad 9$$

and the ranks without signs are

1 2.5 2.5 4 5.5 5.5 7.5 7.5 9 10 11

Using the rules given in Section 2.2.3 for mid-ranks with ties together with the convention mentioned at the end of that section for replacing the rank 1 associated with the zero difference by 0, we get these signed ranks:

0 −2.5 2.5 −4 −5.5 5.5 −7.5 −7.5 −9 −10 11

Our statistic is the sum of the positive ranks $S_+=0+2.5+5.5+11=19$. For the exact permutation test STATXACT gives $\Pr(S_+\leq 19.0)=0.124$. Doubling this value for a two-tail test we immediately see there is no question of rejection of H_0, since $\alpha=0.248$, well in excess of the conventional 0.05.

If we have no facility to calculate this exact tail probability we might use an asymptotic approximation with these rank scores. To calculate the relevant statistic we use (3.1) where $S=19$, $\Sigma|s_i|=65$ and $\Sigma s_i^2=503.5$, whence $Z=-13.5/\sqrt{(503.5/4)}=-1.203$. This equals the asymptotic value given by STATXACT and the relevant two-tail probability is 0.229, comparing reasonably well with the exact probability of 0.248. Also the value 19 for S_+ clearly exceeds the critical value 10 given in Table II for $n=11$ in the no-tie situation. When the test statistic clearly exceeds the no-tie critical value, then even with ties there is virtually no question of rejecting H_0.

Conclusion. We do not reject the hypothesis that the mean improvement may be 10.

Comments. 1. How does one decide if two tests are equally difficult? Might not the improved marks in the second test imply it was easier? The statistician should seek the assurance of the educationalist conducting the test that reasonable precautions have been taken to ensure equal difficulty. Sometimes standard tests that have been tried on large groups of students with results that show convincingly that they are for all practical purposes of equal difficulty, are used in such situations.

2. When $n=11$ in the 'no-tie' case a nominal 95% confidence interval is obtained by rejecting the 10 largest and 10 smallest Walsh averages. This (see Exercise 4.1) gives an interval (4.5, 12). Using a STATXACT exact test for $H_0: \mu=4.5$ gives a lower tail probability of 0.0088. Increasing the lower limit to 4.51 breaks certain ties but the associated lower tail probability only increases to 0.0112. Indeed, not until we increase the lower limit to 5 do we get a further change in the lower tail value and this increases it above 0.025 (indeed to 0.0645 when we test $H_0: \mu=5$). Similarly, at the upper limit we find we would reject $H_0: \mu=12$ in a two-tail test at the 5% significance level since the upper-tail probability is 0.022. However we would accept H_0 for any μ slightly less than 12. We conclude that a nominal 95% confidence interval is $5\leq\mu<12$, and that the exact confidence level associated with this interval is $(1-0.011-0.022)100\%=$

96.7%. The normal theory 95% confidence interval is (5.14, 11.77).

3. It may not be realistic to test simply for a shift in location. Sometimes pupils who are good to start with benefit little from extra tuition; likewise very poor pupils sometimes gain little. Often only those in the mid-ability range show appreciable benefit. The statistician may find evidence of this by looking at the data – and there are indeed tests for such tendencies. Deciding what should be tested or esitmated is often a matter for fruitful discussion between statistician and experimenter.

Example 4.3 is based on data for a group of 77 students made available to me by Nigel Smeeton.

Example 4.3

The problem. The data below are differences *systolic blood pressure after exercise – systolic blood pressure before exercise* measured in mm Hg for a random sample of 24 from the group of 77 students. Obtain 95% confidence limits for the population location difference based on (i) the sign test, (ii) the Wilcoxon signed rank test and (iii) the t-test, and comment on the appropriateness of each. For convenience we have arranged the differences in increasing order.

−5 −5 0 2 10 15 15 15 18 20 20 20 20 22 30 30 34 40 40 40 41 47 80 85

Formulation and assumptions. We use standard methods developed in Chapter 2 and in this chapter.

Procedure. We omit detail of the relevant computations (see Exercise 4.16) since examples of similar computations have already been given.

Conclusion. The relevant nominal 95% confidence intervals are:

Sign test	(15, 40)
Wilcoxon	(17.5. 33.5)
t-test	(16.86, 35.96)

Comments. 1. Not surprisingly, the sign test interval is the longest. There is slight evidence of skewness, but none of the intervals is seriously displaced relative to the others. The Wilcoxon interval being shorter than that based on normal theory may arise from a slightly heavier 'upper tail' (especially the values 80, 85) than one would expect with a sample from a normal distribution. However, Lilliefors' test for normality gives a test statistic value of 0.162, a little short of the value required for significance at the 5% level.

2. The only reason for taking a random sample of 24 from 77 observations was to provide a convenient illustrative example. Almost certainly

one would use all observations in any detailed studies of the effects of exercise on blood pressure (and interest may well extend beyond studies of the mean or median difference).

3. A sidelight of the data set was that there was clearly an error on the computer printout I received. For one student the difference recorded was $15-118=-103$. It is immediately evident to anyone familiar with blood pressure levels that the value 15 mm Hg for systolic blood pressure after exercise is incorrect! There is a strong suspicion that the final digit has been omitted and that the true reading should be between 150 and 159. It is is easy to spot such a discrepancy on a print-out and in practice one would then attempt to track down the source of error (and if possible make the needed correction). Possible sources of error are failure of the printer to reproduce a character, a mistake in entering the original data or in initial recording of the data. With increasing use of computer packages to process data such an error might go undetected. Had this value, -103, been included in our sample of 24 instead of a recorded value of -5 (i.e. if one of our entries of -5 had been replaced by -103) we would have obtained the following 95% confidence intervals:

Sign test	(15, 40)
Wilcoxon	(12.5, 33.5)
t-test	(7.82, 36.84)

The reader who has difficulty explaining differences between these results and those recorded above in terms of the effect of an 'outlier' should refer to Section 2.2.6.

Good experimenters and statisticians check for data errors. Many computer programs include output helpful in detecting outliers; e.g. a print-out of maximum and minimum data values often (but by no means always) highlights a glaring data error.

4.2 A LESS OBVIOUS USE OF THE SIGN TEST

The way the data are presented in Example 4.4 does not make it immediately obvious that a sign test is relevant.

Example 4.4

The problem. Members of a mountaineering club argue interminably about which of two rock climbs is the more difficult. Hoping to settle the argument one member checks the club log book. This records for any climb by a member whether it is successfully completed. The log shows that 108 members have attempted both climbs with the outcomes summarized in Table 4.4. Is there evidence that one climb is more difficult?

Table 4.4 Outcomes of two rock climbs

		First climb	
		Succeeded	Failed
Second climb	Succeeded	73	14
	Failed	9	12

Formulation and assumptions. A moment's reflection shows that if a climber succeeds in both climbs, or fails in both climbs, the outcomes give no information about relative difficulty; such cases are **ties** so far as comparing difficulty is concerned. If we had additional information, for instance about each climber's personal assessment of the difficulty, the situation would be different. As it is, our meaningful comparators of difficulty are numbers who succeed at one climb, but fail at the other. From Table 4.4 we see that 9 succeed at the first climb but fail at the second; we may think of this as a 'plus' for the first climb. Also 14 fail on the first climb but succeed at the second; we may think of this as a 'minus' for the first climb.

This is a sign test situation. If the climbs are of equal difficulty a 'plus' or 'minus' is equally likely for each climber. Thus under H_0 : *climbs are equally difficult* the number of 'plus' signs has a binomial distribution with n equal to the total number of plus and minus signs and $p=½$.

Procedure. There are 9 plus and 14 minus signs so $n=23$. Using the normal approximation to the binomial distribution we find, using (2.2), that $Z=(9.5-11.5)/\sqrt{(23/4)}=-0.83$, or $|Z|=0.83$, well below the magnitude required for significance.

Conclusion. We accept H_0 : *climbs are equally difficult.*

Comments. That 73+12=85 out of 108 observed pairs provide no information on the relative difficulty of the climbs may seem wasted data, but such 'wasted' data give an indication of how big or small any difference is. In some situations we need many observations because we are looking for a small difference which may not be clearly distinguishable when the only relevant criteria are success/failure or failure/success categories.

Computational aspects. STATXACT, DISFREE and other packages include programs for this simple test which may be used for any sign test.

The test just described is one form of a test known as **McNemar's test.** Conover (1980, Section 3.5) presents the test in a more formal way that nevertheless reduces in application to essentially a paired sample sign test.

4.3 FIELDS OF APPLICATION

In most of these applications if a numerical value of the difference for each matched pair is available and these do not appear too skew, the Wilcoxon test (or an analogous test using normal scores) is likely to be appropriate. The matched pairs t-test is appropriate if the differences $d_i=y_i-x_i$ are approximately normally distributed; sometimes this may be the case even when the distributions of X, Y are far from normal.

Laboratory instrument calibration

Two different brands of instrument reputedly measure the same thing (e.g. blood pressure, hormone level, sugar content of urine, bacterial content of sputum), but each is subject to some error. Samples from, say, each of 15 patients might be divided into two sub-samples, the first being analysed with one kind of instrument, the second with the other. A Wilcoxon test is appropriate to test for any systematic difference between instruments. When purporting to measure the same thing a systematic difference from the true values in means or medians is often described as **mean** or **median bias.** The term 'bias' alone is usually taken to imply mean bias.

Biology

The heartbeat rate of rabbits might be measured before and after they are fed a hormone-rich diet. The Wilcoxon test is appropriate to investigate a shift in mean. 'Before' and 'after' measurements are common in medical and biological contexts, including experiments on drugs and other stimuli, which may be either physical or biological (e.g. a rabbit's blood presssure may be measured when on its own and again after it has shared a cage for half an hour with a rabbit of the opposite sex). Confidence intervals for the mean difference are useful both as an indication of the precision of the experiment (Section 1.4.2) and to help in reaching a decision as to whether any statistically significant difference is of practical importance.

Occupational medicine

An instrument called a Vitalograph is used to measure lung capacity. Readings might be taken on workers at the beginning and end of a shift to study any effect on lung capacity of fumes inhaled in some industrial process, or on athletes before and after competing in a 100 metre sprint.

Agriculture

In a pest control experiment each of 10 plots may contain 40 lettuce plants. Each plot is divided into two halves: one half chosen at random is sprayed with one insecticide, the second with another. Differences in numbers of uninfested plants can be used in a Wilcoxon test to compare effects of insecticides. Incidentally, pest control experiments are a situation where a normality assumption is often suspect.

Psychology

Given sets of identical twins, it being known for each pair which was the first born, times are observed for each individual to carry out a manual task to see if there is any indication that the first-born in a pair tends to be quicker. The choice may lie between a Wilcoxon test and a t-test. Confidence intervals for the mean difference will indicate the precision with which any difference is measured and whether it is of practical importance.

Road safety

Drivers' reaction times in dangerous situations may be compared before and after each has consumed 4 pints of beer. (This is a response to stimulus situation of the type mentioned above under the heading **Biology**).

Space research

Potential astronauts may have the enzyme content of their saliva determined before and after they are subjected to a zero gravitational field in a simulator. Such biochemical evidence is important in determining physiological reactions to space travel.

Education

To decide which of two examination questions is the harder each may be given to 150 candidates and records taken of numbers who complete both, neither, only the first, only the second. Numbers in the latter two categories can be used in a McNemar sign test for evidence of unequal difficulty.

4.4 SUMMARY

All matched pair sample location tests for differences considered in this chapter reduce to the analogous single-sample tests considered in Chapters 2 and 3. See in particular the sign test (Section 2.1), the Wilcoxon signed rank test (Section 2.2), raw data scores (Section 3.1), normal, or modified van der Waerden scores (Section 3.2). General tests for distribution of the paired differences include the Kolmogorov test (Section 3.3.1) and Lilliefors' test for normality (Section 3.3.3).

McNemar's test (Section 4.2) is relevant to paired observations to assess changes in attitude or assessments of relative difficulty. It is equivalent to the sign test (Section 2.1).

EXERCISES

4.1 Verify the confidence intervals given in comment 2 on Example 4.2.

4.2 The blood pressures of 11 patients are measured before and after administration of a drug known not to raise blood pressure, but which might lower it. The differences in systolic blood pressure (pressure before − pressure after) for each patient are:

$$7 \quad 5 \quad 12 \quad -3 \quad -5 \quad 2 \quad 14 \quad 18 \quad 19 \quad 21 \quad -1$$

Use an appropriate nonparametric test to see if the sample (assumed random) contradicts the hypothesis of no systematic change in blood pressure.

4.3 Samples of cream from each of 10 dairies (A to J) are divided each into two portions. One portion from each is sent to Laboratory I and the other to Laboratory II for bacterium counts. The counts (thousands bacteria ml^{-1}) are:

Dairy	A	B	C	D	E	F	G	H	I	J
Laboratory I	11.7	12.1	13.3	15.1	15.9	15.3	11.9	16.2	15.1	13.6
Laboratory II	10.9	11.9	13.4	15.4	14.8	14.8	12.3	15.0	14.2	13.1

Use the Wilcoxon signed rank test to test whether the mean difference between laboratories for sub-samples from the same dairy differ significantly. Obtain also nominal 95% and 99% confdience intervals for the mean difference and compare these with the intervals using the optimal method when normality is assumed.

4.4 A hormone is added to one of otherwise identical diets given to each of 40 pairs of twin lambs. Growth differences over a 3-week period are recorded for each pair and signed ranks are allocated to the 40 differences. The lower rank sum was $S_{-}=242$. There was only one rank tie. Investigate the evidence that the hormone may be affecting (increasing or decreasing) growth rate.

4.5 A psychologist interviews both father and mother of each of 17 unrelated but mentally handicapped children, asking each individually a series of questions designed to test how well they understand the problems their child is likely to face in adult life. The psychologist records whether the father (F) or mother (M) shows the better understanding of these potential problems. For the 17 families the findings are

F M M F F F F F F F M F F F M F F

Is the psychologist justified in concluding that fathers show better understanding?

4.6 To test people's ability to induce events by paranormal influence directed at specific scenes, a parapsychologist asks nine subjects who each claim psychic powers to use these to induce carelessness in customers visiting a china shop so as to increase the breakage rate due to accidents while the customers examine stock. Each subject is asked to 'will' customers to break items on a specified (but different) morning (the test day) and the numbers of breakages per 500 customers entering the shop on each test day is compared with the number of breakages per 500 customers entering the shop on the same day of the previous week (the control day) when no such attempt to produce breakages had been made. The results (numbers of items broken per 500 customers) for the nine subjects are:

Subject	1	2	3	4	5	6	7	8	9
Test day	2.5	3.5	1.0	4.6	3.8	5.9	6.1	15.1	2.4
Control day	0	3.7	2.4	3.8	4.1	1.2	1.1	1.3	0

Analyse these results by what you consider the most appropriate parametric or nonparametric methods to determine whether or not they provide acceptable evidence that a paranormal phenomenon is exhibited.

4.7 One hundred travelling salesmen are asked whether or not they think it is in the interests of public safety to prohibit a person drinking any alcohol for up to 3 hours before driving a car. They are then given a lecture and shown a video on the relevance of alcohol as a cause of some serious accidents. They are then again asked their opinion on the desirability of such a ban. Do the results given below indicate a significant change in attitudes as a result of the lecture and video?

		Before lecture	
		In favour	Against
After lecture	In favour	31	16
	Against	8	45

4.8 A canned-soup manufacturer is experimenting with a new-formula tomato soup. A tasting panel of 70 each taste samples of the current product and the new one (without being told which is which). Of the 70, 32 prefer the new-formula product, 25 the current product and the remainder cannot distinguish between the two. Is there enough evidence to reject the hypothesis that consumer preference is likely to be equally divided?

4.9 Do the data in Exercise 4.8 support a claim that as many as 75% of those who have a preference may prefer the new formula?

4.10 To produce high-quality steel one of two hardening agents A, B may be added to the molten metal. The hardness of steel varies from batch to batch, so to test the two agents batches are sub-divided into two portions, for each batch agent A being added to one portion, agent B to the other. To compare hardness, sharpened specimens for each pair are used to make scratches on each other: that making the deeper scratch on the other is the harder specimen. For 40 batch pairs, B is adjudged harder in 24 cases and A in 16. Is this sufficient evidence to reject the hypothesis of equal hardness?

4.11 For a sub-sample of 10 pairs from the steel batches in Exercise 4.10 a more expensive test is used to produce a hardness index. The higher the value of the index, the harder the steel. The indices recorded were:

Batch no.	1	2	3	4	5	6	7	8	9	10
Additive A	22	26	29	22	31	34	31	20	33	34
Additive B	27	25	31	27	29	41	32	27	32	34

Use an appropriate test to determine whether these data support the conclusions reached in Exercise 4.10.

4.12 On the day of the third round of the Open Golf Championship in 1987 before play started a TV commentator said that conditions were such that the average scores of players were likely to be at least three higher than those for the second round. For a random sample of 10 of the 77 players participating in both rounds the scores were:

Round 2	73	73	74	66	71	73	68	72	73	72
Round 3	72	79	79	77	83	78	70	78	78	77

Do these data support the commentator's claim? Consider carefully whether a one- or two-tail test is appropriate.

4.13 Pearson and Sprent (1968) gave data for hearing loss (in decibels below prescribed norms) at various frequencies. The data below show these losses for 10 individuals aged between 46 and 54 at frequencies of 0.125 and 0.25 kc s^{-1}. A negative loss indicates hearing above the norm. Is there an indication of a different loss at the two frequencies?

Subject	A	B	C	D	E	F	G	H	I	J
0.125 kc s^{-1}	2.5	−7.5	11.25	7.5	10.0	5.0	7.5	2.5	5.0	8.75
0.25 kc s^{-1}	2.5	−5.0	6.35	6.25	7.5	3.75	1.25	0.0	2.5	5.0

4.14 Apply a normal scores test to the data in Example 4.2.

4.15 Scott, Smith and Jones (1977) give a table of estimates of the percentages of UK electors predicted to vote Conservative by two opinion polling organizations, A and B, in each month in the years 1965–70. For a random sample of 15 months during that period the percentages were:

A 43.5 51.2 46.8 55.5 45.5 42.0 36.0 49.8 42.5 50.8 36.6 47.6 41.9 48.4 53.5
B 45.5 44.5 45.0 54.5 49.5 43.5 41.0 53.0 48.0 52.5 41.0 47.5 42.5 45.0 52.5

Do these results indicate a significant tendency for one of the organizations to return higher percentages than the other? Obtain an appropriate 95% confidence interval for any location difference between predictions during the period covered.

4.16 Verify the correctness of the confidence intervals and the result for Lilliefors' test quoted in Example 4.3.

5

Methods for two independent samples

5.1 LOCATION TESTS AND ESTIMATES

We now consider independent samples $x_1, x_2, \ldots, x_m$ and $y_1, y_2, \ldots, y_n$ of m, n observations from two populations. For independent samples inferences about location, etc., no longer reduce to one-sample problems.

In Chapters 2 to 4 we showed that several distribution-free test and estimation procedures differed basically only in the scores assigned. Choice of procedure was governed by assumptions and by computational practicality. With asymmetry the sign test was appropriate. The Wilcoxon signed rank test appealed because it was valid for **any** symmetric parent population and was moderately robust against some departures from symmetry, a feature lacking for normal theory parametric inference. Given adequate computational facilities, permutation tests using raw scores (based on the data themselves) had an intuitive appeal, but at least in the one-sample situation, inferences followed closely those using the t-distribution and so lacked robustness against departures from symmetry. Normal scores guaranteed high Pitman efficiencies in a wide range of circumstances.

Given computational facilities for exact permutation distribution tests for any scoring system we may, in theory, deal in one framework with practical complications such as rank ties, observations equal to a hypothesized value of some parameter or censored data with any chosen scores. Nevertheless, simplicities in particular cases make it convenient to describe some specific and widely used counting or ranking procedures in their own right.

Many aspects of the one-sample situation carry over to two independent sample problems. We deal first with commonly used but relatively simple methods that are analogues of some in Chapters 2 and 3, indicating how specific techniques fit into a more general pattern. This approach helps develop a 'feel' for the rationale behind simple permutation distribution theory. Our first method generalizes the sign test.

5.1.1 The median test

If each population has the same median, whether or not they differ in other respects, for each sample the number of values above (or below) that common median has a binomial distribution with $p=½$. The median test we develop does **not** test whether that common median, θ, say, has a particular value, but rather, whether it is reasonable to suppose both populations have the same unknown median. Formally, if the sample of m comes from a population with unknown median θ_1 and the sample of n from a population with unknown median θ_2, we test $H_0 : \theta_1=\theta_2$ against alternatives like $H_1 : \theta_1\neq\theta_2$ (two-tail) or $H_1 : \theta_1>\theta_2$ (one-tail).

If two populations have the same median, then the combined sample median, M, say, provides a point estimate of that common median. The median test examines, for each sample, how many values are above and how many are below M. If both samples come from populations with the same median the distributions of the numbers in each above M will be approximately binomial with parameters $p=½$ and m, n respectively, where the relevant probabilities are now conditional upon the numbers above M in each sample adding to the number above M in the combined sample. If $m+n$ is even and no sample value equals M, in the combined sample the numbers of values above and below M will be $½(m+n)$. If $m+n$ is odd at least one sample value equals M. In that case, or if some values equal M when $m+n$ is even, we suggest (as we did in the sign test) omitting such values and proceeding with a reduced sample. Unless there are many tied values at M this procedure is usually satisfactory. Conditioning on the total numbers above or below M in the combined sample influences the theory for the permutation test, but the situation has much in common with procedures we develop in a more general context in Chapter 9. We illustrate how the test works in practice by an example.

Example 5.1

The problem. Sample 1 consists of 7 observations and sample 2 of 21 observations. The samples are pooled and the median of the combined sample is determined. There are no observed values equal to M, i.e. 14 are greater and 14 are less. In sample 1 there is one value greater than M and in sample 2 there are 13 values greater than M. If the unknown population medians are θ_1, θ_2 respectively, test $H_0 : \theta_1=\theta_2$ against $H_1 : \theta_1<\theta_2$.

Formulation and assumptions. Clearly, in sample 1, the fact that there are 6 values less than M and only 1 greater than M, suggests that if H_0 is not true it is likely that this sample is from a population with a lesser median than that suggested by the pooled sample; similarly, the fact that sample 2 has more values above M than it has below suggests that if H_0 is not true then this sample is likely to have come from a population with a greater median than that suggested by the pooled sample. This suggests any departure from equality is in the direction specified in H_1. It is convenient to set out the information in the form of Table 5.1.

Table 5.1 Numbers of observations above and below M in Example 5.1

	Above M	Below M	Total
Sample 1	1	6	7
Sample 2	13	8	21
	14	14	28

An important feature of this table – often referred to in statistical jargon as a **2×2 contingency table** – is that the row and column totals and the grand total are fixed. The 2×2 refers to the fact that there are two rows and two columns of data. The four data positions are often referred to as cells. That at the intersection of the ith row and jth column is referred to as cell (i, j). In Table 5.1 cell (2, 1) contains the entry 13.

The row totals are determined by the sample sizes m, n and the column totals by the rule that M is the combined sample median. The test is based on the permutation distribution of all **possible** cell entries in Table 5.1 consistent with the marginal totals. A moment's reflection shows that if we know one cell value **all** other values are **automatically** determined to give the correct marginal totals. It is convenient to concentrate on the possible values in cell (1, 1) (i.e. the number of values above M in sample 1). A possible value is 3. To give the correct marginal totals it is clear that the cell entries in the body of a table like Table 5.1 must then be

$$\begin{matrix} 3 & 4 \\ 11 & 10 \end{matrix}$$

Clearly such an outcome is more favourable to H_0 than the one we observed. Indeed, the only outcome less favourable to H_0 than that observed is

$$\begin{matrix} 0 & 7 \\ 14 & 7 \end{matrix} \tag{5.1}$$

This allocation and that in Table 5.1 form an appropriate 'tail' for our test of H_0 against H_1. We need the probability associated with this tail when H_0 is true. Let us assume for the moment that if H_0 is true then M is the value of the common median for the two populations. This may not be true, but it is the 'best' estimate of the common median to be obtained from our data. We see below that this assumption is not critical to our argument. As in the sign test, under H_0 the probability of getting one observation above the median in a sample of 7 is the binomial probability $p_1={}^7C_1(\frac{1}{2})^7$; that of getting 13 above the median in a sample of 21 is $p_2={}^{21}C_{13}(\frac{1}{2})^{21}$. These are independent as they refer to different samples. Turning now to the column margin, in the combined sample of 28 the probability of observing 14 out of 28 observations above M is $p_c={}^{28}C_{14}(\frac{1}{2})^{28}$. We require the probability of observing respectively 1 and 13 observations above the median in samples 1 and 2, conditional upon observing 14 above the median in the combined sample. From the definition of conditional probability this is $P=p_1p_2/p_c$. For the data in Table 5.1, $P={}^7C_1.{}^{21}C_{13}/{}^{28}C_{14}$. The binomial probability $p=\frac{1}{2}$ cancels out in the expression for P. This will still happen if the binomial analogues of p_1, p_2, p_c are written down for any probability p; i.e. P is independent of p. Thus if H_0 is true, we do not need to assume that M is the population median. That P is independent of the binomial parameter p when row and column marginal totals are fixed has far-reaching implications in nonparametric inference. In Chapter 9 we use this result and extensions to it in a variety of situations. To determine a relevant tail probability we calculate P for the configuration in Table 5.1 and for the more extreme configuration (5.1).

Procedure. Evaluation of P is tedious. It is commonly written in an equivalent form. For a general 2×2 contingency table with cell entries

$$\begin{matrix} a & b \\ c & d \end{matrix}$$

this is

$$P=\frac{(a+b)!(c+d)!(a+c)!(b+d)!}{(a+b+c+d)!a!b!c!d!} \tag{5.2}$$

The factorials in the numerator of (5.2) are for the marginal totals and those in the denominator for the grand total and the individual cell entries. For Table 5.1, (5.2) gives $P=[7!21!14!14!]/[28!1!6!13!8!]$. There is appreciable cancellation between numerator and denominator and using a pocket calculator it is easy to verify that $P=0.0355$. Computing P for (5.1) gives $P=0.0029$ (Exercise 5.1). Thus the size, α, of the critical region for testing H_0 against H_1 is $\alpha=0.0029+0.0355=0.0384$.

Conclusion. We reject H_0 at the 3.84% significance level.

Comments 1. For a two-tail test of $H_0 : \theta_1=\theta_2$ against $H_1 : \theta_1\neq\theta_2$ in this case the permutation distribution is symmetric; our upper tail is associated with the two extreme outcomes

$$\begin{matrix} 6 & 1 \\ 8 & 13 \end{matrix} \quad \text{and} \quad \begin{matrix} 7 & 0 \\ 7 & 14 \end{matrix}$$

and these have the same four cell entries (in different order) and marginal totals as those in Table 5.1 and in (5.1). It follows they have the same associated probabilities. Doubling the one-tail probability for a two-tail test, the size of the critical region is 2×0.0384=0.0768, so we would not reject H_0 at the conventional 5% significance level in a two-tail test.

2. Clearly there are marked discontinuities in the permutation distribution, despite there being 28 observations. Discontinuities arise in part

because the observations are split into a relatively large (21) and a relatively small (7) sample. Because fixing the entry in one cell determines all others (to give correct marginal totals) we say the table has **one degree of freedom.** If we denote the number in the first cell by X, then X is a random variable which, in this example, can take only the values 0, 1, 2, 3, 4, 5, 6, 7.

Computational aspects. STATXACT, TESTIMATE and DISFREE have specific programs to carry out this test; so also have a number of standard statistical packages. The test is usually described as **Fisher's (exact) test** and is of wider applicability. STATXACT generates the full permutation distribution corresponding to all possible X values. This is given in Table 5.2, which confirms symmetry of the distribution and illustrates the marked probability discontinuities between successive X values. In Chapters 2 and 3 we saw that discontinuity decreased rapidly with increasing sample size. Because we have 28 observations the level of discontinuity here may surprise. With two or more samples discontinuities are often quite marked if one sample is small relative to the other(s). This may make asymptotic results unreliable.

Table 5.2 Exact permutation distribution for median test in Example 5.1

X	*Probability*
0	0.0029
1	0.0355
2	0.1539
3	0.3077
4	0.3077
5	0.1539
6	0.0355
7	0.0029

A large-sample approximation to the median test uses

$$Z=2(X-\tfrac{1}{2}m)\sqrt{[(n+m)/mn]}. \qquad (5.3)$$

Asymptotically, Z has a standard normal distribution but the approximation may be poor for small n or m even when $n+m$ is large. Some improvement is obtained with a continuity correction analogous to that recommended for the sign test, i.e. replacing X by $X+\frac{1}{2}$ if $X<\frac{1}{2}m$ or by $X-\frac{1}{2}$ if $X>\frac{1}{2}m$. For this example we easily verify (Exercise 5.2) that without a continuity correction $Z=-2.18$ and with the continuity correction $Z=-1.75$. From tables we find $\Pr(Z<-2.18)=0.0146$ and $\Pr(Z<-1.75)= 0.0401$. The latter is reasonably near the exact one-tail probability of 0.0384 found in Example 5.1, but without the continuity correction we greatly overestimate significance. Check when using a computer program that relies on an asymptotic result: some do not use a continuity correction. In Section 9.2.2 we give an alternative asymptotic result useful in more general situations.

The reader may feel that markedly different sample sizes (in Example 5.1 one is three times the other) are unlikely to be met in practice, but they often occur in clinical trials where there may be only a few patients with a rare illness. Then comparison is often between responses for these patients and a much larger available 'control' group without that illness.

5.1.2 Confidence intervals for a median difference

Given full measurement data we may establish a confidence interval for the difference between population medians based on the median test permutation distribution. Writing the true difference $\delta=\theta_2-\theta_1$, a nominal 95% confidence interval for δ is (d_1, d_2) where the end points d_1, d_2 are chosen so that if we change each observation in sample 2 (or sample 1) by more than these amounts we would reject H_0 : *the medians are equal for population 1 and the amended population 2*. This is relevant because increasing or decreasing all sample (or population) values by a constant k increases or decreases the median by k. For given sample sizes m, n it is not difficult to work out, especially if one has the appropriate computer software, the numbers above the median in the amended sample 2 that will just give significance. The procedure is best explained by an example.

Example 5.2

The problem. To determine whether books on management tended to be longer or shorter than books on biology, random samples of 10 books on biology and 12 on management were selected in Dundee University library and the number of pages recorded for each book sampled. The numbers (arranged in order of magnitude within each sample) were

Biology	143	173	226	233	250	287	291	303	634	637		
Management	50	164	198	221	225	302	328	335	426	534	586	618

Determine a nominal 95% confidence interval for the difference between population medians based on the median (Fisher's exact) test procedure.

Formulation and assumptions. The median test will indicate significance at the 5% level in a two-tail test if the contingency table with row totals fixed at 10, 12 (the sample sizes) and column totals 11, 11 (numbers above and below the combined sample median) is such that the entry in row 1 column 1 (the number above the median in sample 1) is in a tail region of nominal size 0.025. If a suitable computer program is available it is not difficult to establish that significance is established if the number above the median in the first sample is 2 or less or 8 or more, the actual level then being $2\times0.0150=0.0300$ (3%). If a computer program is not available this can be established using formula (5.2) for the contingency tables with cell (1, 1) entries 0, 1, 2 respectively. Having established this we determine what constant adjustment is needed to each sample 2 value so that we would just avoid rejecting H_0 in a median test based on sample 1 and the adjusted sample 2 but would reject H_0 for any greater adjustment.

Procedure. Since we reject H_0 if only 2 values in sample 1 exceed the joint sample median this implies we shall reject H_0 if the joint sample median exceeds 303. Since the number above the median in both samples must be 11, this implies that for the adjusted sample 2 if we are not to reject H_0, it must have 8 values above the adjusted median. If the adjusted median is not to exceed 303 this means the adjustment must ensure that there are 8 values at or above 303 in the adjusted second sample. i.e. the eight largest values in the second sample must exceed 303. Clearly this occurs if we add to all values the difference between 303 and the eighth largest sample value in Sample 2. That value is 225, so we must add $303-225=78$ to all second sample values to achieve this. This implies that if $\theta_2-\theta_1>-78$ we would reject H_0. Similarly, since we reject H_0 if 8 or more values exceed the combined sample median, this implies we shall reject H_0 if the joint sample median is less than 226. This means that for the adjusted second sample, if we are not to reject H_0 we must have at least 3 values at or above 226 in that adjusted second sample. Clearly this occurs if we subtract $426-226=200$ from all second sample values. This implies that if $\theta_2-\theta_1>200$ we reject H_0.

Conclusion. A nominal 95% confidence interval for the median difference $\theta_2-\theta_1$ is $(-78, 200)$.

Comments. 1. As usual, at the limits of this interval we have ties at the combined sample medians for the first sample and the adjusted second sample. To check that we would accept H_0 at the end points we must in effect perform a test with allowance for ties at the joint median. Exercise 5.3 asks the reader to check that we would accept H_0 in such a test.

2. The actual confidence level of the above interval is 97% since the actual rejection level of H_0 is 3% (see formulation and assumptions above).

3. Assuming normality, the t-based 95% confidence interval for the mean difference is $(-142.94, 172.04)$. This is a wider interval than that based on median test theory. Indeed, normal theory is not very appropriate for these data because the distribution of numbers of pages is skew. While a majority of books (as on most subjects) are between 150 and 350 pages in length there are a few works of 600 or more pages and a relatively small number below 150 pages – in practice usually virtually no books are less than 50 pages if we exclude pamphlets and a few special tracts. Minimal restrictions about the nature of the populations make the median test

reasonable in circumstances where the precise population distribution is in doubt.

Computational aspects. Given a program for the Fisher exact test for a 2×2 contingency table one can easily determine for any given m, n the number above the median in the first sample which just gives significance, and, because there is only one degree of freedom, all other entries in the 2×2 table. It is then, as indicated in the above example, relatively simple to determine confidence limits by appropriate additions or subtractions from all second-sample observations.

5.1.3 The Wilcoxon–Mann–Whitney test

The literature refers to equivalent tests formulated in different ways as the **Wilcoxon rank sum test** and the **Mann–Whitney test.** The formulations were developed independently by Wilcoxon (1945) and Mann and Whitney (1947). We refer to the two versions jointly as the **Wilcoxon–Mann–Whitney test** or, for brevity, as the **WMW test.** Example 1.4 gave a specific case of the Wilcoxon formulation that reflected the basic theory directly. The Mann–Whitney approach is easy to apply and leads more directly to confidence intervals for differences between population means or medians.

The Wilcoxon formulation needs a joint ranking of observations from the two samples, and we sum the ranks associated with one sample. As indicated in Example 1.4, if both samples come from the same population (which may be of any continuous form and need not be symmetric) we expect a mix of low, medium and high ranks in each sample. If the alternative to the null hypothesis of identical populations is that the population distributions differ only in location (i.e. mean or median), then under that alternative we expect lower ranks to dominate in one population and higher ranks in the other. A shift in location epitomizes the concept of an 'additive' treatment effect, or a 'constant' difference between two treatments. The test is also relevant when we sample from two distributions with cumulative distribution functions $F(u)$ and $G(v)$ identical under H_0, but under H_1, for all x, either $F(x) \leq G(x)$ or $F(x) \geq G(x)$ with strict inequality for at least some x; a moment's reflection shows that under H_1 low or high ranks should dominate in one sample, as opposed to a fairly even distribution of ranks under H_0. Given the permutation distribution of rank sums under H_0, critical regions may be determined in the way described for the particular case in Example 1.4.

Example 5.3

The problem. Given the data on page numbers for books on biology and management in Example 5.2, test the hypothesis that the medians do not differ against a two-sided alternative. The data are

Biology	143	173	226	233	250	287	291	303	634	637		
Management	50	164	198	221	225	302	328	335	426	534	586	618

Formulation and assumptions. We require the sum of ranks associated with (either) sample in a joint ranking of all data. If this is sufficiently small or large to lie in an appropriate critical region we reject H_0 : *the population locations are identical.*

Procedure. Since the data within each sample are in ascending order it is easy to write down the $m+n$ ranks (here $m=10$, $n=12$) associated with the combined samples. These ranks are:

Biology	2	4	8	9	10	11	12	14	21	22		
Management	1	3	5	6	7	13	15	16	17	18	19	20

Denoting by S_m, S_n the sums of ranks associated with each sample we find

$$S_m=2+4+8+9+10+11+12+14+21+22=113$$

Similarly, $S_n=140$. We may compare the lesser of these with tabulated values, although the more common tables give not a relevant value of S_m itself but that of a related statistic $U_m=S_m-\frac{1}{2}m(m+1)$ (or the equivalent $U_n=S_n-\frac{1}{2}n(n+1)$). Here $U_m=113-\frac{1}{2}\times 10\times 11=58$; $U_n=140-\frac{1}{2}\times 12\times 13=62$. Tables V and VI give for various m, n the value of the lesser of U_m, U_n which must not be exceeded for significance at nominal 1% and 5% levels in one- and two-tail tests. Table VI gives 29 as the maximum value of the lesser of U_m, U_n for significance in a two-tail test with samples of 10 and 12. We found $U_m=58$.

Conclusion. Since U_m exceeds 29 we do not reject H_0.

Comments. More extensive tables of critical values for the Wilcoxon–Mann–Whitney statistic are given by Neave (1981, p. 30). Conover (1980, Table A7) gives various quantiles for S_m, S_n. Actual, rather than nominal, significance levels may be obtained from a computer program giving the exact permutation distribution.

Computational aspects. STATXACT, TESTIMATE and DISFREE give exact tail probabilities corresponding to observed S or U. Many general statistical packages compute S or U but leave the user to consult tables or give an asymptotic result which may be unsatisfactory if, for example, one sample is large but the other small, or if there are many tied ranks. For our data STATXACT indicates $\Pr(U_m\leq 58)=0.4614$. It also gives the exact significance level for a two-tail test corresponding to $U\leq 29$ as 4.26%.

We only need to compute *ab initio* one of S_m or S_n, for the sum of ranks from 1 to $m+n$ is $\frac{1}{2}(m+n)(m+n+1)= S_m+S_n$. Since $U_m=S_m-\frac{1}{2}m(m+1)$

and $U_n = S_n - \frac{1}{2}n(n+1)$ we easily see that each has minimum value zero and verify that

$$U_m = mn - U_n \tag{5.4}$$

so that again only one of U_m, U_n need be computed *ab initio*. Either may be used in a test. We chose the lesser in Example 5.3 because most tables are relevant to this.

5.1.4 The Mann–Whitney formulation

The statistic U_m or U_n is calculated directly in the Mann–Whitney approach. Ranks are not needed, and the procedure eases calculation if no computer program is available for the test. It also forms the basis for determining a confidence interval for location difference. To obtain U_m or U_n we count the number of observations in one sample exceeding each member of the other sample.

Example 5.4

The problem. Recalculate the test statistic for the data in Example 5.3 using the Mann–Whitney approach.

Formulation and assumptions. We inspect the observations in each sample; they need not be ordered, but counting is easier if they are. We count the number of times each observation in sample 1 (biology books) is exceeded by an observation in sample 2. This gives U_n.

Procedure. The numbers of pages are:

Biology	143	173	226	233	250	287	291	303	634	637		
Management	50	164	198	221	225	302	328	335	426	534	586	618

Clearly the first observation 143 in sample 1 is exceeded by 11 observations in sample 2 (i.e. by all observations except 50). Similarly 173 in sample 1 is exceeded by 10 observations in sample 2. Proceeding in this way we find the numbers of observations in sample 2 exceeding each observation in sample 1 and add these, viz. $U_n = 11+10+7+7+7+7+7+6+0+0 = 62$, as found in Example 5.3. Using (5.4), $U_m = 120 - 62 = 58$.

Conclusion. As in Example 5.4.

Comment. Equivalence of the Wilcoxon and Mann–Whitney formulations is general.

5.1.5 Ties

As with the signed rank test (Section 2.2.3) we use mid-ranks for ties, and for only a few ties basing significance tests on the appropriate critical values for the 'no-tie' case is unlikely to be seriously misleading. If m, n are both reasonably large (say 15 or greater), the normal approximation we develop in Section 5.1.7 may be used with reasonable confidence, adjustment being essential only when there are quite a few ties. In Example 5.6 we meet a situation where ties dominate.

In the Mann–Whitney formulation, if an observation in the second sample equals an observation in the first sample it is scored as ½ in counting the number of observations in the second sample exceeding that observation in the first.

For small to medium-sized samples ties present no difficulty if a computer program is available to generate exact probabilities for the WMW test under the appropriate permutation distribution.

Example 5.5

The problem. In Example 2.1 we used samples of numbers of references in papers published in the journal *Biometrics* in the years 1956–7. We give below numbers of references in a different sample of 12 for those years and also for a sample of 14 for the same journal for the year 1990. If θ_1, θ_2 represent the population medians, test $H_0 : \theta_1 = \theta_2$ against $H_1 : \theta_2 > \theta_1$.

1956–7 2 3 6 6 6 9 9 10 14 15 16 72
1990 6 8 9 10 16 18 20 22 22 23 23 26 28 59

Formulation and assumptions. We use the WMW test with mid-ranks for ties. A one-tail test is appropriate in the light of general trends evident in many journals for more references in later issues, especially in areas of high research activity. Some skewness is evident in the samples.

Procedure. As the data are ordered it is easy to allocate ranks and mid-ranks. These are

1956–7 1 2 4.5 4.5 4.5 9 9 11.5 13 14 15.5 26
1990 4.5 7 9 11.5 15.5 17 18 19.5 19.5 21.5 21.5 23 24 25

However, unless we need S_n there is no need to allocate ranks as we may get U_n by simply counting for each first sample value the number of observations exceeding it in the second sample (scoring one half for ties) and summing these. For example, for *each* value 6 in the 1956–7 sample there is one tied value, scored as ½ and 13 values exceeding 6 in the second sample, giving a contribution of 13.5 to U_n. Proceeding in this way we find

$$U_n = 14 + 14 + 13.5 + 13.5 + 13.5 + 11.5 + 11.5 + 10.5 + 10 + 10 + 9.5 + 0 = 131.5$$

From (5.4), $U_m = 168 - 131.5 = 36.5$. Table VI indicates that in a one-tail test U, the lesser of U_m, U_n, must not exceed 38 for significance at the 1% level, so there is a clear indication of significance although the presence of ties means reservations about claiming significance at the nominal 1% level. The exact permutation test in STATXACT, which allows for ties, indicates $\Pr(U \leq 36.5) = 0.0065$.

Conclusion. We reject H_0 at an actual 0.65% significance level.

Comments. 1. This one example suggests that a few ties do not seriously upset conclusions based on 'no-tie' critical values for moderate sample sizes.

2. The situation is less satisfactory for ties in unbalanced samples. For example, if we had a sample of 3 with values 1, 2, 2, and a sample of 13 with values 1, 1, 4, 5, 5, 5, 7, 8, 9, 9, 9, 9, 10 it is easily verified that $U=5$. This would not indicate significance in a no-tie situation two-tail test since then $\Pr(U \leq 5) = 0.0571$. However the exact permutation test allowing for ties gives $\Pr(U \leq 5) = 0.0464$.

3. As we demonstrated in Section 2.2.3, if ties result from rounding and more accuracy can be introduced to break ties, the precise nature of the tie breaks may appreciably alter our conclusions regarding significance.

Computational aspects. Software to determine exact permutation probabilities asociated with the test statistics U or S is particularly valuable when there are ties. STATXACT, TESTIMATE and DISFREE allow this. Some general programs such as MINITAB take ties into account by using mid-ranks but provide only an asymptotic test (which may be unreliable for small $m+n$ or when one of m, n, is small). Alternatively, one may resort to tables appropriate to the no-tie situation but nominal significant levels can no longer be guaranteed. Daniel (1990, p.125) lists a number of packages that include WMW and other two-sample location tests.

A common tie situation is one where we are not given precise measurements, but only grouped data. For example, instead of the complete sample values in Example 5.5 we may be given only numbers of papers in which there are 0 to 9 references, 10 to 19 references and so on. We may still calculate U or S making allowances for ties, but it may now be misleading to use tabulated critical values for the Mann–Whitney statistic.

Example 5.6

The problem. Instead of the data in Example 5.6 we are given only the numbers of papers with 0–9, 10–19, 20–29 references, etc. so that our data are now:

No. of references	*0–9*	*10–19*	*20–29*	*≥30*
1956–7	7	4	0	1
1990	3	3	7	1

Perform the hypothesis test in Example 5.5 using this reduced information.

Formulation and assumptions. We use a WMW test based on mid-ranks for ties, using Table VI for the critical value of U but realizing this may not be completely satisfactory. A program giving exact permutation probabilities using mid-ranks is more appropriate.

Procedure. Calculation of U_n is reasonably straightforward but needs care. For *each* of the 7 ties in 0–9 for 1956–7 there are 3 ties (scoring 1.5 in all) and $3+7+1=11$ greater values in 1990, thus each scores $1.5+11=12.5$ Similarly each of the 4 ties in 10–19 for 1956–7 scores 9.5 and so on. The reader should verify that $U_n=7\times12.5+4\times9.5+0.5=126$. From (5.4) $U_m=168-126=42$. Thus our test statistic is $U=42$. Table VI indicates that in the no-tie case the critical value for significance at the 5% level for a one-tail test is $U\leq51$. The exact permutation probability that $U\leq42$ given by STATXACT is 0.01.

Conclusion. We reject the hypothesis of equal medians at an actual 1% significance level.

Comments. 1. With no ties significance at the 1% level requires $U\leq38$.
2. The median test in Section 5.1.1 can be formulated as a WMW test with ties. Effectively, we score all values below the combined sample median as ties within each sample; similarly, another set of ties is formed by all values above the combined sample median in each set. This is one of many situations we meet where two different approaches have a common meeting ground. Also, as indicated in Section 5.1.1, the median test is a special case of the more general Fisher exact test.

5.1.6 Wilcoxon–Mann–Whitney confidence intervals

The Mann–Whitney statistic compares all differences $d_{ij}=x_i-y_j$ between sample values. In computing U_m we allocate a score of 1 if d_{ij} is positive, zero if d_{ij} is negative and ½ if d_{ij} is zero. To calculate U_n we reverse these scores, i.e. score 1 if d_{ij} is negative, zero if d_{ij} is positive and ½ if d_{ij} is zero. To calculate a confidence interval for location difference we need the actual values of some or all d_{ij}. If c is the value of the Mann–Whitney statistic U required for significance at the $100\alpha\%$ significance level, the $100(1-\alpha)\%$ confidence limits are given by the $c+1$ smallest and $c+1$ largest d_{ij}.

Example 5.7

The problem. Obtain nominal 95% confidence limits for a population location shift based on the WMW method using the data in Example 5.3.

Formulation and assumptions. Table VI indicates the critical value $U=29$ for significance at the nominal 5% level for samples of 10, 12. Denoting the sample values by $x_1, x_2, \ldots, x_{10}$ and $y_1, y_2, \ldots, y_{12}$ we require the 30 largest and smallest $d_{ij}=x_i-y_j$. It is not essential to compute all d_{ij}, but for completeness we give these in Table 5.3.

Procedure. It eases computation if the sample values are ordered. We write those for the first sample in the top row and those for the second sample in the first (left) column. The entries in the body of Table 5.3 are the differences between the data entry at the top of that column and at the left of that row, e.g. the first entry of 93 is 143−50. Note that the difference between each entry in any pair of rows is a constant equal to the difference between the corresponding entries in the left-hand data column; there is an analogous constant difference between entries in pairs of columns. The largest entries appear in the top right of the table and entries decrease and eventually change sign as we move towards the bottom left.

A count shows 62 negative and 58 positive values implying $U_n=62$, $U_m=58$, as found in Example 5.4. The critical value is $U=29$ for significance at the nominal 5% level so we find the lower limit for a 95% confidence interval for the difference $\theta_1-\theta_2$ by eliminating the 29 largest negative differences; the next largest negative difference is the required lower limit. Inspection of Table 5.3 shows this lower limit is −185. Similarly, elimination of the 29 largest positive differences gives the upper limit 103.

Table 5.3 Paired differences for numbers of pages in Biology and Mangement books

	143	173	226	233	250	287	291	303	634	637
50	93	123	176	183	200	237	241	253	584	587
164	−21	9	62	69	86	123	127	139	470	473
198	−55	−25	28	35	52	89	93	105	436	439
221	−78	−48	5	12	29	66	70	82	413	416
225	−82	−52	1	8	25	62	66	78	409	412
302	−159	−129	−76	−69	−52	−15	−11	1	332	335
328	−185	−155	−102	−95	−78	−41	−37	−25	306	309
335	−192	−162	−109	−102	−85	−48	−44	−32	299	302
426	−283	−253	−200	−193	−176	−139	−135	−123	208	211
534	−391	−361	−308	−301	−284	−247	−243	−231	100	103
586	−443	−413	−360	−353	−336	−299	−295	−283	48	51
618	−475	−445	−392	−385	−368	−331	−327	−315	16	19

Conclusion. A nominal 95% confidence interval for the difference $\theta_1 - \theta_2$ is $(-185, 103)$ or equivalently, for the difference $\theta_2 - \theta_1$ the interval is $(-103, 185)$.

Comments. 1. In Example 5.2, using the median test method a nominal 95% confidence interval for the difference $\theta_2 - \theta_1$ was $(-78, 200)$. The current interval is slightly longer and shifted to the left, but is shorter than the interval $(-142.94, 172.04)$ based on an assumption of normality.

2. Confidence intervals at other levels can be obtained. For example, a 99% confidence interval is obtained (see Table VI) by rejecting the 21 extreme differences and is $(-283, 200)$ for the difference $\theta_1 - \theta_2$.

3. The computational process is reminiscent of that with Walsh averages in Section 2.2.4.

Computational aspects. As for the Wilcoxon signed rank test, there are few completely satisfactory programs for confidence intervals. A program with similar limitations to that provided for the signed rank test is included in MINITAB, and Neave and Worthington (1988) give a BASIC program listing for confidence intervals that parallels that given by them for the signed-rank procedure. If one wishes to explore further the exact confidence level one may examine results of hypothesis testing situations at or near the end points of the interval, much as we did for the Wilcoxon signed rank test. This is achieved in the case of a 95% confidence interval by an appropriate addition or subtraction to all observations in one sample, as we did for median test confidence intervals in Example 5.2, so that we in fact test the hypothesis $H_0 : \theta_2 = \theta_1 - 103$ or $\theta_2 = \theta_1 + 185$ (see Exercise 5.4).

5.1.7 A large-sample normal approximation

There are simplified formulae for asymptotic results for the WMW test without ties, but it is convenient to introduce asymptotic results valid for any scoring system. These cover not only mid-ranks for ties but also apply to the rank transformations that we consider in Sections 5.2 and 5.3. In all these cases we denote the score for the ith ordered observation in the joint sample by s_i. If the sample sizes are m, n and T denotes the sum-of-scores statistic based on the sample of m, then the general theory for the relevant permutation distribution gives, after some straightforward but tedious applications of standard results, the mean value $\mathrm{E}(T)$ and the variance $\mathrm{Var}(T)$ of T to be

$$\mathrm{E}(T) = m(\Sigma s_j)/(m+n) \qquad (5.5)$$

$$\mathrm{Var}(T) = \frac{mn}{(m+n)(m+n-1)}\left[\Sigma s_j^2 - \frac{1}{m+n}(\Sigma s_j)^2\right] \qquad (5.6)$$

where summations are over all j from 1 to $m+n$. For large m, n

$$Z=[T-\mathrm{E}(T)]/\sqrt{\mathrm{Var}(T)} \tag{5.7}$$

has approximately a standard normal distribution. For the WMW test with no ties (5.7) reduces to

$$Z=\frac{S_m-\frac{1}{2}m(m+n+1)}{\sqrt{[mn(m+n+1)/12]}} \tag{5.8}$$

A continuity correction in the 'no-tie' case has little practical effect for large m and n, but if used the correction is to add ½ to S_m if Z is negative and to subtract ½ if Z is positive. If one prefers to use S_n the only alteration needed in (5.8) is to interchange m and n throughout. If one uses the Mann–Whitney statistic the numerator of (5.8) is modified, giving

$$Z=\frac{U_m-\frac{1}{2}mn}{\sqrt{[mn(m+n+1)/12]}} \tag{5.9}$$

If there are ties and we use mid-rank scores the numerators, but not the denominators, of (5.8) and (5.9) are unaltered. Some writers (e.g. Daniel, 1990, p. 94) give rules for modifying the denominator, but with modern calculators or computers it is easy to use (5.6) directly where the s_j are mid-rank scores. Unless there are many ties the effect of these is small.

Example 5.8

The problem. Ages at death for members of two more clans (Section A7) – McBeta and McGamma – in the Badenscallie burial ground (arranged in ascending order) are:

McBeta	0	19	22	30	31	37	55	56	66	66	67	67
	68	71	73	75	75	78	79	82	83	83	88	96
McGamma	13	13	22	26	33	33	59	72	72	72	77	78
	78	80	81	82	85	85	85	86	88			

Use the WMW large-sample approximation to test H_0 : *the populations are identical* against H_1 : *the populations differ in location.*

Formulation and assumptions, With sample sizes 24, 21 a large-sample approximation will suffice. To adjust for ties we require mid-ranks.

Procedure. The appropriate mid-ranks are:

McBeta	1	4	5.5	8	9	12	13	14	16.5
	16.5	18.5	18.5	20	21	25	26.5	26.5	30
	32	35.5	37.5	37.5	43.5	45			
McGamma	2.5	2.5	5.5	7	10.5	10.5	15	23	23
	23	28	30	30	33	34	35.5	40	40
	40	42	43.5						

While one may use either S or U, once the ranks are recorded it is probably as easy to use S. For the McBeta's we easily find $S_m=516.5$. Ignoring the effect of ties and using (5.8) gives

$$Z=(516.5-552)/\sqrt{1932})=-35.5/43.95=-0.81.$$

We leave it as an exercise (Exercise 5.5) to show that using (5.6) to compute the variance in formula (5.7) reduces the denominator above from 43.95 to 43.92, demonstrating the trivial effect of only a few ties.

Conclusion. Since 0.81 is well short of the critical value $|Z|=1.96$ required for significance in a two-tail test at the 5% level we do not reject the hypothesis of identical populations.

Comment. For both m, n greater than 20 one expects asymptotic results to agree closely with those for the exact permutation test. Tables show $\Pr(|Z|\geq 0.81)=0.418$; the exact WMW permutation test two-tail probability in this case is 0.427. See also comment 1 on Example 5.9.

Computational aspects. Even with efficient algorithms computation of the exact permutation distribution tail probability in this case takes minutes on most PCs, whereas the asymptotic result is available in seconds. While computational times are software- and processor-dependent, there is little point in calculating exact probabilities when an asymptotic result suffices. Of particular interest when using an asymptotic result is whether a reasonable fit in the tails is obtained. For the case $m=24$, $n=21$, if $S_m=394$ the exact test one-tail probability is 0.0001 and the asymptotic probability is 0.0002; if $S_m=478$ the corresponding probabilities are 0.0317 and 0.0311.

If an approximate 95% confidence interval is required for large m, n the value of U giving the number of extreme differences x_i-y_j in a table similar to Table 5.3 is obtained by setting $Z=-1.96$ and substituting the values of m, n in (5.9). In practice, for large m, n this value of U will be large and a computer program will be almost essential to get the required limits. MINITAB and several other packages provide such a program.

5.1.8 The Gehan–Wilcoxon test for censored data

We met a one-sample example of censored data in Example 1.2. Censoring is common in medical studies that involve long-term follow-up of patients after treatment and also in some industrial contexts. Davis and Lawrance (1989) give an example involving tyre failures. Tyres that had not failed at the end of the experimental period were regarded as censored observations. In a medical context censoring may arise because patients can no longer be traced due to factors that may be similar for all treatments being compared.

In a study of survival times patients may be lost due to their leaving the district or refusing to attend follow-up clinics and so on, and many may still be alive when the study is terminated. Gehan (1965a; 1965b) proposed a generalized Wilcoxon test for both single- and two-sample problems and various types of censoring. We consider one two-sample case. The scoring procedure involves a modified ranking for censored observations that takes account of censoring time and an average life expectancy under the null hypothesis of no treatment difference (i.e. the hypothesis that the two samples come from the same population).

All $m+n$ observations, including censored observations, are arranged in ascending order. If the ith ordered observation is uncensored it is given the score $s_i=i$ (the usual WMW rank). If the ith ordered observation is censored it is given the score $s_i^*=\frac{1}{2}i+\frac{1}{2}(m+n+1)$. The term $\frac{1}{2}(m+n+1)$ is the average rank of all $m+n$ observations. The term $\frac{1}{2}i$ may be looked upon as an additional credit to life expectation for the time survived prior to censorsing. In particular, this means that a person dropping out before the first death is given the average ranking plus ½ (corresponding broadly to an average or median life expectancy score). A person who is censored as the last survivor would receive the rank $m+n+\frac{1}{2}$, compared with the rank $m+n$ if he or she were the last to die. All survivors at the time of the ith death are ensured a higher score than that accorded to the ith death. The range of possible scores is similar to that for the Wilcoxon ranks, and a reasonable assumption is made about expected survival times with the appealing property that expected total time of survival score increases the longer one has survived. We illustrate the method with an example, discussing 'ties' under *procedure*.

Example 5.9

The problem. We indicated in Example 1.2 that Dinse (1982) gave survival data for 28 asymptomatic cases as well as for the 10 symptomatic cases in that example. The complete data, an asterisk denoting censoring, are

Symptomatic	49	58	75	110	112	132	151	276	281	362*
Asymptomatic	50	58	96	139	152	159	189	225	239	242
	257	262	292	294	300*	301	306*	329*	342*	346*
	349*	354*	359	360*	365*	378*	381*	388*		

We test H_0 : *the survival time distributions are identical* against H_1 : $G(x)\geq F(x)$ *with strict inequality for some* x, where $F(u)$, $G(v)$ are survial time cdfs for symptomatic and asymptomatic cases.

Formulation and assumptions. A one-tail test is appropriate since H_1 is one-sided. Because of censorship Gehan–Wilcoxon scores are appropriate.

Procedure. Scores are allocated as described above. If a death and censored observation coincide this is **not** a tie; the censored observation is, as is intuitively reasonable, the greater. If there are ties for time to death or ties for times to censorship a mid-score is given to such ties. Bearing these rules in mind, since the smallest censored observation occurs at 300, the 23 lesser observations are ranked 1 to 23 in the usual way (with a mid-rank of 3.5 for the tie at 58). Since 300 is the first censored observation and there are 23 lesser values we calculate the appropriate score as $s_{24}^* = \frac{1}{2} \times 24 + \frac{1}{2}(10+28+1) = 31.5$. The entry 301 (uncensored) is the 25th ordered observation and thus $s_{25}=25$. Proceeding in this way the remaining observations are given the ranks recorded below.

Symptomatic	1 3.5 5 7 8 9 11 20 21 36.5
Asymptomatic	2 3.5 6 10 12 13 14 15 16 17 18 19 22 23 31.5 25 32.5 33 33.5 34 34.5 35 32 36 37 37.5 38 38.5

We compute $S_m = 122$. The STATXACT program for calculating the exact Gehan–Wilcoxon one-tail probability gives $\alpha=0.0046$. The asymptotic value given by (5.7) is 0.0053.

Conclusion. We reject H_0 at an actual 0.46% significance level.

Comments. 1. We expressed H_1 in terms of one distribution 'dominating' the other rather than as a location shift. This is often appropriate for survival data; an efficient treatment tends to increase life expectancy not by a constant amount for all subjects, but by amounts which vary from subject to subject, often depending on whether there is complete recovery or merely arrested development of a disease. Similar arguments might be advanced on genetical grounds in Example 5.8 for expressing H_1 as a hypothesis of dominance rather than one of shift in location.

2. All censored values in this example are at least 300; for the asymtomatic cases survival times are sufficiently long for many patients to survive this time, whereas only one symptomatic case does so.

Computational aspects. STATXACT and DISFREE contain programs for the Gehan–Wilcoxon test for censored data. In STATXACT the weights are different from those given here, being a linear transformation of those we use, but they lead to an equivalent test. They are calculated by a formula given in the STATXACT manual.

5.2 TESTS BASED ON TRANSFORMATION OF RANKS

Permutation tests for location or dominance of one population distribution over the other may be based on transformations of ranks that preserve order, including the observations themselves – the raw scores. Using these, the permutation distribution must be calculated afresh for each problem, a difficulty that is not serious with packages such as STATXACT or TESTIMATE, but as for single samples the approach lacks robustness. If we regard the combined-sample ordered observations as 'scores' relevant asymptotic tail probabilities are given by (5.7). We do not pursue this approach, but STATXACT, TESTIMATE and DISFREE provide programs to calculate exact probabilities for small samples as well as giving asymptotic approximations.

Normal scores provide a more viable approach. Van der Waerden scores use the $r/(m+n+1)$th standard normal quantiles ($r=1, 2, \ldots, m+n$) in place of ranks. Expected normal scores may also be used, and in either case we get samples like those from a normal distribution. Symmetry implies that the mean of the $m+n$ van der Waerden scores is zero if there are no ties. Thus (5.5) and (5.6) reduce respectively to $\mathrm{E}(T)=0$ and $\mathrm{Var}(T)=mn(\Sigma s_i^2)/[(m+n-1)(m+n)]$ where summation is over the squares of all $m+n$ van der Waerden scores s_i. The test statistic $Z=T/\sqrt{[\mathrm{var}(T)]}$ is asymptotically standard normal. This approximation is good even for moderately small samples in view of the near-equivalence of the scores to an ideal sample from a normal distribution.

The mean of all van der Waerden scores is not exactly zero with ties if we use quantiles based on mid-ranks; for a few ties the effect is negligible as adjacent quantile scores are replaced by scores close to their mean.

Example 5.10

The problem. For the data in Example 5.3 test H_0 : *the distributions are identical* against that of a location difference using van der Waerden scores.

Formulation and assumptions. There are $10+12=22$ observations, so we replace rank i by the $i/23$rd quantile, e.g. when $i=1$ we require the 1/23rd quantile, given by tables of the standard normal distribution as -1.71. It suffices to record these scores to two decimal places.

Procedure. In Example 5.3 we allocated the following ranks:

Biology	2	4	8	9	10	11	12	14	21	22		
Management	1	3	5	6	7	13	15	16	17	18	19	20

The corresponding van der Waerden scores for Biology are

Biology -1.36 -0.94 -0.39 -0.28 -0.16 -0.05 0.05 0.28 1.36 1.71.

The sum of these scores is 0.22. Since the sum of all scores is zero, it follows that the sum associated with management books is $T=-0.22$. The sum of squares of all the 22 scores is easily found to be 16.8968 whence

$$Z=-0.22/\sqrt{[10\times 12\times 16.8968/(21\times 22)]}=-0.105$$

Tables of the standard normal distribution give $\Pr(Z\leq -0.105)=0.458$.

Conclusion. We accept the hypothesis of no population difference.

Comments. Under computational aspects in Example 5.3 we pointed out that the permutation test probability associated with the WMW test was 0.462. It is a common experience that normal score tests give similar results to rank tests over a wide range of conditions.

Computational aspects. Exact probabilities can be calculated using these scores in a general permutation test such as that in STATXACT.

A number of alternatives to the WMW procedure have been proposed which are dependent on ranks or a transformation of ranks. Many references are given by Conover, (1980, pp. 226–7).

5.3 TESTS FOR EQUALITY OF VARIANCE

We may want to test for heterogeneity of variance, i.e. to test H_0 : *samples are from identical distributions* against H_1 : *samples are from distributions with different variances.* Some relevant tests use ranks or transformation of ranks much as in the WMW and related location tests except that ranks are allocated differently.

5.3.1 The Siegel–Tukey test

This test is easy to carry out but not very powerful. The basic idea behind it is that if two samples come from populations differing in variance, the sample from the population with greater variance will be more spread out. If there is a location difference we first align the populations by shifting the median (or mean) of one sample to coincide with that of the other sample. This requires an appropriate addition to or subtraction of a constant for all observations in one sample. The sample variances are unaltered by this change. If we now arrange the combined samples in order and allocate the rank 1 to the smallest observation, 2 to the largest, 3 to the next largest, 4 and 5 to the next two smallest, 6 and 7 to the next two largest, and so on, the sum of the ranks for the sample from the population having the greater variance should be smaller than if there were no difference in variance.

Example 5.11

The problem. Davis and Lawrance (1989) give (as part of a larger data set collected for different purposes) the time in hours to two different types of tyre failure under similar test conditions. Failure type A is rubber chunking on shoulder and failure type B is cracking of the side wall. Do the following data suggest the population variances may differ?

Type A 177 200 227 230 232 268 272 297
Type B 47 105 126 142 158 172 197 220 225 230 262 270

Formulation and assumptions. We align the medians of the two samples, then allocate ranks as described above. The lesser sum of ranks associated with one sample is calculated and tested for significance as in the WMW test. Our null hypothesis is that the variances are identical. Without further information a two-tail test is appropriate.

Procedure. The first sample median (see Section A1.2) is ½(230+232) =231 and the second sample median is 184.5. We align the samples for location by subtracting 231−184.5=46.5 from all first-sample values and work with the adjusted samples

Type A (adjusted) 130.5 153.5 180.5 183.5 185.5 221.5 225.5 250.5
Type B 47 105 126 142 158 172 197 220 225 230 262 270

To allocate ranks using the scheme outlined above it is conveninent to arrange all sample values in ascending order before allocation. The result of this operation is given in Table 5.4 where type A (adjusted) values are indicated in **bold.**

Table 5.4 Allocation of ranks for Siegel–Tukey test; type A (adjusted) in **bold**

Value	47	105	126	**130.5**	142	**153.5**	158	172	**180.5**	**183.5**
Rank	1	4	5	**8**	9	**12**	13	16	**17**	**20**
Value	**185.5**	197	220	**221.5**	225	**225.5**	230	**250.5**	262	270
Rank	**19**	18	15	**14**	11	**10**	7	**6**	3	2

The sum of the ranks for type A faults is S_m=106, giving U_m=70 and U_n=96−70=26. Table VI gives U=22 as the critical value for significance at a 5% level in a two-tail test for samples of 8 and 12 observations.

Conclusion. We do not reject H_0 at a nominal 5% significance level.

Comments. 1. The critical value for significance at a nominal 5% level in a one-tail test is U=26, so had a one-tail test been appropriate we would have rejected H_0.

2. We 'relocated' the type A data by equating sample medians. Since variance is based on squared deviations from the mean, equating means has intuitive appeal. In practice, which we choose usually makes little difference. We chose the median because it is simple to calculate. In Exercise 5.6 we seek the consequences of equating means.

Computational aspects. After ranks are allocated any program that computes exact probabilities associated with the permutation distribution for the WMW test may be used.

5.3.2 The Conover squared rank test for variance

If the means of X, Y are respectively μ_x, μ_y then equality of variance implies $E[(X-\mu_x)^2]=E[(Y-\mu_y)^2]$, where $E[X]$ is the expectation of X.

Conover (1980) proposed a test for equality of variance based on the joint squared ranks of squared deviations from the means, $(x_i-\mu_x)^2$, $(y_i-\mu_y)^2$. The population means are seldom known but it is reasonable to replace them by their sample estimates m_x, m_y. We do not need to square the deviations to obtain the required rankings because the same order is achieved by ranking the absolute deviations $|x_i-m_x|$, $|y_i-m_y|$. We rank these deviations and use as scores squares of the ranks. Our test statistic T is the sum of the scores of these ranks for one of the samples. For large samples Z given by (5.7) has a standard normal distribution. Conover (1980) gives quantiles of T for a range of sample sizes in a 'no-tie' situation. Programs generating the permutation distribution for arbitrary scores may also be used if exact tail probabilities are required.

Example 5.12

The problem. For the data in Example 5.11 use the squared rank test for equality of variance.

Formulation and assumptions. We require deviations of each observation from its sample mean. The absolute deviations are then ranked, the ranks squared and the statistic T is calculated.

Procedure. Denoting type A sample values by x_i and type B sample values by y_i we find $m_x=238$ and $m_y=180$ (it suffices to express these to the same order of accuracy as the data, here to the nearest integer). We compute the absolute deviations for each sample value, e.g. for the type A observation 227, the deviation is $227-238=-11$, giving an absolute deviation of 11. Table 5.5 gives the ordered absolute deviations for the combined samples together with squared ranks (squared mid-ranks for ties). Type A deviations are in **bold.**

The sum of the bold squared ranks is $T=706.5$. With squared ranks as scores we find from (5.6) that $\text{Var}(T)=78\ 522.49$. Also (5.5) gives $E(T)=1147.6$, whence from (5.7), $Z=(706.5-1147.6)/280.219=-1.574$.

Conclusion. Since $|Z|<1.64$ we do not reject the hypothesis of equal variance at the 5% significance level even in a one-tail test.

Table 5.5 Absolute deviations and squared ranks (sqr), Example 5.12. Type A in **bold**

Adev	**6**	**8**	8	**11**	17	22	**30**	**34**	**38**	38
Sqr	**1**	**6.25**	6.25	**16**	25	36	**49**	**64**	**90.25**	90.25
Adev	40	45	50	54	**59**	**61**	75	82	90	133
Sqr	121	144	169	196	**225**	**256**	289	324	361	400

Comments. 1. Calculating sample means to 1 decimal place would avoid data ties, but this makes little difference to our calculated Z.

2. Whereas the Siegel–Tukey test would establish significance in a one-tail test if that were appropriate, the Conover squared rank test fails to do so. While the test appears in this case to be less powerful than the Siegel–Tukey test, it is reasonably robust.

3. We could have used ranks of absolute deviations rather than squared ranks. In that case we essentially have a WMW test (see Exercise 5.7). It is not clear whether use of absolute deviation ranks or of absolute deviation squared ranks provides the more powerful test.

Computational aspects. If an exact tail probability is of interest the STATXACT program for the exact permutation distribution may be used with the squared rank scores above. For the asymptotic test $\Pr(|Z| \geq 1.574) = 0.1156$. The exact permutation test probability is 0.1152.

Alternative tests for equality of variance have been proposed by Mood (1954), Ansari and Bradley (1960), Klotz (1962), Moses (1963) and others; details are given by Daniel (1990, Section 3.2) and Marascuilo and McSweeney (1977, Sections 11.11, 11.12).

5.4 TESTS FOR A COMMON DISTRIBUTION

In Section 3.3 we developed tests of hypotheses that a single sample was drawn from some specified distribution. For two independent samples we may want to know if each can be supposed to come from the same unspecified distribution. The Smirnov test has similarities to the Kolmogorov test developed in Section 3.3.1.

5.4.1 Smirnov test for a common distribution

The hypothesis H_0 : *two samples come from the same distribution* may be tested against H_1 : *the distributions have different cumulative distribution functions (cdfs).* We do not specify further the nature of the difference: they might have the same mean but different variances; one may be skew, the other symmetric, etc. We compare the sample cdfs; the test statistic is the difference of greatest magnitude between these two functions.

Example 5.13

The problem. Use the Smirnov test to decide if it is reasonable to conclude that the samples in Example 5.11 are from the same poplulations.

Formulation and assumptions. We compute the sample cumulative distribution functions $S_1(x)$, $S_2(y)$ at each sample value and at each of these values we also compute and record the difference $S_1(x)-S_2(y)$. For samples of size m, n respectively $S_1(x)$, $S_2(y)$ are step functions with steps $1/m$, $1/n$ respectively at each sample value (or multiple steps at ties).

Procedure. Table 5.6 gives in successive columns the sample values and corresponding values of $S_1(x)$, $S_2(y)$ and $S_1(x)-S_2(y)$ at each sample point. The difference of greatest magnitude (final column) is 0.5, occurring twice. Tables VII and VIII give critical values for various sample sizes for this test and indicate that with a two-tail test at a nominal 5% significance level we cannot reject H_0. With a one-tail test we are exactly at the critical value at the 5% level. Here a one-tail test is essentially a test of whether the function for type B failures is above that for type A failures for at least one common x, y value, against the alternative that it is everywhere at or below.

Conclusion. There is an indication that the cumulative distribution for type B failures is strictly above that for type A for at least some common x and y.

Comments. 1. Sometimes a test for location or variance establishes a difference when the Smirnov test indicates no overall distributional difference, because Kolmogorov–Smirnov type tests are often less powerful than tests for specific characteristics such as differences in location.

2. Like the Kolmogorov test, the Smirnov test may appear not to be making full use of the data by using only maximum difference. As pointed out in Section 3.3.1, the statistic uses cumulative information. In Section 5.4.2 we see that power may be increased by considering all differences.

3. The nominal significance levels in Table VII and VIII may differ from exact levels because $|S_1(x)-S_2(y)|$ when $x=y$ is a step function with steps at each x_i, y_j and the size of possible steps depends upon the sample size. Many published tables for the Smirnov test give not critical values but quantiles which must be exceeded for significance at a given level. Tables VII and VIII give critical values for maximum$|S_1(x)-S_2(y)|$. Neave (1981, p.31) gives values for the equivalent $mn[\max|S_1(x)-S_2(y)|]$ for significance for a wide range of sample sizes.

Computational aspects. Several standard packages give programs for the Smirnov test (often referred to as the Kolmogorov–Smirnov test). For example, that in STATGRAPHICS gives an approximate significance level but little information about how this is obtained. Usefully, it also gives a plot of the sample cdfs. Neave and Worthington (1988) give a program listing in BASIC to calculate the test statistic but one needs tables to determine whether the result is significant.

Table 5.6 Calculation of the Smirnov test statistic

Type A x_i	Type B y_j	$S_1(x)$	$S_2(y)$	$S_1(x)-S_2(y)$
	47	0	0.083	−0.083
	105	0	0.167	−0.167
	126	0	0.250	−0.250
	142	0	0.333	−0.333
	158	0	0.416	−0.416
	172	0	0.500	−0.500
177		0.125	0.500	−0.375
	192	0.125	0.583	−0.458
200		0.250	0.583	−0.333
	220	0.250	0.667	−0.417
	225	0.250	0.750	−0.500
227		0.375	0.750	−0.375
230	230	0.500	0.833	−0.333
232		0.625	0.833	−0.208
	262	0.625	0.917	−0.292
268		0.750	0.917	−0.167
	270	0.750	1.000	−0.250
272		0.875	1.000	−0.125
297		1.000	1.000	0.000

5.4.2 The Cramér–Von Mises test for identical populations

The differences $S_1(x)-S_2(y)$ at each sample value are not independent; this makes it difficulty to work out the distribution for statistics that take account of all differences. However, for one such test, the Cramér–Von Mises test, an approximate theory gives simple significance tests that, except for very small samples, are virtually independent of sample size.

The test statistic is a function of the sum of squares of the differences $S_1(x)-S_2(y)$ at all sample points. Denoting this sum of squares by S_d^2 the test statistic is $T=mnS_d^2/(m+n)^2$. In a two-tail test for significance at the 5% level T must exceed 0.461; for significance at the 1% level T must exceed 0.743.

Example 5.14

The problem. Perform the Cramér–Von Mises test for the data used in Example 5.11.

Formulation and assumptions. We square and add the differences in the last column of Table 5.6, then form the statistic T.

Procedure. From the last column of Table 5.6 we find the sum of squares of the differences is 2.0024, whence $T=96\times2.0024/(20\times20)=0.48$.

Conclusion. Since $T>0.461$ we conclude that the population cumulative distribution functions differ for at least some x.

Comments. The Cramér–Von Mises test is often more powerful than the Smirnov test and is easy to use because of the approximation. Accurate tables exist for some values of m, n. The only labour additional to that for the Smirnov test is calculation of the sums of squares of differences.

5.5 FIELDS OF APPLICATION

Little imagination is needed to think of realistic situations where one might wish to compare medians or means of two populations (i.e. look for location or 'treatment' differences) on the basis of two independent samples. Here are some relevant situations.

Medicine

For comparing the efficacy of two drugs for reducing hypertension, blood cholesterol levels, relief of headaches or other conditions, independent samples are often needed because of 'interaction' or 'hangover' effects if each drug is given to the same patients (Section 4.1); for this or ethical reasons it may be inappropriate to give both drugs to any one person even after a considerable time lapse.

Sociology

To explore the possibility that town and country children may attain different levels of physical fitness, samples of each might be scored (e.g. by ranking) in a fitness test and the results compared nonparametrically.

Mineral exploration

A mining company has options on two sites but only wishes to develop one. Sample test borings are taken on each and the percentage of the mineral of interest in each boring is determined; these are the basis for a test for population differences in mean or median levels; if there is evidence of one site being richer the company may want an estimate of the difference because development costs and other factors may well differ between sites.

Manufacturing

New and cheaper production methods are often tried. Manufacturers may compare products using a new process or raw material with existing ones to assess quality and durability. Interest here is often not only in 'average' quality, but also in differences in variability.

Psychology

Children with learning difficulties may be given treatment that it is hoped will encourage them to respond to commands. Sixty commands are given to a sample of 10 treated children and to a further sample of 12 untreated children. The number of favourable reponses is recorded for each child. Interest will lie in whether there is a response level difference and a confidence interval for any response shift is likely to be informative.

5.6 SUMMARY

The median test. A test for location based on numbers in each of two samples above and below the combined sample median. It is equivalent to the more general Fisher exact test (Section 5.1.1). An asymptotic approximation is given by (5.3). Confidence intervals (Section 5.1.2) are available if we have full measurement data. The only assumption about the population distributions is that of identical medians under H_0.

Wilcoxon–Mann–Whitney test. A rank test for location differences or domination of one distribution over the other. The test statistic is the rank sum associated with either sample. An equivalent formulation was given by Mann and Whitney. The two formulations are described in Sections 5.1.3 and 5.1.4. Adjustments for ties are given in Section 5.1.5 and confidence intervals for measurement data are obtained in Section 5.1.6. Asymptotic results are given by (5.7), (5.8) and (5.9) in Section 5.1.7. The **Gehan** modification for censored data is given in Section 5.1.8.

Van der Waerden scores. Among tests based on transformation of ranks that using van der Waerden scores is easy to use, but results are often similar to those given by the WMW test (Section 5.2).

Equality of variance. The **Siegel–Tukey test** (Section 5.3.1) is often less powerful and less robust than the **Conover squared rank test** (Section 5.3.2) or a similar test using absolute deviations.

Comparison of distribution functions. The **Smirnov test** (Section 5.4.1) has analogies with the one-sample Kolmogorov test (Section 3.3.1). The alternative **Cramér–Von Mises test** (Section 5.4.2) is easy to use as the significance levels are virtually independent of sample size except for very small samples.

EXERCISES

5.1 Using (5.2) calculate P for the data in (5.1).

5.2 For Example 5.1 verify the values obtained for Z using (5.3) with and without a continuity correction.

5.3 In Example 5.2 verify that we accept H_0 at the end points of the nominal 95% confidence interval computed in that example for the population median difference. Verify also that for any point outside that interval we would reject H_0.

5.4 For the data in Example 5.7 test the hypothesis $H_0 : \theta_2 = \theta_1 - 103$ and $H_0 : \theta_2 = \theta_1 + 185$ and use your results to determine whether or not the end points of the confidence interval (−103, 185) for the difference $\theta_2 - \theta_1$ should be included in that interval.

5.5 Use (5.6) to compute the exact variance using mid-ranks in Example 5.8 and compare it with the result given in that example.

5.6 Perform a Siegel–Tukey test on the data in Example 5.11 after shifting one set of sample values to align the sample means.

5.7 Using the data in Example 5.11 carry out a test analogous to the Conover squared rank test but using absolute ranks instead of squared ranks.

5.8 To compare two different keyboard layouts on a pocket calculator it is designing, a company divided 21 staff volunteers randomly into groups of 10(A) and 11(B). Each group is asked to carry out the same set of calculations, group A using the first type of keyboard, group B the second. Total times in minutes for each individual to complete the calculations are:

Group A	23	18	17	25	22	19	31	24	28	32	
Group B	24	28	32	28	41	27	35	34	27	35	35

Use the Wilcoxon–Mann–Whitney procedure to obtain a nominal 95% confidence interval for median time difference for the two layouts. Is there evidence that one layout is preferable?

5.9 An alloy is composed of zinc, copper and tin. It may be made at one of two temperatures H (higher) or L (lower). We wish to know if one temperature produces a harder alloy. A sample is taken from each of 9 batches at L and 7 at H. To arrange them in ascending order of hardness, all specimens are scraped against one another to see which makes a deeper scratch (a deeper scratch indicates a softer specimen). On this basis the specimens are ranked 1 (softest) to 16 (hardest) with the results given below. Should we reject the hypothesis that hardness is unaffected by temperature? State any assumptions needed for validity of the test you use.

Temperature	H	L	H	H	H	L	H	L	L	H	H	L	L	L	L	L
Rank	1	2	3	4	5	6	7	8	9	10	11	12	13	14	15	16

5.10 Hotpot stoves use a standard oven insulation. To test its effectiveness they take random samples from the production line and heat the ovens selected to 400°C, noting the time taken to cool to 350° after switching off. For a sample of 8 ovens the times in minutes are:

15.7 14.8 14.2 16.1 15.3 13.9 17.2 14.9

They decide to explore a cheaper insulation, and using this on a sample of 9 the times taken for the same temperature drop are:

13.7 14.1 14.7 15.4 15.6 14.4 12.9 15.1 14.0

Are the firm justified in asserting there is no firm evidence of a different rate of heat loss? Obtain a 95% confidence limit for the difference in median heat loss (a) with and (b) without a normality assumption. Comment critically on any difference in your conclusions.

5.11 A psychologist wants to know whether men or women are more upset by delays in being admitted to hospital for routine surgery. He devises an anxiety index measured on patients one week before scheduled admission and records it for 17 men and 23 women. These are ranked 1 to 40 on a scale of increasing anxiety. The sum of the ranks for the 17 men is 428. Is there evidence that anxiety is sex-dependent? If there is, which sex appears to show the greater anxiety?

5.12 A psychologist notes total time (in seconds) needed to perform a series of simple manual tasks for each of 8 children with learning difficulties and 7 children without learning difficulties. The times are:

Without difficulties	204	218	197	183	227	233	191	
With difficulties	243	228	261	202	343	242	220	239

Use a Smirnov test to find whether the psychologist is justified in asserting these samples are likely to be from different populations. Do you consider a one- or a two-tail test appropriate? Perform also a Cramér–Von Mises test. Does it lead you to a different conclusion? If you think the psychologist should have tested more specific aspects of any difference, perform the appropriate tests.

5.13 Apply the Smirnov test for different population distributions to the oven cooling data in Exercise 5.10.

5.14 Suppose we are given the data for response times in LVF and RVF in Table 4.1, but the information that they are paired is omitted. In these circumstances we might analyse them as independent samples. Would we then conclude that reponses in the two fields differed? Does your conclusion agree with that found in Example 4.1? If not, why not?

5.15 The numbers of words in the first complete sentence on each of 10 pages selected at random is counted in each of the books by Conover (1980) and Bradley (1968). The results were:

Conover	21	20	17	25	29	21	32	18	32	31
Bradley	45	14	13	31	35	20	58	41	64	25

Perform tests to determine whether there is evidence that in these books

(i) sentence lengths show a location difference;

(ii) the variances of sentence lengths differ between authors;

(iii) the distributions of sentence lengths differ in an unspecified way;

(iv) the sentence lengths for either author are not normally distributed.

5.16 Lindsey, Herzberg and Watts (1987) give data for widths of first joint of the second tarsus for two species of the insect *Chaetocnema*. Do these indicate population differences between the width distributions for the two species?

Species A	131	134	137	127	128	118	134	129	131	115
Species B	107	122	144	131	108	118	122	127	125	124

5.17 Carter and Hubert (1985) give data for percentage variation in blood sugar over one-hour periods for rabbits given two different dose levels of a drug, Is there evidence of a response difference between levels?

Dose I	0.21	−16.20	−10.10	−8.67	−11.13
Dose II	1.59	2.66	−6.27	−2.32	−10.87
Dose I	1.96	−10.19	−15.87	−12.81	
Dose II	7.23	−3.76	3.02	15.01	

5.18 The journal *Biometrics* published data on the numbers of completed months between receipt of a manuscript for publication and the first reply to the authors for each of the years 1979 and 1983. The data are summarized below. Is there evidence of a difference in average waiting times between 1979 and 1983?

Completed months		0	1	2	3	4	5	>6
Number of authors	1979	26	28	34	48	21	22	34
	1983	28	27	42	44	17	6	16

5.19 Hill and Padmanabhan (1984) give body weights (g) of diabetic and normal mice. Is there evidence of a significant difference in mean body weight? Obtain the Hodges–Lehmann estimate of the difference together with a 95% confidence interval. Compare this interval with that based on the t-distribution.

Diabetic	42	44	38	52	48	46	34	44	38					
Normal	34	43	35	33	34	26	30	31	27	28	27	30	37	38
	32	32	36	32	32	38	42	36	44	33	38			

5.20 The data below are numbers of words with various numbers of letters in 200-word sample passages from the presidential addresses to the Royal Statistical Society by W.F. Bodmer (1985) and J. Durbin (1987). Is there acceptable evidence of a difference between the average lengths of words used by the two presidents?

Numbers of letters	1–3	4–6	7–9	10 or more
Bodmer	91	61	24	24
Durbin	87	49	32	32

5.21 Nigel Smeeton provided the following data on DMF scores for 34 male and 54 female first year dental students. The DMF score is the total of the numbers of decayed+missing+filled teeth.

Males	8	6	4	2	10	5	6	6	19	4	10	4	10	12	7	2	5
	1	8	2	0	7	6	4	4	11	2	16	8	7	8	4	0	2
Females	4	7	13	4	8	8	4	14	5	6	4	12	9	9	9	8	12
	4	8	8	4	11	6	15	9	8	14	9	8	9	7	12	11	7
	4	10	7	8	8	7	9	10	16	14	15	10	4	6	3	9	3
	10	3	8														

Use an asymptotic WMW test to determine whether the DMF score differs significantly between males and females. Do tied ranks have much influence on the appropriate test statistic?

6

Three or more samples

6.1 COMPARISONS WITH PARAMETRIC METHODS

This section assumes some familiarity with experimental design and the analysis of variance. As it only gives background information readers unfamiliar with these topics may wish to proceed directly to Section 6.2.

One- or two-sample analyses of continuous data using normal theory test and estimation procedures are modified for more than two samples. Emphasis shifts from the t-distribution and related test and estimation procedures to the analysis of variance and tests based on the F-distribution. The t-distribution is a special case of the F-distribution in the sense that when a statistic t has a t-distribution with υ degrees of freedom then t^2 has an F-distribution with 1, υ degrees of freedom.

Given p independent samples from normal distributions all with the same variance and means respectively $\mu_1, \mu_2, \ldots, \mu_p$, the basic overall significance test is that of $H_0 : \mu_1 = \mu_2 = \ldots = \mu_p$ against H_1 : *not all* μ_i *are equal*. In analysis of variance terminology large values of the statistic F=(*between samples mean square/within samples mean square*) indicate rejection of H_0. Under H_0, F has an F-distribution with $p-1$ and $N-p$ degrees of freedom, where N is the total number of observations. Where the context makes this appropriate, H_0 is sometimes expressed as the hypothesis of *no difference between treatments*.

The F-test above is usually a preliminary to more specific tests and estimation procedures concerning possible differences within chosen subsets (often pairs) of the μ_i, $i=1, 2, \ldots, p$. New concepts such as least significant differences and multiple comparison tests are then needed even in the independent sample or 'one-way classification' situation.

The two-dependent-samples situation in Chapter 4 generalizes to designed experiments. For the overall parametric test of H_0 : *no difference between treatments* against H_1 : *some treatments differ* (in location) testing and estimation procedures depend on the experimental design. A well-known design is that of randomized blocks. The nature of the treatment structure also influences the analysis; factorial experiments not only allow us to study the basic effect of two or more factors, but also whether the factors 'interact' with one another.

In normal theory inference these topics are included in the framework of **linear models**, a framework that also includes linear regression analysis.

Nonparametric analyses parallel many aspects of the normal theory linear model, but relaxing assumptions means that some sophisticated linear model techniques have no direct nonparametric analogue.

In this chapter we first develop overall tests for location differences, analogous to those in analysis of variance, that generalize from some of the tests in Chapters 4 and 5. Detailed testing of particular aspects of data subsets is deferred to Section 6.4.

More samples often imply more observations in total, and asymptotic results may be used with fewer reservations, unless a few samples are very small. Reliance on asymptotic results reflects a paucity of tables giving even nominal significance levels for permutation tests. Programs like STATXACT, TESTIMATE and DISFREE alleviate this problem.

In this and the following four chapters we show that the same nonparametric technique may often be applied to problems that appear, at least superficially, to be different. We have already met examples of such equivalences; e.g. the sign test and McNemar's test (Section 4.2), also that between Wilcoxon's signed rank test in a specific highly tied situation and the sign test (p.47). We presented the median test (Section 5.1.1) as a special case of the the widely applicable Fisher's exact test (Section 9.2.1).

6.2 LOCATION TESTS FOR INDEPENDENT SAMPLES

Tests described in this section are generalizations of some in Chapter 5.

6.2.1 The median test for several samples

The median test in Section 5.1.1 generalizes easily from two to three or more samples except that our alternative hypothesis is one of a difference between population medians without specifying which populations differ in location, how many differences there are, or their direction. Thus our test is a generalization of the two-tail test $H_1 : \theta_1 \neq \theta_2$. As in the two-sample case, each sample may come from **any** unspecified population (they need not all have the same distribution). Given samples from p populations with unknown medians $\theta_1, \theta_2, \ldots, \theta_p$, we test $H_0 : \theta_1 = \theta_2 = \ldots = \theta_p$ against H_1 : *not all* θ_i *are equal*. We extend the procedure in Section 5.1.1. If M is the combined sample median for all observations, in each sample we count the numbers of observations above and below M. As in Section 5.1.1, we reject sample values equal to M and then work with reduced samples. Assuming sample values that equal M have already been dropped, suppose that of the n_i observations in sample i there are $a_i > M$ and $b_i < M$, $i = 1, 2, \ldots, p$. We record these numbers above and below M in what is called a $p \times 2$ contingency table of p rows and 2 columns. Table 6.1 is an example: it is an obvious extension of Table 5.1 and we explain the test in Example 6.1.

Example 6.1

The problem. Six samples contain respectively 4, 7, 5, 4, 2, 6 observations. The median, M, of all 28 observations is obtained. No sample values equal M and the numbers above and below M in each sample are given in Table 6.1. Use a median test to determine whether to accept H_0 : *all six samples come from populations with the same median.*

Table 6.1 Numbers of observations above and below M in Example 6.1

Sample no.	Above M	Below M	Total
I	4	0	4
II	2	5	7
III	2	3	5
IV	3	1	4
V	2	0	2
VI	1	5	6
	14	14	28

Formulation and assumptions. Generalizing the argument in Example 5.1, for any sample of n, the probability of getting r observations above the population median is $p={}^nC_r(\frac{1}{2})^n$. Denoting this probability by p_i for the ith sample we require the probability of obtaining the cell values in the body of Table 6.1 conditional upon observing 14 from 28 above the population median in the combined sample; this is $P={}^{28}C_{14}(\frac{1}{2})^{28}$. The required conditional probability is

$$P_c=p_1\,p_2\,p_3\,p_4\,p_5\,p_6\,/P$$

As in Example 5.1, this is independent of the binomial probability $p=\frac{1}{2}$. Our critical region consists of all configurations of the $p\times2$ contingency table having the given marginal totals which, under H_0, have the same or a lesser P_c than that observed. Unlike the 2×2 table in Example 5.1, the more extreme configurations are not obvious, nor can we attribute these to a particular tail; i.e. our alternative hypothesis H_1: *not all medians are equal* is essentially a two-tail situation.

Procedure. In a general p-sample situation if we write a_i for the number above M and b_i for the number below M for sample i of n_i observations, then the a_i, b_i are constrained to give the correct row totals n_i and column totals A, B, say, in the $p\times2$ table. Table 6.2 generalizes Table 6.1. If no observations equal M, then $A=B=\frac{1}{2}N$: unless there are many data ties at M, A and B will be nearly equal.
The corresponding P_c is given by a generalization of (5.2):

Table 6.2 A general contingency table for the median test

Above M	Below M	Total
a_1	b_1	n_1
a_2	b_2	n_2
.	.	.
.	.	.
a_p	b_p	n_p
A	B	N

$$P_c = \frac{A!B!\Pi_i(n_i!)}{N!\Pi_i(a_i!)\,\Pi_i(b_i!)}$$

where $\Pi_i(x_i!)$ is the product $(x_1!)\times(x_2!)\times \ldots \times(x_p!)$. For Table 6.1

$$P_c = \frac{14!14!4!7!5!4!2!6!}{28!4!2!2!3!2!1!0!5!3!1!0!5!}$$

where, by definition, $0!=1$. Despite cancellations between numerator and denominator, even with a pocket calculator it takes time to verify that $P_c=0.000126$. The labour of calculating P_c for all other possible sets of cell values that might give rise to the same or lower values of P_c calls for either a computer program or some suitable approximation. An appropriate program gives, in this case, $\alpha=0.046$ as the size of the critical region.

Conclusion. We reject H_0 at the 4.6% significance level.

Comments. 1. Individual sample sizes here are all small (particularly sample V). Inspecting Table 6.1, the main differences seem to be that sample I probably has a median above M and sample VI (also perhaps sample II) a median below M. If we were to performed separate sign tests of $H_0 : \theta_i=M$ for each individual sample we would not reject H_0 (even in a one-tail test) for any sample because of the small sample sizes. However, the collective information confirms unequal population medians.

2. Difficulty of carrying out an exact test without a suitable computer program makes an asymptotic test of interest. We discuss one below.

3. A situation where we may be interested in testing for a common median when samples come from distributions differing in other respects (e.g. spread, skewness, etc.) arises if we compare alternative components that may be used in an industrial process. We may measure a characteristic such as time to failure for a number of replicates of each type of component. A preliminary step may then be to test whether all have the same median time to failure. If there is clear evidence they do not, one might reject from further consideration those that appear to have a lower

median time. Further choice may depend on preference for a component with little variability in its failure time distribution, or on non-statistical factors, such as cost.

Computational aspects. The STATXACT or TESTIMATE programs for Fisher's exact test (Section 9.2.1) may be used for the median test.

In the notation of Table 6.2 an asymptotic test uses the statistic

$$T=\Sigma_i[\{(a_i-n_iA/N)^2\}/\{n_iA/N\}]+\Sigma_i[\{(b_i-n_iB/N)^2\}/\{n_iB/N\}] \quad (6.1)$$

Readers familiar with the chi-squared test for independence in contingency tables will recognize (6.1) as the statistic often described as the sum of [(*observed*−*expected value*)^2/*expected value*] for all cells, which we discuss in Section 9.2.2, where an alternative form easier to compute is given in (9.8). For a $p\times 2$ table T has asymptotically a chi-squared distribution with $p-1$ degrees of freedom (Section A5). Chi-squared tables (e.g. Neave, 1981, p. 21) give minimal critical values for significance at conventional levels.

Example 6.2

The problem. For the data in Example 6.1 test for equality of population medians using (6.1).

Formulation and assumptions. We calculate all quantities in (6.1) from Table 6.1 and compare the resulting T with the tabulated critical value.

Procedure. The 'expected numbers' (n_1A/N, etc.) are calculated for each cell in (6.1) and are in the order given in that table,

2	2
3.5	3.5
2.5	2.5
2	2
1	1
3	3

Substituting these and the observed cell values from Table 6.1 in (6.1), we obtain $T=2\times(2^2/2+1.5^2/3.5+0.5^2/2.5+1^2/2+1^2/1+2^2/3)=11.15$. When $p=6$, tables give the critical value of chi-squared with $p-1=5$ degrees of freedom for significance at the 5% level to be $T\geq 11.07$.

Conclusion. We reject H_0 at a nominal 5% level.

Comments. 1. All six samples are small, but exact and asymptotic results agree well. For the chi-squared distribution with 5 degrees of freedom

$\Pr(T \geq 11.15) = 0.048$, close to the exact test probability of 0.046.

2. The alternative hypothesis that not all medians are equal is essentially two-tailed and the critical region includes all $p \times 2$ tables giving a probability equal to or less than that observed; however, the asymptotic test uses a single tail of the chi-squared distribution. This is because we use the squares of the discrepancies *observed−expected* and these are necessarily positive, no matter what is the sign of the actual discrepancy.

Computational aspects. If STATXACT is used for the Fisher exact test it gives an asymptotic test statistic that differs from T given by (6.1). This is because it uses a statistic, called the Fisher statistic, which also has an asymptotic chi-squared distribution. For this particular example the Fisher statistic takes the value 10.23 and $\Pr(\chi^2 \geq 10.23) = 0.069$. The reason for this discrepancy is explored further in Section 9.2.3. For larger samples the two asymptotic statistics are usually in reasonable agreement.

In practice, we may be interested in whether there are differences in medians for some subset of the p populations. One might take pairwise samples and compute, say, 95% confidence limits for the median differences for each pair using the method given in Section 5.1.2. Since these confidence intervals will generally be determined with varying precision for different pairs (due to differences in sample size and perhaps differences in the population distributions) direct comparisons involving these intervals sometimes lead to what at first sight appear bizarre consequences. For example, if zero is included in the 95% confidence interval for the difference $\theta_2 - \theta_1$ and also in the 95% confidence interval for the difference $\theta_3 - \theta_1$ we accept the null hypotheses that $\theta_2 = \theta_1$ and that $\theta_3 = \theta_1$. It is tempting to conclude that this implies we should accept $\theta_3 = \theta_2$. Yet one may find that the 95% confidence interval for the difference $\theta_3 - \theta_2$ does not include zero. This is not a logical contradiction because acceptance of a null hypothesis does not prove it true. We would accept *any* value in a relevant 95% confidence interval in a hypothesis test at the 5% significance level. Similar outcomes may occur with parametric analyses if we compare samples in pairs.

6.2.2 The Kruskal−Wallis test

Kruskal and Wallis (1952) proposed an extension of the Wilcoxon−Mann−Whitney test using the Wilcoxon formulation. It is relevant as an overall test for equality of population means or medians with otherwise identical and continuous distributions. It also has good power for testing equality of population cumulative distribution functions against an alternative that one or more pairs of these cumulative distributions satisfies the inequality $F_r(x) \leq F_s(x)$ (or the reverse inequality) with strict inequality for at least some x. As in the two-sample case a shift in location is often referred to as an additive treatment effect in experimental design terminology.

Suppose we have p random samples, the ith sample ($i=1, 2, \ldots, p$) consisting of n_i observations, the jth of these being x_{ij}, ($j=1, 2, \ldots, n_i$). The total number of observations is $N=\Sigma_i n_i$. We rank all N observations from smallest (rank 1) to largest (rank N), tied ranks being given mid-rank values. Let r_{ij} be the rank allotted to x_{ij} and $s_i=\Sigma_j r_{ij}$ be the sum of the ranks for the ith sample. We compute $S_p=\Sigma_i(s_i^2/n_i)$ and $S_r=\Sigma_{i,j} r_{ij}^2$. Readers familiar with the analyisis of variance will recognize these as uncorrected **treatment** and **total** sums of squares for ranks. From each we subtract an appropriate correction for the mean, namely $C=\frac{1}{4}N(N+1)^2$. If there are no ties $S_r=N(N+1)(2N+1)/6$. The test statistic is

$$T=\frac{(N-1)(S_p-C)}{S_r-C} \tag{6.2}$$

With no ties T simplifies to

$$T=12S_p/[N(N+1)]-3(N+1) \tag{6.3}$$

Tables giving critical values of T for small sample sizes (strictly relevant to the 'no-tie' situation) are given by Neave (1981, pp. 32–4) and elsewhere. For N moderate to large, T has approximately a chi-squared distribution with $p-1$ degrees of freedom if there are no treatment differences. Computation of exact permutation probabilities under H_0 is time consuming except for very small samples. Good estimates of exact probabilities are available using the Monte Carlo facilities in STATXACT or DISFREE.

Example 6.3

The problem. In uniform editions by each of three writers of detective fiction the numbers of sentences per page on randomly selected pages in a work by each are

C.E. Vulliamy	13	27	26	22	26		
Ellery Queen	43	35	47	32	31	37	
Helen McCloy	33	37	33	26	44	33	54

Use the Kruskal–Wallis test to examine the validity of the hypothesis that these may be samples from identical populations.

Formulation and assumptions. We consider the alternative hypothesis that samples are from populations having at least one location difference.

Procedure. Although not essential, it helps to avoid mistakes in ranking if we arrange values in ascending order within each sample. This is done in Table 6.3, where the rank is shown beneath each observed value.

Table 6.3 Sentence numbers (N) and ranks (R) in samples for three different authors

Vulliamy	Number	13	22	26	26	27		
	Rank	1	2	4	4	6		
Queen	Number	31	32	35	37	43	47	
	Rank	7	8	12	13.5	15	17	
McCloy	Number	26	33	33	33	37	44	54
	Rank	4	10	10	10	13.5	16	18

From Table 6.3 we find $s_1=1+2+4+4+6=17$, $s_2=72.5$ and $s_3=81.5$, whence $S_p=(17)^2/5+(72.5)^2/6+(81.5)^2/7=1882.73$. The sum of squares of all mid-ranks is $S_r=2104.5$ and $C=18\times(19)^2/4=1624.5$ since $N=18$. We compute $T=17(1882.73-1624.5)/(2104.5-1624.5)=9.146$. Neave's tables show that $T\geq 8.157$ indicates significance at the 1% level.

Conclusion. We reject the hypothesis of no difference in sentence length. Inspection of the data indicates that Vulliamy uses longer sentences (fewer sentences per page).

Comments. 1. Use of uniform editions avoids difficulties that may arise with different page sizes or type fonts.

2. $N=18$ may seem small for an asymptotic test and $T=9.146$ is short of the value 9.21 required for significance at the 1% level.

3. For practical reasons (e.g. availability of tables) we use T as a test statistic. It would suffice to use S_p since all other quantities in T are invariant in the exact permutation distribution.

Computational aspects. STATXACT gives $\Pr(T\geq 9.146)=0.0103$ for a chi-squared distribution with 2 degrees of freedom. The exact permutation probability is not available from STATXACT for samples of this size, but Monte Carlo estimates indicate that the result almost certainly falls in an exact critical region of size less than $\alpha=0.007$. Packages such as STATGRAPHICS and recent releases of MINITAB include the Kruskal–Wallis test, computing T but giving only asymptotic critical values. Neave and Worthington (1988, p. 262) give a BASIC listing for calculating T.

For the above example most statisticians would have few qualms about using a parametric analysis of variance. This leads to basically similar conclusions (Exercise 6.1).

We defer to Section 6.4 the study of specific contrasts in the context of a Kruskal–Wallis analysis.

6.2.3 Tests based on transformation of ranks

A modification of the Kruskal–Wallis test is to transform ranks to normal scores. For van der Waerden scores we replace the rank or mid-rank r by the $r/(N+1)$th quartile of a standard normal distribution. Ignoring a slight discrepancy in the case of ties, the mean of these scores is zero, and if we evaluate quantities corresponding to S_p and S_r in Section 6.2.2 with ranks replaced by van der Waerden scores and set $C=0$, we may compute a T as in (6.2) for the van der Waerden scores. Again, asymptotically T has a chi-squared distribution with $p-1$ degrees of freedom and the asymptotic result is usually reasonable for all but very small samples. In Exercise 6.2 we ask you to reanalyse the data in Example 6.3 using van der Waerden scores. Expected normal scores may also be used.

Other transformations are possible, including the use of raw data scores. Arguments given in earlier chapters expressing reservations about analyses based on raw score permutation tests also apply in the situations presently under discussion. Pitman (1938) and Welch (1937) discussed some permutation tests using raw data in analysis of variance contexts; their aim was to justify parametric tests as approximations to these.

6.2.4 The Jonckheere–Terpstra test

The Kruskal–Wallis test is an omnibus test for location or other differences. If treatments represent steadily increasing doses we may want to test hypotheses about means or medians of the form H_0 : *all* θ_i *are equal* against $H_1 : \theta_1 \leq \theta_2 \leq \theta_3 \leq \ldots \leq \theta_p$, where at least one of the inequalities is strict, or against $H_1 : F_1(x) \leq F_2(x) \leq \ldots \leq F_p(x)$, again at least one inequality being strict for some x. This is a 'one-tail' test. Reversal of all inequalities gives an analogous test in the opposite tail. For such ordered alternatives a test was devised by Jonckheere (1954), though conceived independently by Terpstra (1952). Its application is an extension of the Wilcoxon–Mann–Whitney test using the Mann–Whitney formulation. The procedure is easily illustrated by an example. Before carrying out the test the samples must be ordered in the sequences specified in H_1. Exact permutation tests are more readily available for small samples than they are for the Kruskal–Wallis test, and an asymptotic test is also available. We discuss this after the example.

Example 6.4

The problem. Hinkley (1989) gives braking distances taken by motorists to stop when travelling at various speeds. A subset of his data is:

Speed (mph)	*Breaking distances (feet)*
20	48
25	33 59 48 56
30	60 101 67
35	85 107

Use the Jonckheere–Terpstra test to determine if there is evidence of a tendency for braking distance to increase as speed increases.

Formulation and assumptions. We test H_0 : *braking distance is independent of initial speed* against H_1 : *braking distance increases with speed.* The samples are already arranged in order of increasing speed implicit as the natural order under H_1.

Procedure. If there are p samples we calculate the sum, U, of all Mann–Whitney statistics U_{rs} relevant to the rth sample ($r=1, 2, \ldots, p-1$) and any sample s for which $s>r$. For the four samples above we calculate U_{12}, U_{13}, U_{14}, U_{23}, U_{24} and U_{34}. For example, U_{12} is the sum of the number of sample 2 values that exceeds each sample 1 value. Clearly here $U_{12}=2.5$ since there is only one sample 1 value (48) and this is equal to one value in sample 2 and exceeded by two others (56, 59); ties are scored as ½ as in the Mann–Whitney statistic. Similarly $U_{13}=3$, $U_{14}=2$, $U_{23}=12$, $U_{24}=8$ and $U_{34}=5$. Adding all U_{rs} gives $U=32.5$. To obtain the exact permutation distribution of ranks is tedious without a suitable computer program. In an exact permutation test $\Pr(U \geq 32.5) = 0.0011$, indicating significance in a one-tail test (relevant here) at an actual 0.11% significance level.

Conclusion. There is clear evidence (not surprisingly!) that increasing speed influences braking distance.

Comments. When relevant, the Jonckheere–Terpstra test is generally more powerful than Kruskal–Wallis. If a Kruskal–Wallis test is used with these data, an asymptotic test would accept H_0 at the nominal 5% significance level. Monte Carlo simulations of the exact permutation distribution indicate we should reject H_0 using the Kruskal–Wallis test at approximately a 1.2% level. A parametric analysis of variance using these data gives a variance ratio $F=4.45$ with 3, 6 degrees of freedom which is not significant at the 5% level.

Computational aspects. STATXACT computes exact permutation test tail probabilities for somewhat larger samples than it does for the Kruskal–Wallis test. It is illuminating to compare these with tail probabilities for the asymptotic test given below. In this example the asymptotic result would only indicate significance in a one-tail test at the

0.23% significance level; although twice as large as the exact test level, the indication of significance is still clear.

The asymptotic Jonckheere–Terpstra test is based on the fact that U as defined in Example 6.4 has a mean $\mathrm{E}(U)=\frac{1}{4}(N^2-\Sigma_i n_i^2)$ and variance $\mathrm{Var}(U)=\{N^2(2N+3)-\Sigma_i\,[n_i^2(2n_i^2+3)]\}/72$. For large N and the individual n_i not too small the distribution of

$$Z=(U-\mathrm{E}(U)/\surd[\mathrm{Var}(U)] \qquad (6.4)$$

is approximately standard normal. Although the sample sizes in Example 6.4 are too small for an asymptotic result to inspire confidence, calculations sought in Exercise 6.3 give, for this example, $Z=2.83$, corresponding, as indicated above to a one-tail significance level of 0.23%.

The expression for Var(U) given above requires adjustment for ties. The adjustment is trivial for relatively few ties but in Section 9.3.2 we show that the existence of many ties may have a dramatic effect. A formula for adjusting Var(U) for ties is given by Lehmann (1975, p. 235) and a modified version is given in Section 9.3.2.

Exercise 6.21 gives an example with $H_1 : F_1(x)\geq F_2(x)\geq \ldots \geq F_p(x)$.

6.3 LOCATION COMPARISONS FOR RELATED SAMPLES

If we have blocks of two units and allocate two treatments to the units in each block at random we have the matched-pair situation discussed in Chapter 4. This is a simple case of a randomized block design. Using matched pairs allows us to make treatment comparisons by analysing differences between responses within blocks, thus eliminating differences **between** blocks. We do this if we expect results for each member of a pair to be more homogeneous under a null hypothesis of no treatment effect than would be the case if treatments were applied randomly to large sets of less homogeneous units.

To compare several (say t) treatments we might replace our homogeneous pairs by blocks of t units, blocks being chosen so that units within each block are as homogeneous as possible. For example, to compare the effects of five different feed regimes on pigs, each block may be a litter of 5 pigs. In each litter the 5 diets are allocated, one to each pig, chosen at random.

To compare three different cake recipes (treatments) we might make batches of each recipe, divide each batch into four samples and cook one sample of each batch in four different ovens (each oven is a block). We do this because ovens may operate at slightly different temperatures and have varying heat efficiencies. To assess the merits of the recipes we might ask experts to rank them in order of preference for taste. Comparison between cakes baked in the same oven is desirable because, although all mixtures may produce poor cakes in a particular oven and all rather nice cakes in another, one hopes that the relative ordering for products from each oven may be reasonably consistent from oven to oven. We have so far not considered randomization within ovens. If all mixtures are cooked at the same time it would be wise to allocate the three different recipes to shelf positions at random as these may affect the cakes produced. If the cakes in each oven are to be cooked one after another the time order could sensibly be randomized separately for each oven as the end product might be affected by the time the mixture stands before cooking, or there may be some carry-over effect on flavour depending upon which cake is cooked first. These are matters reflecting principles of good experimental design and are relevant to all methods of analysis – parametric or nonparametric.

6.3.1. A generalization of the sign test

In Section 4.1.1 we applied the sign test to matched pairs, effectively observing whether a treatment difference had a plus or minus sign. An equivalent analysis uses ranks of treatments within each block of two. If we rank the lower response 1 and the higher response 2 and sum over blocks all ranks allocated to each treatment we may deduce the numbers of plus and minus signs and the difference between them, as shown by a simple numerical example. The data in Table 6.4 represent observations on 9 matched pairs numbered 1 to 9.

Table 6.4 Observations in matched pairs

Pair	1	2	3	4	5	6	7	8	9
Treatment A	17	11	15	14	22	41	7	2	8
Treatment B	15	9	18	11	17	43	5	4	3

The difference *treatment A−treatment B* gives 6 positive and 3 negative values, the respective numbers of plus and minus signs for the sign test. If we rank the observations lower (1) and higher (2) in each pair we get Table 6.5.

Table 6.5 Rank order of responses in Table 6.4

Pair	1	2	3	4	5	6	7	8	9
Treatment A	2	2	1	2	2	1	2	1	2
Treatment B	1	1	2	1	1	2	1	2	1

Clearly in Table 6.5 the number of positive signs is equal to the number of 2s allocated to treatment A (or the number of 1s allocated to treatment B). These results generalize to any number of pairs, so the rankings for either treatment immediately lead to the sign-test statistic. Also the difference between the rank sums for each treatment, 15−12=3, equals the difference between the number of plus and minus signs in a sign test.

We generalize this idea to more than two treatments when we have a randomized block design where the number of units in each block equals the number of treatments.

In parametric analysis of variance of continuous observations we remove block differences as a source of variability before making treatment comparisons (but do not worry about how this is done if you are unfamiliar with the analysis of variance). Friedman (1937) proposed the test described below which automatically removes block differences by the expedient of replacing observations by their ranks, doing this separately within each block. The test is immediately applicable if the basic data are only ranks within each block (see Exercise 6.11); in these circumstances it provides a test for consistency of ranks rather than one for location and as such was developed independently by M.G. Kendall, and we discuss it further in Section 7.2. In Example 6.7 also the initial data are ranks.

If t treatments are each applied to 1 of t units in each of b blocks in a randomized block design we denote by x_{ij} the response (observation) for treatment i in block j; here i runs from 1 to t and j from 1 to b. We replace the observations in **each** block by ranks 1 to t, this ranking being carried out separately for each block, using mid-ranks for ties. The sum of the ranks for treatment i is denoted by s_i, $i=1, 2, \ldots, t$. We calculate $S_t=\Sigma_i (s_i^2)/b$ (equivalent to the uncorrected treatment sum of squares in an analysis of variance). If r_{ij} is the rank or mid-rank corresponding to x_{ij} we also calculate $S_r=\Sigma_{i,j} r_{ij}^2$. If there are no ties $S_r=bt(t+1)(2t+1)/6$. The correction factor, analogous to that in the Kruskal–Wallis test, is $C=\frac{1}{4}bt(t+1)^2$. A commonly used form of the Friedman statistic is

$$T=b(t-1)(S_t-C)/(S_r-C) \tag{6.5}$$

If b, t are not too small T has approximately a chi-squared distribution with $t-1$ degrees of freedom. For $t\leq 6$, Neave (1981, p. 34) gives tables of critical values of T strictly valid for no ties. Iman and Davenport (1980) suggest that a better approximation is

$$T_1=(b-1)(S_t-C)/(S_r-S_t) \tag{6.6}$$

which, under the null hypothesis of no treatment difference, has approximately an F-distribution with $t-1$ and $(b-1)(t-1)$ degrees of freedom. If the ranking is identical in all blocks the denominator of T_1 is zero. Iman and Davenport show that this may be interpreted as a result significant at the $100(1/t)^{b-1}\%$ significance level.

Example 6.5

The problem. Pearce (1965, p. 37) quotes results of a greenhouse experiment carried out by J. I. Sprent (unpublished). The data, given in Table 6.6, are the numbers of nodes to first initiated flower summed over four plants in each experimental unit (pot) for the pea variety Greenfeast subjected to six treatments – one an untreated control, the others various growth substances. There were four blocks allowing for differences in light intensities and temperature gradients depending on proximity to greenhouse glass. The blocks were arranged to make these conditions as like as possible for all units (pots of four plants) in any one block. Use the Friedman statistic to test for differences between treatments in node of flower initiation.

Formulation and assumptions. Within each block we replace each observation by its rank and calculate the Friedman statistic for these ranks to test H_0 : *no difference between treatments* against H_1 : *at least one treatment has a different location from the others.* We may test this using (6.5) or (6.6).

Table 6.6 Nodes to first flower, total for four plants

Block	I	II	III	IV
Treatment				
Control	60	62	61	60
Gibberellic acid	65	65	68	65
Kinetin	63	61	61	60
Indole acetic acid	64	67	63	61
Adenine sulphate	62	65	62	64
Maelic hydrazide	61	62	62	65

Procedure. Relevant ranks are given in Table 6.7 and we append a further column giving rank totals s_i for each treatment.

Table 6.7 Nodes to first flower, ranks within blocks

Block	I	II	III	IV	Total
Treatment					
Control	1	2.5	1.5	1.5	6.5
Gibberellic acid	6	4.5	6	5.5	22
Kinetin	4	1	1.5	1.5	8
Indole acetic acid	5	6	5	3	19
Adenine sulphate	3	4.5	3.5	4	15
Maelic hydrazide	2	2.5	3.5	5.5	13.5

Squaring each rank and adding we get $S_r=361$. The uncorrected treatment sum of squares is

$$S_t=(6.5^2+22^2+8^2+19^2+15^2+13.5^2)/4=339.625$$

and $C=4\times6\times7^2/4=294$; using (6.5) $T=4\times5(339.625-294)/(361-294)=13.62$, and (6.6) gives $T_1=3(339.625-294)/(361-339.625)=6.40$. Using T_1, as recommended by Iman and Davenport, we find that with 5, 15 degrees of freedom critical values for significance at the 5%, 1% and 0.1% significance levels are respectively 2.90, 4.56 and 7.57 so we would judge our result significant at the 1% level. If instead we compare T with the critical values of chi-squared with 5 degrees of feeedom the result is significant at the 5% but not the 1% level (for the chi-squared distribution with 5 degrees of freedom, $\Pr(T \geq 13.62 = 0.0182)$. Comparing T with the critical values given by Neave for 6 treatments and 4 blocks we find significance is indicated at the 1% level if $T \geq 12.71$, so again significance is indicated at the 1% level, but remember that Neave's tables apply strictly to the 'no-tie' situation.

Conclusion. There is clearly a significant difference between treatments, almost certainly at a nominal 1% significance level.

Comments. 1. It would be helpful to base our test on a relevant exact permutation distribution. DISFREE provides a program to calculate exact probabilities; however, the computation time is prohibitively long on most PCs for this data set. The simulation option in DISFREE using 20 000 simulations gave an estimate of 0.00485 for the upper-tail probability with a 95% confidence interval (0, 0.01178) for this probability. Overall, then, it would seem reasonable to accept significance at the 1% level.

2. While for practical reasons we use T or T_1 as our test statistic it is worth noting that it would suffice to use S_t as both S_r, C remain unaltered in the permutation distribution which involves only permuting ranks **within** blocks.

3. A parametric randomized blocks analysis of variance of the original data gives $F=4.56$, which just indicates significance at the 1% level.

4. Whereas analysis of variance introduces a sum of squares reflecting difference between block totals there is no such term in the Friedman analysis because the sum of ranks in all blocks is the same, namely $\frac{1}{2}t(t+1)$.

Computational aspects. DISFREE provides facilities described in comment 1. The test statistic, and in some cases asymptotic tail probabilities, are available in some general statistical packages. A BASIC program listing is given by Neave and Worthington (1988, p.282).

It is feasible to replace ranks within blocks by normal scores. Experience with the Friedman method suggests that such rescoring has few advantages. Indeed, if there are ties, the block differences removed by ranking may be reintroduced, though usually not dramatically.

Ranking within blocks is robust against many forms of heterogeneity of variance, in that it removes any inequalities of variance between blocks (see Exercise 6.14).

6.3.2 Ordered alternatives

Page (1963) proposed an analogue to the Jonckheere–Terpstra test for ordered alternatives applicable to blocked data. If the treatments are arranged in the order specified in the alternative hypothesis and s_i is the sum of ranks for treatment i (as in the Friedman test) the Page statistic is $P=s_1+2s_2+3s_3+ \ldots +ts_t$. Asymptotically, if H_0 : *no treatment difference* holds, then P has a normal distribution with mean $tb(t+1)^2/4$ and variance $b(t^3-t)^2/[144(t-1)]$. Tables (Daniel, 1990, Table A17) give critical values of P for small b, t. The test is discussed in more detail by Daniel (1990, Section 7.3) and by Marascuilo and McSweeney (1977, Section 14.12).

6.3.3 Extension of Wilcoxon type tests

The Friedman test is a natural extension of the sign test. Several tests for blocked data that extend the Wilcoxon rank sum test have been proposed.

Hora and Conover (1984) rank all observations simultaneously without regard to treatments or blocks and carry out an analysis of variance on the ranks (or normal scores derived from these). For ranks the procedure is described by Iman, Hora and Conover (1984).

When applied to the data in Example 6.5 the relevant F-statistic for treatment difference is $F=4.96$ (compared with 4.56 for a parametric test and 6.40 using (6.6)). The reader familiar with standard analysis of variance may verify these results (see Exercises 6.4 and 6.13).

6.4 MORE DETAILED TREATMENT COMPARISONS

We pointed out in Section 6.1 that after preliminary parametric analysis of variance interest usually shifts to specific treatment comparisons. In factorial experiments, for example, interest may centre on main effects and interactions. In more general contexts multiple comparisons may be important. These concepts have their analogues in nonparametric analyses but complications arise as some of the simplicity associated with the **linear** model aspects of the parametric analysis is lost. It is beyond the scope of this book to discuss subtle aspects of these generalizations, but we indicate some analogies between parametric and nonparametric approaches. We give a number of references for those wanting to follow up these aspects in detail.

6.4.1 Multiple comparisons

As in parametric analyses, nonparametric multiple comparisons require care in selecting appropriate tests or estimation procedures. In the absence of exact tests we are highly dependent on large-sample approximations for establishing significance. In this section we consider tests equivalent to using what are known as **least significant differences.** Logically they can only be justified if applied to pre-selected comparisons when an overall Kruskal–Wallis, Friedman or similar test indicates significance. The significance level used in multiple comparison tests should be no less stringent than that in the overall test.

When using the Kruskal–Wallis test the criteria for accepting a location difference between the *i*th and *j*th sample are that

1. the overall test indicates significance and
2. if $m_i=s_i/n_i$, $m_j=s_j/n_j$ are the mean ranks for these samples, then

$$|m_i-m_j|>t_{N-t,\alpha}\sqrt{[(S_r-C)(N-1-T)(n_i+n_j)/\{n_in_j(N-t)(N-1)\}]} \quad (6.7)$$

where T is given by (6.3) and $t_{N-t,\alpha}$ is the t value required for significance at the $100\alpha\%$ significance level in a t-test with $N-t$ degrees of freedom, the other quantities being as defined in Section 6.2.

An analogous result holds for van der Waerden scores with sample rank means replaced by sample score means and C now zero and T being the form of the statistic appropriate for these scores.

Nominated comparisons for the Friedman test may be based on either treatment rank means or totals since we have equal replication. The algebraic expression in terms of rank totals is simpler. An analogue of (6.7) gives the requirement for a least significant difference between the treatment rank totals s_i, s_j:

$$|s_i-s_j|>t_{(b-1)(t-1),\alpha}\sqrt{[2b(S_r-S_t)/(b-1)(t-1)]} \quad (6.8)$$

in the notation of Section 6.3.1.

Most statisticians agree with Pearce (1965, pp. 21–2) that it is reasonable to compare a sub-set of all samples (or treatments) if that sub-set is selected before the data are obtained (or at least before they are inspected) using formulae like (6.7) or (6.8) if one could sensibly anticipate that the specified pair are likely to show a difference in location.

Example 6.6

The problem. In Example 6.3 interest attached to differences in sentence length betweeen Vulliamy and Queen, the former being an English author with an academic background and limited fictional output and Queen a popular and prolific American writer. Use a least significant difference test to determine whether the difference is significant.

Formulation and assumptions. Relevant quantities are mostly available from the solution to Example 6.3, and these are subsitituted in (6.7).

Procedure. The relevant rank means are $m_1=17/5=3.4$ (Vulliamy) and $m_2=72.5/6=12.1$ (Queen). Also $S_r-C=2104.5-1624.5=480.0$, $T=9.146$,

$N=18$, $n_1=5$, $n_2=6$, $t=3$ and from tables we find the critical t-value at the 1% significance level with 15 degrees of freedom is 2.95. Thus the left-hand side of (6.7) is $12.1-3.4=8.7$ and the right-hand side reduces to $2.95\times\sqrt{[480\times(17-9.164)\times 11/(30\times 17\times 15)]}=6.86$.

Conclusion. Since $8.7>6.86$ the difference is significant at the 1% level.

Comment. We use the 1% level since the overall test indicated significance at this level.

The description of the experiment and the data used in Example 6.7 were kindly provided by Chris Theobald.

Example 6.7

The problem. As part of a study of the feeding habits of the larvae of the Blue-tailed Damsel Fly, 6 of the larvae were collected along with six members of each of 7 species of prey on which they usually fed. Each larva was placed on a cocktail stick in a glass beaker which contained water and and one each of the 7 types of prey. Records were kept of the order in which each larva ate the prey. The results are shown in Table 6.8. Some ties denoted by mid-ranks arose due to failure to observe the order when a larva ate two prey in quick succession. Is there evidence that the larvae prefer to eat some species of prey before others?

The cyclops was usually the last to be eaten. If the experimenter had prior reason to anticipate such an outcome, how might you appropriately examine whether the tendency was statistically significant?

Formulation and assumptions. Evidence of preference requires a Friedman test analogous to that in Example 6.5. A multiple comparison test using (6.8) is one approach to examining the situation regarding cyclops.

Table 6.7 Preferences of Blue-tailed Damsel Fly larvae for various prey

	Larva						
Prey	1	2	3	4	5	6	Total
Anopheles	1	2	1	1	3	1	9
Cyclops	7	7	7	6.5	6.5	7	41
Ostracod	5.5	5.5	5	5	5	2.5	28.5
Simocephalus (wild)	3	4	4	6.5	1.5	4	23
Simocephalus (domestic)	2	1	6	4	6.5	6	25.5
Daphnia magna	4	3	2	2	1.5	2.5	15
Daphnia longspina	5.5	5.5	3	3	4	5	26

Procedure. The Friedman test follows exactly the lines in Example 6.5. The reader should confirm (Exercise 6.20) that $t=7$, $b=6$, $S_r=837$, $S_t=775.75$, $C=672$, $T=22.64$ and $T_1=8.47$, indicating significance at the 1% (and indeed at the 0.1%) level. Setting $\alpha=0.01$ we find the appropriate t-value with $(b-1)(t-1)=30$ degrees of freedom is 2.75, so when using (6.8) the least significant difference between treatment rank totals given in the last column of Table 6.8 is $2.75\times\sqrt{[2\times6(837-775.75)/30]}=13.61$.

Conclusion. The difference 13.61 is exceeded between totals for cyclops and all other prey except ostracod, so there is strong evidence that there is a low preference for cyclops.

Comment. We have used the only relevant test given in this book to clarify the situation regarding cyclops but it is possible to devise a test comparing cyclops with the average rank response for all other prey which would confirm the conclusion that cyclops was less favoured. A more sophisticated approach to multiple comparisons is given by Leach (1979, Section 6.2). A recent valuable paper on this topic is that by Shirley (1987), and an important earlier paper is Rosenthal and Ferguson (1965).

6.4.2 Factorial treatment structures

Many papers have recently appeared on the nonparametric analysis of factorial treatment structures. The detail is beyond the scope of this book and indeed any fuller discussion would only be meaningful to readers already familiar with the equivalent analysis of variance. Key references to work in this area include Grizzle, Starmer and Koch (1969), Iman (1974), Conover and Iman (1976), Scheirer, Ray and Hare (1976), Mack and Skillings (1980), de Kroon and van der Laan (1981), Groggel and Skillings (1986) and the paper by Shirley (1987) referred to in Section 6.4.1. Further references, together with a critical review of some problems in analysis of data expressed as scores, are given by Thomas and Kiwanga (1991).

6.4.3 Commonsense analysis

In this section we show how simple nonparametric methods may be adapted to deal with a practical complication. This example indicates the versatility that goes with common-sense applications of nonparametric methods in what is often termed **exploratory data analysis.**

Example 6.8

The problem. Suppliers of word processing programs are testing an updated version designed to make it easier to prepare technical reports containing graphs, mathematical formulae, etc. To see whether the updated version (package A) shows an advantage over the current version (package B) four people chosen at random from nine are asked to prepare a specimen report using package A and the remaining five to prepare the same report using package B. All participants had similar experience in word processing. The numbers of mistakes made by each operator in their first drafts of the report were:

Package A: 2 11 24 26
Package B: 17 25 28 31 63

A Wilcoxon–Mann–Whitney (WMW) test shows no significant difference in a one-tail test (appropriate because improvements in package A should not tend to increase mistakes). The relevant Wilcoxon statistic (Exercise 6.5) is $S_m=13$ and $\Pr(S_m \leq 13)=0.0556$.

It is now disclosed that some participants in this test had a training in the technology which was the subject of the report whilst others did not. The results could now be displayed separately for each group:

With knowledge of technology (K)	Package A	2	11	
	Package B	17	25	
No knowledge of technology (NK)	Package A	24	26	
	Package B	28	31	63

We now see clearly that whichever package is used, those with no knowledge of the technology make more mistakes than those using the same package who have a knowledge. Explore the use of this additional information in establishing whether there is convincing evidence that package A reduces mistakes, all other factors being equal.

Formulation and assumptions. A simplistic approach is to apply the WMW test separately to the two groups K and NK but the samples are now too small to show significance no matter what outcome! We might regard the results as representing samples from four populations and apply a Kruskal–Wallis test. If this gave a significant result we would need to follow it by multiple comparison tests comparing packages A and B with technical knowledge (K), and packages A and B with no technical knowledge (NK). For the Kruskal–Wallis test (Exercise 6.6) $T=7.1333$. In an asymptotic test this is not significant at the 5% level, but it proves significant in an exact test at a nominal 1% level (Neave, 1981, p. 33). Tests based on least significant differences (Exercise 6.6) indicate significance at the 5% level between packages A and B for K but not for NK.

Alternatively, we might look upon knowledge (K) as a factor (often called a covariate) likely to reduce errors in a similar way for each package and allow for its effect by considering separately permutation distributions for K and NK, and then combine these results in an appropriate manner.

Procedure. We rank all observations as for the Kruskal–Wallis test, i.e.

With knowledge of technology (K)	Package A	1	2	
	Package B	3	5	
No knowledge of technology (NK)	Package A	4	6	
	Package B	7	8	9

We carry out separate permutation distribution calculations for K, NK respectively using these scores. With K the possible scores for package A are clearly (1, 2) (1, 3) (1, 5) (2, 3) (2, 5) (3, 5). The associated sums are S_1 = 3, 4, 6, 5, 7, 8 each with probability 1/6. Similarly, for NK possible scores for package A are (4, 6) (4, 7) (4, 8) (4, 9) (6, 7) (6, 8) (6, 9) (7, 8) (7, 9) (8, 9) with sums S_2 = 10, 11, 12, 13, 13, 14, 15, 15, 16, 17 each with probability 1/10 under H_0. The samples are independent, so each sum $S=S_1+S_2$ has associated probability 1/60 and the tail probability relevant to a one-tail test is $\Pr(S \leq k)$, where k is the value we observe. Symmetry implies doubling this probability for a two-tail test. To work out the relevant probabilities we proceed, in a way reminiscent of that used in Section 1.3.2, to work out the number of ways each sum can be obtained. In Table 6.9 each entry is the sum of the S_1 value at the top and the S_2 value at the left. We repeat the values 13 and 15 for S_2 as each occurs twice. This ensures equal probabilities of 1/60 for all 60 sums in the body of the table. If a particular sum occurs r times the associated probability is r/60. We easily deduce the probabilities associated with each sum: these are given in Table 6.10.

Table 6.9 Equiprobable sums under separate permutation of sub-samples in Example 6.8

	S_1	3	4	5	6	7	8
S_2							
10		13	14	15	16	17	18
11		14	15	16	17	18	19
12		15	16	17	18	19	20
13		16	17	18	19	20	21
13		16	17	18	19	20	21
14		17	18	19	20	21	22
15		18	19	20	21	22	23
15		18	19	20	21	22	23
16		19	20	21	22	23	24
17		20	21	22	23	24	25

Table 6.10 Probability of each rank sum in Table 6.8

Value k of S	$\Pr(S=k)$	Value k of S	$\Pr(S=k)$
13	1/60	20	8/60
14	2/60	21	7/60
15	3/60	22	5/60
16	5/60	23	4/60
17	6/60	24	2/60
18	8/60	25	1/60
19	8/60		

In our example adding the ranks associated with package A for K and NK we find $S=1+2+4+6=13$. This is the same as the WMW statistic. Can you see why? From Table 6.9, $\Pr(S \leq 13)=1/60$.

Conclusion. Since for a one-tail test the critical region is of size $\alpha=1/60=0.0167$, we reject the hypothesis of no difference betwen packages and accept that package A is superior.

Comments. 1. In this somewhat *ad hoc* analysis the assumption that if one package is superior that superiority should show for both K and NK is critical. In practice, one might have found that one group preferred one package and one the other – an effect known as an **interaction** in the terminology of factorial experiments.

2. A parametric analysis of variance of the raw data when broken into the 4 samples does not indicate a significant difference (see Exercise 6.7).

Computational aspects. The method outlined under *procedure* would not be feasible for larger data sets without suitable computer programs. STATXACT gives the requisite program for what are termed WMW tests with stratification. The method extends to more than two covariate levels. Monte Carlo estimates for the Kruskal–Wallis test using STATXACT confirm significance at the 1% level.

6.4.4 Binary responses

Sometimes each experimental unit may show one of two responses – win or lose, succeed or fail, live or die, subject is male or female. For analytic purposes we usually score such responses as 0 or 1. For example, five members A,B,C,D,E of a mountaineering club may each attempt three rock climbs at each of which they succeed or fail. If a success is recorded as 1 and a failure as 0, the outcomes may be summarized as follows:

Member	A	B	C	D	E
Climb 1	1	1	0	0	1
Climb 2	1	0	0	1	0
Climb 3	0	1	1	1	1

Cochran (1950) proposed a method applicable to such situations to test the hypothesis H_0 : *all climbs are equally difficult* against H_1: *the climbs vary in difficulty*. In conventional terms the climbs are 'treatments' and the 'climbers' are blocks. If we have t treatments in b blocks and binary (i.e. 0, 1) responses the appropriate test statistic is

$$Q=\frac{t(t-1)\Sigma_i T_i^2-(t-1)N^2}{tN-\Sigma_j B_j^2} \tag{6.9}$$

where T_i is the total (of 1s and 0s) for treatment i, B_i is the total for block j and N is the grand total. The exact permutation distribution of Q is not easily obtainable but for large samples Q has approximately a chi-squared distribution with $t-1$ degrees of freedom. Although not immediately obvious, for two treatments the test reduces to McNemar's test. Indeed, we introduced the McNemar test with an example comparing two climbs in Section 4.2 and in Section 9.5 we give an alternative form of the McNemar test which is exactly equivalent to Q given above.

Cochran's test is discussed more fully by Conover (1980, Section 4.6) and also by Leach (1979, Section 6.3), Daniel (1990, Section 7.5) and by Marascuilo and McSweeney (1977, Section 7.7).

6.5 TESTS FOR HETEROGENEITY OF VARIANCE

The squared rank test given in Section 5.3.2 extends to several independent samples. The procedure parallels that for the Kruskal–Wallis test with squared ranks of absolute deviations replacing data ranks as scores. The absolute deviations of all sample values from their sample mean are ranked over all samples and the squares of these ranks are noted. The statistic T is identical to (6.2) except that now s_i is the sum of the squared rank deviations from sample i and C is the square of the sum of squared ranks divided by N: symbolically, denoting a squared rank by r_{ij}^2, $C=[\Sigma_{i,j} r_{ij}^2]^2/N$. S_r in (6.2) becomes the sum of squares of the the squared ranks, i.e. $S_r=\Sigma_{i,j} r_{ij}^4$. If all population variances are equal, for a reasonably large sample T has a chi-squared distribution with $t-1$ degrees of freedom. If there are no rank ties the sum of squared ranks is $N(N+1)(2N+1)/6$. Also, for no ties the denominator of T reduces to

$$S_r-C=(N-1)N(N+1)(2N+1)(8N+11)/180$$

Example 6.9

The problem. As well as increasing with speed, as in Example 6.4 it is possible that braking distance becomes more variable from driver to driver as speed increases. Use the squared rank test to test for heterogeneity of population variance on the basis of the following sub-sample from Hinkley (1989) of initial speeds (mph) and braking distance (feet).

Speed	*Braking distance*			
5	2	8	8	4
10	8	7	14	
25	33	59	48	56
30	60	101	67	

Formulation and assumptions. We obtain the absolute deviations $|x_{ij} - m_i|$ of each observation x_{ij} from its sample mean m_i. These deviations are then ranked over combined samples and the statistic T is calculated and compared to the relevant chi-squared critical value.

Procedure. The means at each speed are $m_1=5.5$, $m_2=9.67$, $m_3=49$, $m_4=76$. The absolute deviations together with their ranks (1 for least, 14 for greatest, with mid-ranks for ties) are as follows:

5 mph	Absolute deviation	3.5	2.5	2.5	1.5
	Overall rank	7	4.5	4.5	2
10 mph	Absolute deviation	1.67	2.67	4.33	
	Overall rank	3	6	8	
25 mph	Absolute deviation	16	10	1	7
	Overall rank	12.5	11	1	9
30 mph	Absolute deviation	16	25	9	
	Overall rank	12.5	14	10	

The sum of squared ranks at 5 mph is $s_1=7^2+4.5^2+4.5^2+2^2=93.5$. Similarly, $s_2=109$, $s_3=359.25$, $s_4=452.25$, so the sum of all squared ranks is 1014 and $S_p=(93.5)^2/4+(109)^2/3+(359.25)^2/4+(452.25)^2/3=106\,587.724$. It is left as an exercise for the reader to verify that $C=73\,442.571$ and $S_r=7^4+4.5^4+\ldots+10^4=127\,157.25$, whence $T=8.02$. Tables of the chi-squared distribution indicate that the critical value for significance at the 5% level with 3 degress of freedom is $T=7.815$.

Conclusion. There is evidence of heterogeneity of variance significant at a nominal 5% level.

Comments. 1. One has reservations about an asymptotic result with such small samples, but we can do little better except in extremely small samples where we might calculate exact permutation distributions. It is important to align the samples for location by taking absolute deviations from the mean. This does not affect variance. For the original data there may of course be differences in location whether or not we accept H_0 : *no variance difference* in a squared rank test.

2. An alternative would be a Kruskal–Wallis test using overall ranks of absolute deviations.

6.6 FIELDS OF APPLICATION

Parametric analysis of variance of designed experiments with historical origins in agriculture soon spread to the life and natural sciences, medicine, industry and, more recently, to business and the social sciences. Development of nonparametric analogues was stimulated by a realization that data often clearly violated normal theory assumptions and, more importantly, to provide a tool for analysis of ordinal data expressed only as ranks or preferences (see e.g. Example 6.7). Ranking or preference scores often combine assessment of a number of factors that are given different weightings by individuals. It is thus of considerable practical interest to see if there is still consistency between the way individuals rank the same objects despite the fact that they give different weights to different factors. Our first example illustrates this point.

Preferences for washing machines

Consumers' preferences for washing machines are influenced by their assessment of several factors, e.g. price, reliability, power consumption, washing time, load capacity, ease of operation and clarity of instructions. Individuals weigh such factors differently; a farmer's wife offering bed and breakfast to tourists will rate ability to wash bed linen highly; a parent with a large young family the ability to remove sundry stains from children's clothing; to another low price and running economy may be key factors. No machine is likely to get top ratings on all points, but a manufacturer will be keen to achieve a high overall rating for his product from a wide range of consumers. A number of people may be asked to state preferences (e.g. ranks) for that manufacturer's machine and those of several competitors; the manufacturer wants to know if there is consistency in rankings – whether most people give his machine a preferred rating – or whether there is inconsistency or general dislike for his machine. Each consumer is a block, each machine a treatment, in the context of a Friedman test. The hypothesis under test is H_0 : *no consistency in rankings* against H_1 : *some consistency in rankings*. We discuss this type of situation further in Section 7.2.

Literary discrimination

A professor of English asserts that short stories by a certain writer (A) are excellent, those by a second writer (B) are good, and those by a third (C) are inferior. To test his claim and judgement he is given 20 short stories to read on typescripts that do not identify the authors and asked to rank them 1 to 20 (1 for best, 20 for worst). In fact 6 are by A, 7 by B and 7 by C. Rankings given by the professor when checked against authors are

Excellent author	1	2	4	8	11	17	
Good author	3	5	6	12	16	18	19
Inferior author	7	9	10	13	14	15	20

Do these results justify his claim of discriminatory ability? A Kruskal–Wallis test could be used to test the hypothesis that the ranks indicate the samples are all from the same population (i.e. no discriminatory ability) but as the samples have a natural ordering – excellent, good, inferior – a Jonckheere–Terpstra test is appropriate (see Exercise 6.9).

Assimilation and recall

A list of 12 names is read out to students. It contains in random order, 4 names of well-known sporting personalities, 4 of national and international political figures, and 4 prominent in local affairs. The students are later asked which names they can recall and a record is made of how many names each student recalls in each of the three categories. By ranking the results we may test whether recall ability differs systematically between categories, e.g. do people recall names of sporting personalities more easily than those of people prominent in local affairs? See Exercise 6.8.

Tasting tests

A panel of tasters may be asked to rank different varieties of raspberry in order of preference. A Friedman test is useful to detect any pattern in taste preference. See Exercise 6.11.

Quantal responses

Four doses of a drug are given to batches of rats, groups of four rats from the same litter forming a block. The only observation is whether each rat is dead or alive 24 hours later. Cochran's test is appropriate to test for different survival rates at the four doses.

6.7 SUMMARY

Location tests for several independent samples include the following:

The median test (Section 6.2.1) is applicable to any populations (these may differ from one another in aspects other than the median). The test is essentially a special case of the Fisher exact test, or asymptotically a Pearson chi-squared test applied to a $p \times 2$ contingency table calculating the statistic using (6.1).

The Kruskal–Wallis test (Section 6.2.2) is a rank analogue of the one-way classification analysis of variance. The test statistics commonly used are (6.2) or, if there are no ties, (6.3). If p is the number of samples (6.2) and (6.3) have asymptotically a chi-squared distribution with $p-1$ degrees of freedom. An alternative is to transform ranks to van der Waerden scores.

The Jonckheere–Terpstra test (Section 6.2.4) is appropriate if H_1 orders the treatments. The test statistic is a sum of Mann–Whitney statistics. An asymptotic approximation (6.4) is widely used, but care is needed with ties.

For related samples the **Friedman test** (Section 6.3.1) is applicable to randomized block designs where ranks are allocated within blocks. One test statistic is given by (6.5) and asymptotically it has a chi-squared distribution. An alternative statistic is given by (6.6). The **Page test** (Section 6.3.2) is an analogue of the Jonckeere–Terpstra test applicable in a randomized block context if H_1 orders treatments.

Multiple comparisons (Section 6.4.1) analogous to least significant differences for both the Kruskal–Wallis and Friedman test situations are given in (6.7) and (6.8).

The Cochran test (Section 6.4.4) is applicable to blocked binary response data and the test statistic (6.9) has an asymptotic chi-squared distribution.

The squared rank test for heterogeneity of variance (Section 6.5) extends from that for two samples.

EXERCISES

Readers unfamiliar with (parametric) analysis of variance may ignore questions on that topic.

6.1 Perform a parametric analysis of variance on the data in Example 6.3, comparing your result with those for the Kruskal–Wallis test.
6.2 Reanalyse the data in Example 6.3 using van der Waerden scores.
6.3 Carry out the asymptotic Jonkheere–Terpstra test (6.4) for the data in Example 6.4.
6.4 Analyse the data in Example 6.5 using parametric analysis of variance.
6.5 Verify for Example 6.8 that the relevant Wilcoxon statistic for the WMW test is 13.
6.6 Perform the Kruskal–Wallis test on the four-sample data in Example 6.8 to confirm the results quoted in that example, including computation of least significant

differences between package A and package B (i) for those with technical knowledge and (ii) for those without technical knowledge.

6.7 Perform an analysis of variance of the data on numbers of mistakes in Example 6.8.

6.8 At the beginning of a session 12 names are read out in random order to 10 students. Four are names of prominent sporting personalities (Group A), four of national and international politicians (Group B) and four of local dignitaries (Group C). At the the end of the session students are asked to recall as many of the names as possible. The results are:

Student	I	II	III	IV	V	VI	VII	VIII	IX	X
Group A	3	1	2	4	3	1	3	3	2	4
Group B	2	1	3	3	2	0	2	2	2	3
Group C	0	0	1	2	2	0	4	1	0	2

Rank the data within each block (student) and use a Friedman test to assess evidence of a difference between recall rates for the three groups. In particular, is the recall rate for group B and/or group C significantly lower than that for group A?
Carry out an analysis of variance on the given data. Do the conclusions agree with the Friedman test? If not, why not?

6.9 Use the ranks given in the literary discrimination example in Section 6.6 to assess validity of the professor's claim to discriminate between the work of authors.

6.10 A sergeant major orders 34 men to parade tallest on the right, shortest on the left, numbered 1 (tallest) to 34 (shortest). Each man is then asked whether he smokes or drinks and the rank numbers of men in the various categories are as follows:

Drinker and smoker	3 8 11 13 14 19 21 22 26 27 28 31 33
Smoker, non-drinker	2 12 25 32 34
Drinker, non-smoker	1 7 15 20 23 24 30
Non-smoker, non-drinker	4 5 6 9 10 16 17 18 29

Is there evidence of association between height and drinking and smoking habits? Would you reach the same conclusion if ranks were replaced by van der Waerden scores? (In this basic analysis ignore the 'factorial' nature of the treatment structure).

6.11 Five tasters rank four varieties of raspberry in order of preference. Ties are allowed and specified by mid-ranks. Do the results indicate a consistent taste preference?

Taster	1	2	3	4	5
Variety:					
Malling Enterprise	3	3	1	3	4
Malling Jewel	2	1.5	4	2	2
Glen Clova	1	1.5	2	1	2
Norfolk Giant	4	4	3	4	2

6.12 Four share tipsters are each asked to predict on 10 randomly selected days whether the London FTSE Index (commonly known as Footsie) will rise or fall on the following day. If they predict correctly this is scored as 1, if incorrectly as 0. Do the scores indicated below indicate differences in tipsters' ability to predict accurately?

Day	1	2	3	4	5	6	7	8	9	10
Tipster 1	1	0	0	1	1	1	1	0	1	1
Tipster 2	1	1	1	1	0	1	1	0	0	0
Tipster 3	1	1	0	1	1	1	1	1	0	1
Tipster 4	1	1	0	0	0	1	1	1	0	1

6.13 Replace the data in Table 6.6 by ranks 1 to 24 (using mid-ranks for ties where appropriate) and carry out an ordinary randomized block analysis of variance of these ranks to confirm the F-value quoted in Section 6.3.3.

6.14 Berry (1987) gives the following data for numbers of premature ventricular contractions per hour for 12 patients with cardiac arrhythmias when each is treated with 3 drugs A, B, C.

Patient	1	2	3	4	5	6	7	8	9	10	11	12
A	170	19	187	10	216	49	7	474	0.4	1.4	27	29
B	7	1.4	205	0.3	0.2	33	37	9	0.6	63	145	0
C	0	6	18	1	22	30	3	5	0	36	26	0

Use a Friedman test for differences in response between drugs. In particular, is there evidence of a difference in response between drug A and drug B?
Note the obvious heterogeneity of variance between drugs. Carry out an ordinary randomized block analysis of variance on these data. Do you consider it to be valid? Is the Friedman analysis to be preferred? Why?

6.15 Cohen (1983) gives data for numbers of births in Israel for each day in 1975. We give below data for numbers of births on each day in the 10th, 20th, 30th and 40th weeks of the year.

Day	Mon	Tue	Wed	Thu	Fri	Sat	Sun
Week							
10	108	106	100	85	85	92	96
20	82	99	89	125	74	85	100
30	96	101	108	103	108	96	110
40	124	106	111	115	99	96	111

Perform Friedman analyses to determine whether the data indicate (i) a difference in birth rate between days of the week that shows consistency over the four selected weeks and (ii) any differences between rates in the 10th, 20th, 30th and 40th weeks.

6.16 Snee (1985) gives data on average liver weights per bird for chicks given three levels of growth promoter (none, low, high). Blocks correspond to different bird houses. Use a Friedman test to see if there is evidence of an effect of promoter.

Block	1	2	3	4	5	6	7	8
None	3.93	3.78	3.88	3.93	3.84	3.75	3.98	3.84
Low dose	3.99	3.96	3.96	4.03	4.10	4.02	4.06	3.92
High dose	4.08	3.94	4.02	4.06	3.94	4.09	4.17	4.12

(As the dose levels are ordered, a Page test (Section 6.3.2) is appropriate. Try this.)

6.17 Lubischew (1962) gives measurements of maximum head width in units of 0.01 mm for three species of *Chaetocnema*. Part of his data is given below. Use a Kruskal–Wallis test to see if there is a species difference in head widths.

Species 1	53	50	52	50	49	47	54	51	52	57	
Species 2	49	49	47	54	43	51	49	51	50	46	49
Species 3	58	51	45	53	49	51	50	51			

6.18 Biggins, Loynes and Walker (1987) considered various ways of combining examination marks where all candidates sat the same number of papers but different candidates selected different options from all those available. The data below are the marks awarded by 4 different methods of combining results for each of 12 candidates. Do the schemes give consistent ranking of the candidates? Is there any evidence that any one scheme treats some candidates strikingly differently than the way they are treated by other schemes so far as rank order is concerned? Is there any evidence of a consistent difference between marks awarded by the various schemes?

Cand.	1	2	3	4	5	6	7	8	9	10	11	12
A	54.3	30.7	36.0	55.7	36.7	52.0	54.3	46.3	40.7	43.7	46.0	48.3
B	60.6	35.1	34.1	55.1	38.0	47.8	51.5	44.8	39.8	43.2	44.9	47.6
C	59.5	33.7	34.3	55.8	37.0	49.0	51.6	45.6	40.3	43.7	45.5	48.2
D	61.6	35.7	34.0	55.1	38.3	46.9	51.3	44.8	39.7	43.2	44.8	47.5

6.19 The pea node data in Example 6.5 include a control treatment given no growth substance because the experimenter wished to compare all other treatments with this as a base. Regarding each such as a nominated comparison check whether any exceed the least significant difference.

6.20 Confirm the numerical values quoted in Example 6.7 that are relevant for the Friedman test for the data in that example.

6.21 Chris Theobald supplied the following data from a study of 40 patients suffering from a form of cirrhosis of the liver. One purpose was to examine whether there was evidence of association between spleen size and blood platelet count. Blood platelets form in bone marrow and are destroyed in the spleen, so it was thought that an enlarged spleen might lead to more platelets being eliminated and hence to a lower platelet count. The spleen size of each patient was found using a scan and scored from 0 to 3 on an arbitrary scale, 0 representing a normal spleen and 3 a grossly enlarged spleen. The platelet count was determined as the number in a fixed volume of blood. Do these data indicate an association between spleen size and platelet count in the direction anticipated by the experimenter?

Spleen size	*Platelet count*												
0	156	181	220	238	295	334	342	359	365	374	391	395	481
1	65	105	121	150	158	170	214	235	238	255	265	390	
2	33	70	87	109	114	132	150	179	184	241	323		
3	79	84	94	259									

7

Correlation and concordance

7.1 CORRELATION IN BIVARIATE DATA

Nonparametric correlation is concerned largely with paired observations consisting of ranks. The ranks may be the primary data or they may be derived from continuous measurements. In a parametric context with measurement data, the most widely used indicator of correlation is the **Pearson product moment correlation coefficient.** For n paired observations $(x_1, y_1), (x_2, y_2), \ldots, (x_n, y_n)$ this coefficient is

$$r = c_{xy} / [\sqrt{(c_{xx} c_{yy})}] \tag{7.1}$$

where $c_{xy} = \Sigma(x_i y_i) - (\Sigma x_i)(\Sigma y_i)/n$, $c_{xx} = \Sigma(x_i^2) - (\Sigma x_i)^2/n$, $c_{yy} = \Sigma(y_i^2) - (\Sigma y_i)^2/n$, all summations being over the subscript i. The Pearson coefficient is particularly relevant to samples from a bivariate normal distribution where it is an appropriate estimate of the population correlation coefficient ρ, but it is used in practice in a wider context as a measure of linear association in the sense that r takes the value $+1$ or -1 if there is a straight line relationship between x and y. Generally values of r close to $+1$ or -1 imply a near-linear relationship between continuous x and y. If both x and y have a normal distribution, values near zero imply independence. In general, if x and y are independent r will be close to zero, but the converse is not true; there may well be some nonlinear relationship between x and y.

Desirable properties of any correlation coefficient are that its values should be confined to the interval $(-1, 1)$ and that lack of association implies a value near zero. Values near $+1$ imply a strong positive association (i.e. high values of y are associated with high values of x and low values of y with low values of x) and values near -1 imply a strong negative association (i.e. high values of y are associated with low values of x and low values of y are associated with high values of x). For the Pearson coefficient, $r = \pm 1$ implies linearity, but for rank coefficients values of ± 1 need not imply linearity in continuous data from which these ranks may have been derived. They do imply a property called **monotonicity.** This means that as x increases y increases (monotonic increasing) or as x increases y decreases (monotonic decreasing). For rank correlation the value 1 implies strictly increasing monotonicity, the value -1 strictly decreasing monotonicity. Values near ± 1 imply a situation approaching monotonicity.

Rank correlation is also relevant where there is no underlying continuous measurement scale but the ranks simply indicate order of preference expressed by two assessors for a group of objects (e.g. contestants in a diving contest or different brands of tomato soup in a tasting trial).

To illustrate many of the basic concepts in nonparametric correlation we shall refer to the data set in Table 7.1. Figure 7.1 is a scatter diagram for these data.

Table 7.1 A data set to illustrate some facets of nonparametric correlation

x	1	3	4	5	6	8	10	11	13	14	16	17
y	13	15	18	16	23	31	39	56	45	43	37	0

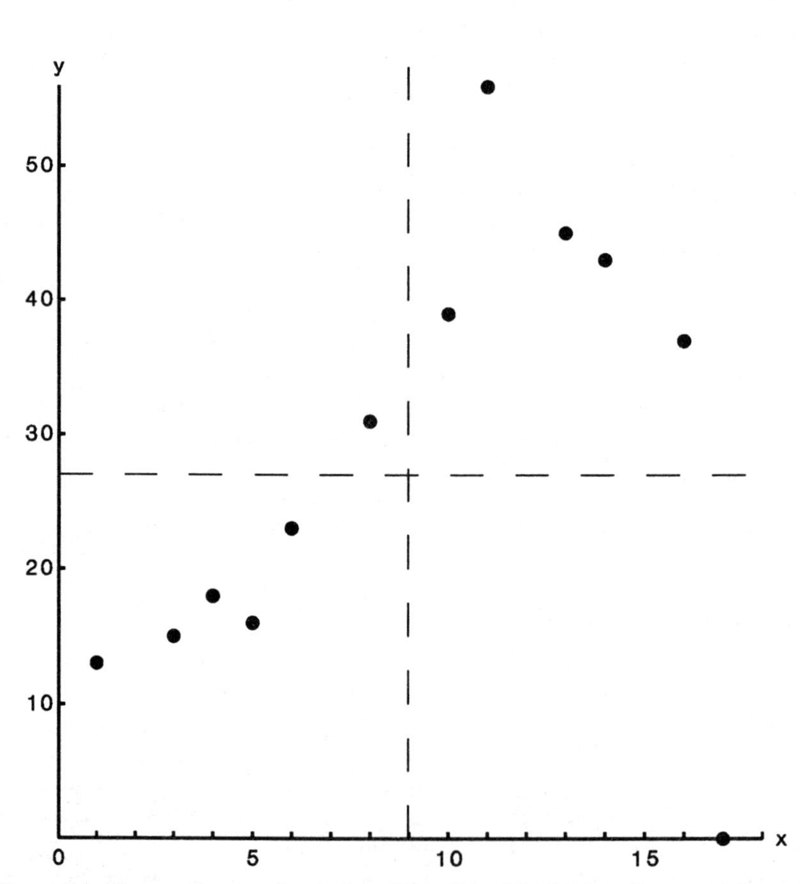

Figure 7.1 Scatter diagram for data in Table 7.1. The broken lines at right angles represent axes when the origin is transferrred to the x, y sample medians.

The data in Table 7.1 are artificial; their general pattern is not unrealistic and is chosen to illustrate some key aspects of the behaviour of correlation coefficients. Figure 7.1 exhibits a tendency for y to increase as x increases, with a suggestion that this becomes erratic for high x; the final observation $x=17$, $y=0$ is a glaring departure from the prevalent pattern. Although numerical values of x and y may be very different in specific applications, the pattern is common. For example, in a football or baseball league table if the observations x, y are the total points gained by a team in the years 1991, 1992 respectively, there is likely to be close agreement between the ordering of x and y for most teams with only small juxtapositions between years, though one or two top performers in the first year may suffer a dramatic change in fortune in the second year or vice versa. Similarly, in a biological context a measured reponse often increases broadly with increasing level of a stimulus, although the rate of increase may fall off or be more erratic at higher levels, with sometimes a subject showing no response even at high levels of stimulus, perhaps as a result of a toxicity effect. For the data in Table 7.1, the Pearson coefficient is $r=0.3729$. This is not significantly different from zero at the 5% level even in a one-tail test if we assume the sample is from a bivariate normal distribution (an assumption that is hardly realistic for these data). Nonsignificance of the Pearson coefficient goes against an intuitive feeling that there is evidence of a fairly high degree of association in the data.

In Sections 7.1.1 to 7.1.3 we consider several nonparametric measures of correlation and see how each responds to the data in Table 7.1. We present the coefficients in a way that shows parallels between these measures of correlation and concepts introduced in earlier chapters.

As in earlier chapters, exact tests are based on appropriate permutations of data, ranks or counts. For correlation tests based on ranks, for example, the appropriate permutation procedure is to fix the rank order associated with one variable, conventionally x, and to calculate the value of the chosen coefficient under all equally likely permutations of the other variable, y. The probabilities of the observed or more extreme values of the chosen coefficient (or some linear function of it) determine the size of the relevant critical region. Example 7.4 gives an illustration.

7.1.1 A median test for correlation

We consider first a correlation coefficent proposed by Blomqvist (1950; 1951) that embodies similar concepts to the median test considered in Section 5.1.1. Suppose the medians of the marginal distribution of X, Y are respectively θ_x, θ_y. In a sample we expect about half the observed x to be below θ_x and about half the y to be below θ_y. If X and Y are independent we expect a good mix of high, medium and low values of Y to be associated with the various values of X. If we knew θ_x and θ_y we could shift the origin in a scatter diagram of the observed (x_i, y_i) to (θ_x, θ_y). Then under $H_0 : X$, *Y are independent* we would expect about one quarter of all points to be in each of the four quadrants determined by the new axes. Usually we do not know θ_x, θ_y, so we replace them by their estimates the sample medians M_x, M_y. In Figure 7.1 axes with a new origin at (M_x, M_y) are shown by broken lines, and working clockwise from the first (or top right) quadrant the numbers of points in the respective quadrants are 5, 1, 5, 1. The concentration of points in the first and third quadrants suggests dependence between X and Y. We may formalize a test of H_0 against a one-sided alternative of either *positive* or *negative* association, or a two-sided alternative of *some association*, following closely the procedure in Section 5.1.1. In general, if n is even, denoting the sample medians for X, Y by M_x, M_y and assuming no values coincide with these medians, we count the numbers of pairs a, b, c, d in each of the four quadrants determined after this shift of the scatter plot origin to (M_x, M_y). Because of sample median properties we get a 2×2 contingency table as in Table 7.2.

Table 7.2 Contingency table for the median test for correlation

	Above M_x	Above M_x	Row total
Above M_y	a	b	$\frac{1}{2}n$
Below M_y	c	d	$\frac{1}{2}n$
Column total	$\frac{1}{2}n$	$\frac{1}{2}n$	n

The marginal totals follow from the properties of the sample median and our restriction to even n with no obervation at either M_x or M_y. If n is odd at least one sample value for x and one sample value for y will coincide with the medians. This results in either one or two sample points lying at the median or on axes through the sample medians; such points are omitted from the count. This necessitates modifications that we give at the end of this section.

In the case envisaged in Table 7.2, if X, Y are independent then clearly a, b, c and d each has expected value $\frac{1}{4}n$. Marked departures from this value indicate an association between X, Y. The appropriate test procedure uses Fisher's exact test. We illustrate this in Example 7.1, but first we seek an appropriate correlation measure having the desirable properties that it takes values in the interval $(-1, 1)$ with values near ± 1 indicating near-monotonic dependence, and independence resulting in values near zero (though the converse may not be true).

Two possible coefficients are

$$r_{\mathrm{d}} = (a+d-b-c)/n \tag{7.2}$$

and

$$r_{\mathrm{m}} = 4(ad-bc)/n^2 \tag{7.3}$$

It is easily verified that in the special circumstances of Table 7.2, where the marginal totals imply that $a=d$, $b=c$ and $a+b=\frac{1}{2}n$, both (7.2) and (7.3) have the desirable properties for a correlation coefficient and indeed are equivalent, each taking the same set of $\frac{1}{2}n+1$ possible values with the same associated probabilities under the null hypothesis of independence. While (7.2) has arithmetic simplicity, (7.3) generalizes more readily to other contexts, including the case n odd, so is usually preferred.

Example 7.1

The problem. Calculate r_m (or r_d) for the data in Table 7.1 and test the hypothesis of independence against that of positive association between X and Y.

Formulation and assumptions. We find the sample medians M_x, M_y and deduce the entries in Table 7.2. The relevant coefficient values are computed using (7.2) or (7.3) and the test of significance is based on Fisher's exact test as in Example 5.1. In this case a one-tail test is appropriate since the alternative is in a specified (positive) direction.

Procedure. For the data in Table 7.1 the sample medians are $M_x=9$, $M_y=27$, from which we deduce the values of a, b, c, d in Table 7.2 (most easily done by inspection of Figure 7.1) to be $a=d=5$ and $b=c=1$. Since $n=12$ we easily calculate $r_m=2/3=r_d$. We use Fisher's exact test as in Section 5.1, the relevant tail probabilities being associated with the observed value $a=5$ and the more extreme $a=6$. Calculating these probabilities using (5.2) is not difficult even if an appropriate computer program is not available. The relevant probability is 0.0400.

Conclusion. We reject the hypothesis of independence in favour of that of positive association at an actual 4% significance level in a one-tail test. The numerical value of the appropriate median correlation coefficient is $r_m=0.667$.

Comments. 1. The Pearson correlation coefficient was not significant for these data; it is not robust against departures from normality of the type implicit in the observation (17, 0). However, r_m is robust and indeed we would reach the same conclusion if the point (17, 0) were replaced by any point in the second quadrant (positive x, negative y) relative to axes with origin at the sample medians. In general, the median correlation test is robust against a few quite major departures from a monotone or near-monotone trend. Unfortunately, it has low Pitman efficiency compared to the Pearson coefficient test when the relevant normality assumption holds.

2. An observation such as (17, 0) in these data, well out of line with the general pattern, is an example of an outlier. Many analyses are strongly

influenced by the presence of outliers, a matter we discuss in a wider context in Chapter 11.

3. The test is applicable no matter what the marginal distributions of X, Y may be.

Computational aspects. For all but very small samples a computer program such as that in STATXACT or TESTIMATE is desirable for Fisher's exact test. The asymptotic result (5.3) is generally fairly reliable for larger samples.

If n is odd, or generally if there are values equal to one or both of the sample medians, the marginal totals in a table like Table 7.2 will no longer be $\frac{1}{4}n$. It is then appropriate to use the Fisher exact test with the marginal totals actually observed, and if a correlation coefficient is required to use a modification of r_m, namely $r_m=(ad-bc)/\sqrt{[(a+b)(c+d)(a+c)(b+d)]}$, where a, b, c, d are the observed cell values. This reduces to (7.2) when all marginal totals are $\frac{1}{2}n$.

The coefficient r_m is easy to calculate, but in practice it is overshadowed by the rank correlation coefficients described in Sections 7.1.2 and 7.1.3.

7.1.2 The Kendall rank correlation coefficient

We may adapt methods reminiscent of Wilcoxon–Mann–Whitney procedures to test for monotonic association. The test we develop is based on a coefficient commonly known as **Kendall's correlation coefficient,** or **Kendall's tau**, and it is closely related to the Jonckheere–Terpstra test statistic.

In practice, as in the Jonckheere–Terpstra test, we need not replace continuous data by ranks, but we illustrate the basic ideas of the test using ranks. Computation is simplified if we assume the x data arranged in ascending rank order. For illustrative purposes we use ranks corresponding to the data in Table 7.1, where the x are already arranged in ascending order. Allocating ranks to y, we obtain the paired rankings in Table 7.3. The labels 'x-ranks' and 'y-ranks', referring repectively to the ranks associated with the first and second members of a bivariate pair (x, y), are used for convenience in this chapter even when our basic data are ranks only, with no specific underlying continuous variable (e.g. in the case of taste preferences or two judges ranking a number of competitors in a sporting event like ice skating or high diving).

Table 7.3 Paired ranks corresponding to data in Table 7.1

x-rank	1	2	3	4	5	6	7	8	9	10	11	12
y-rank	2	3	5	4	6	7	9	12	11	10	8	1

Note the general tendency in Table 7.3 for the *y*-ranks to increase as those of *x* increase.

Denoting the ranks of x_i, y_i by r_i, s_i respectively, Kendall's coefficient of correlation is based on the principle that if there is association between the ranks of *x* and *y*, then if we arrange the *x*-ranks in ascending order (i.e. so that $r_i=i$) the s_i should show an increasing trend if there is positive association and a decreasing trend if there is negative association. Kendall (1938) therefore proposed that after arranging observations in increasing order of *x*-ranks, we score each paired difference s_j-s_i for $i=1, 2, \ldots, n-1$ and $j>i$ as $+1$ if this difference is positive and as -1 if negative. Kendall called a positive difference a **concordance** and a negative difference a **discordance**. Denoting the numbers of concordances and discordances by n_c, n_d respectively, Kendall's tau, which we shall denote by *t*, although historically it has usually been denoted by the Greek letter τ (tau), is

$$t=\frac{n_c-n_d}{\frac{1}{2}n(n-1)} \tag{7.4}$$

We distinguish between this sample estimate *t* and the corresponding population value τ, but do not confuse *t* here with the *t* in a Student's *t*-test. If we are comparing ranks assigned by two judges to a finite set of objects such as ten different brands of tomato soup this is essentially the entire population of interest, so our calculated coefficient is a measure of agreement between the judges in relation to that population. On the other hand, if we had a sample of *n* bivariate observations from some continuous distribution (of measurements, say) and calculate *t* using (7.4) this is an estimate of some underlying τ that is a measure of monotonic association between the variables *X*, *Y* in the population. Since there are $\frac{1}{2}n(n-1)$ pairs s_j-s_i, if all are concordances $n_c=\frac{1}{2}n(n-1)$ and $n_d=0$, whence $t=1$. Similarly, if all are discordances, clearly $t=-1$. If the rankings of *x* and *y* are independent we expect a fair mix of concordances and discordances and *t* should be close to zero. If there are no ties in the rankings $n_c+n_d=\frac{1}{2}n(n-1)$. We consider ties in Section 7.1.4.

Example 7.2

The problem. Compute Kendall's rank correlation coefficient for the data in Table 7.1.

Formulation and assumptions. We compute n_c from the *y*-ranks in Table 7.3. We may also obtain n_d this way, or deduce it from $n_d=\frac{1}{2}n(n-1)-n_c$.

Procedure. To count the number of concordances we inspect the *y*-ranks in Table 7.3, noting for each successive rank the number of succeeding ranks that are greater. The first *y*-rank is 2 and this is succeeded by 10 greater ranks, namely 3, 5, 4, 6, 7, 9, 12, 11, 10, 8. Similarly, the next y-rank, 3, gives 9 concordances. Proceeding this way it is easily verified that the total number of concordances is

$$n_c = 10+9+7+7+6+5+3+0+0+0+0=47.$$

Since $n=12$ we easily deduce that $n_d = \frac{1}{2} \times 12 \times 11 - 47 = 19$. This may be verified by counting discordances directly from Table 7.3.

Conclusion. For the given data $t=(47-19)/66=0.4242$.

Comments. 1. Since the x are ordered in Table 7.1, the order of the corresponding y values is an ordered transformation of the ranks in Table 7.3. Thus we could equally well have computed the number of concordances and discordances by noting the signs of all differences $y_j - y_i$ in Table 7.1 as these correspond to the signs of the $s_j - s_i$ above.

2. The counting procedure for concordances is reminiscent of the Mann–Whitney counts for the Wilcoxon–Mann–Whitney test, and we show below equivalence to the Jonckheere–Terpstra test.

3. We discuss a test of significance for $H_0 : \tau=0$ in Example 7.3.

The x_i are in ascending order in Table 7.1; though not essential, this simplifies computation. Generally, the number of concordances is the number of positive b_{ij} among all $\frac{1}{2}n(n-1)$ of the $b_{ij}=(y_j-y_i)/(x_j-x_i)$, $i=1, 2, \ldots, n-1$, $j>i$. The number of negative b_{ij} is the number of discordances.

Equivalence to a Jonckheere–Terpstra test situation is evident if we arrange the x in ascending order as in Table 7.1 and regard the x_i (or the associated ranks) as 'indexing' n ordered samples each of 1 observation which is the corresponding y value. It is clear that the number of concordances is the sum of the Mann–Whitney sample pairwise statistics used in the Jonckheere–Terpstra test in Section 6.2.4. Thus, in theory at least, any program for the exact Jonckheere–Terpstra test may be used for Kendall's coefficient to test H_0 : *no association between x and y ranks* against either a one- or two-sided alternative. In practice, however, most programs for the Jonckheere–Terpstra test will, in this situation, only give exact tail probabilities for very small n. Asymptotic results, or Monte Carlo estimates, of the exact tail probability must be used with this approach if the samples are moderate to large. We consider more specific asymptotic results in Section 7.1.5. Programs in general statistical packages often calculate t directly, some indicating significance levels (usually asymptotic), others leaving the user to determine these at a nominal level from tables. Extensive tables are given, for example, by Neave (1981, p. 40): we give a short table as Table IX. Some tables give equivalent critical values of $n_c - n_d$ for various n. However, calculating the t value gives a feel for the degree of association implied by any correlation coefficient. When there are no ties tests may be based on n_c only, for it is clear from (7.4) that if n is fixed, t is a linear function of n_c.

Example 7.3.

The problem. For the data in Table 7.1 use Kendall's coefficient to test H_0 : *no association* against H_1 : *positive association between x and y.*

Formulation and assumptions. The appropriate test is a one-tail test of $H_0 : \tau=0$ against $H_1 : \tau>0$.

Procedure. In Example 7.2 we found $t=0.4242$, also showing that $n_c-n_d=47-19=28$. Table IX or Neave's tables give nominal 5% and 1% critical values for significance when $n=12$ in a one-tail test as 0.3929 and 0.5455. Corresponding values for n_c-n_d are 26 and 36.

Conclusion. We reject H_0 at a nominal 5% significance level and accept a positive association between the ranks of x, y.

Comments. 1. The result is consistent with that obtained with the median correlation test in Section 7.1.1.

2. Had our alternative been $H_1 : \tau\neq 0$ a two-tail test would have been appropriate and we would not have rejected H_0 at the 5% significance level.

Computational aspects. As indicated above, most general statistical programs do not lead to exact significance levels for relevant hypothesis tests. Since extensive tables are widely available for small and moderate n this is no bar to the use of the coefficient in testing, especially as it is relatively easy to calculate.

For small samples it is easy to calculate the exact distribution of t.

Example 7.4

The problem. Calculate the exact permutation distribution of Kendall's t when $n=4$, under the null hypothesis of no association.

Formulation and assumptions. If the x-ranks are in ascending order, then for each of the $4!=24$ equally likely permutations of the y-ranks we count the number of concordances, n_c. When $n=4$, $n_c+n_d=6$ whence, given n_c, we easily find n_d. We then compute t using (7.4) and count the number of times each value of t occurs. Division of each number of occurrences by 24 gives the probability of observing each value.

Procedure. In Exercise 7.1 we ask for computation of t for each of the 24 possible permutations of 1, 2, 3, 4. For example, for the permutation 4, 2, 1, 3 we easily see that $n_c=2$, whence, by (7.4), $t=(2-4)/6=-0.333$. It is easily verified that 4 other permutations give the same value of n_c, whence $\Pr(t=-0.333)=5/24$.

Conclusion. Completing Exercise 7.1 gives the following probabilities associated with each of the 7 possible values of t:

t	1	0.67	0.33	0	−0.33	−0.67	−1
Probability	1/24	3/24	5/24	6/24	5/24	3/24	1/24

Comments. 1. The distribution of t is symmetric.

2. When $n=4$ we can only demonstrate significance at a nominal 5% level in a one-tail test. The size of the critical region is $\alpha=1/24=0.042$.

Computational aspects. We may calculate the exact tail probabilities or the complete permutation distribution for samples of size 5 or less using the Jonckheere–Terpstra program in STATXACT. The Monte Carlo option in that program may be used to estimate (very accurately) tail probabilities for larger samples.

The use and calculation of t when there are data or rank ties is discussed in Section 7.1.4. Some asymptotic properties are given in Section 7.1.5.

7.1.3 Spearman's rank correlation coefficient

Another intuitively reasonable coefficient is the Pearson coefficient calculated for ranks (instead of the original continuous data if these are given). This coefficient was first proposed by Spearman (1904) and is often denoted by the Greek letter ρ (rho). To avoid confusion with this symbol when it is used for the correlation coefficient in bivariate normal distributions (estimated by the Pearson coefficient r) we shall denote an estimate of the **Spearman coefficient** by r_s and the corresponding population value by ρ_S. The formula (7.1) may be used to calculate r_s if (x_i, y_i) are replaced by their ranks (r_i, s_i). If there are no ties a simplified formula can be obtained by straightforward algebraic manipulation making use of properties of sums and sums of squares of ranks. The formula is

$$r_s=1-\frac{6T}{n(n^2-1)} \tag{7.5}$$

where $T=\Sigma_i(r_i-s_i)^2$, i.e. T is the sum of the squares of the difference between ranks for each pair. With perfect matching of ranks $T=0$, so $r_s=1$. If there is a complete reversal of ranks, tedious elementary algebra establishes that $r_s=-1$. If there is no correlation between ranks it can be shown that $E(T)=n(n^2-1)/6$, so that r_s has expectation zero. If our observations are a random sample from a bivariate distribution with X, Y independent we expect near-zero rank correlation.

Table X gives critical values at nominal 5% and 1% significance levels for testing $H_0 : \rho_s=0$ against one- and two-sided alternatives. Since, for fixed n, r_s is a monotonic function of T we may use T itself as a test statistic and this has been tabulated for small n. As with the Kendall coefficient, we prefer to use r_s; it gives a more easily appreciated indication of the level of association.

Unlike Kendall's t, which may be computed from the original data, we need the ranks to compute r_s. Both coefficients are often used to test whether there is broad agreement between ranks assigned by different assessors. Do two examiners agree in their ranking of candidates? Do two judges agree in the placings in a diving contest? Do job applicants' ranks for manual skill on the basis of some psychological test show any relationship to their rankings in a further test for mathematical skills?

These coefficients are also appropriate for tests of trend. Does Y increase (or decrease) as X increases?. Such tests are often more powerful than the Cox–Stuart test described in Section 2.1.6.

Example 7.5

The problem. Compute the Spearman correlation coefficient for the ranked data in Table 7.3 and use your result to test $H_0 : \rho_s=0$ against the one-sided alternative $H_1 : \rho_s>0$.

Formulation and assumptions. The coefficient is computed using (7.5); critical values are given in Table X.

Procedure. T is computed from Table 7.3 by taking the difference between ranks in each column, squaring each difference, and adding. Thus

$$T=(1-2)^2+(2-3)^2+(3-5)^2+ \ldots +(12-1)^2=1+1+4+ \ldots +121=162$$

whence $r_s=1-(6\times162)/(12\times143)=0.4336$. One-tail critical values (Table X) for significance at nominal 5% and 1% levels are 0.503 and 0.678.

Conclusion. Because $r_s=0.4336$ is less than the value 0.503 required for significance at a nominal 5% level we do not reject H_0.

Comments. We rejected H_0 using an equivalent test with Kendall's coefficient in Example 7.3. In practice, it is unusual for tests using either statistic to lead to different conclusions though this can occasionally, as here, happen. The reason is that Spearman's coefficient, being conceptually related to the Pearson product-moment correlation coefficient, is less robust against certain fairly extreme situations such as that associated with an odd complete, or almost complete, rank reversal. We have such a reversal for the last data pair in Table 7.3.

Computational aspects. Many general statistical packages have programs that compute r_s or T or an equivalent statistic, usually leaving one to look up tables or use an asymptotic result to establish significance. DISFREE includes a small-sample procedure which uses the permutation distribution to establish appropriate tail probabilities; however, the manual correctly warns that computation time may be appreciable and suggests using simulations. STATXACT has no explicit program for Spearman's coefficient but the linear-by-linear association program (see Section 10.2.2) may be used with appropriate weights as described in the manual. For this example 5000 Monte Carlo simulations with either of these programs clearly indicated that we do not attain significance at a nominal 5% level even in a one-tail test. In view of the ease with which r_s may be calculated, and the wide availability of tables, a dearth of computer programs for exact tests is not too serious for the 'no-ties' situation.

The exact permutation distribution of r_s for small samples may be computed in an analogous manner to that demonstrated in Example 7.4 for Kendall's t (see Exercise 7.2). Generally speaking, for a given n the Spearman coefficient may take a greater number of possible values than does Kendall's coefficient.

We consider ties in Section 7.1.4 and asymptotic results in Section 7.1.5.

Research workers often ask which of the rank correlation coefficients – Kendall's or Spearman's – is to be preferred. There is no clear-cut answer. They seldom lead to markedly different conclusions, though Examples 7.3 and 7.5 show this is possible when one pair of ranks is clearly out of line with the general trend. For instance, in Example 7.5 one pair of ranks, i.e. (12, 1), contributes 121 to the total T value of 162, suggesting r_s may be sensitive to such an outlier.

An alternative way of calculating the coefficients clarifies the relationship between them. We assume there are no ties and that the pairs of ranks are arranged in ascending order of x-ranks as in Table 7.3. We define $s_{ij}=s_j-s_i$, $i=1, 2, 3, \ldots, n-1$ for all $j>i$. We easily see that the number of concordances in Kendall's t equals the number of positive s_{ij}, and the number of discordances equals the number of negative s_{ij}.

We do not prove it, but if we denote by n_{cs} the sum of the values of the positive s_{ij} (the sum of the values of differences that are concordant) and by n_{ds} the sum of the values of the negative s_{ij} (the sum of the values of differences that are discordant), then in the 'no-tie' case $n_{cs}+n_{ds}=n(n^2-1)/6$ and r_s is given by

$$r_s=\frac{n_{cs}-n_{ds}}{n(n^2-1)/6} \tag{7.6}$$

Note a formal similarity to (7.4). It is instructive to write the s_{ij} as an upper triangular matrix

$$\begin{array}{ccccccc}
s_2-s_1 & s_3-s_1 & s_4-s_1 & \cdot & \cdot & \cdot & s_n-s_1 \\
 & s_3-s_2 & s_4-s_2 & \cdot & \cdot & \cdot & s_n-s_2 \\
 & & \cdot & \cdot & \cdot & \cdot & \cdot \\
 & & & \cdot & \cdot & \cdot & \cdot \\
 & & & & & s_{n-1}-s_{n-2} & s_n-s_{n-2} \\
 & & & & & & s_n-s_{n-1}
\end{array}$$

For the data in Table 7.3 the matrix (see Exercise 7.3) is

$$\begin{array}{rrrrrrrrrrr}
1 & 3 & 2 & 4 & 5 & 7 & 10 & 9 & 8 & 6 & -1 \\
 & 2 & 1 & 3 & 4 & 6 & 9 & 8 & 7 & 5 & -2 \\
 & & -1 & 1 & 2 & 4 & 7 & 6 & 5 & 3 & -4 \\
 & & & 2 & 3 & 5 & 8 & 7 & 6 & 4 & -3 \\
 & & & & 1 & 3 & 6 & 5 & 4 & 2 & -5 \\
 & & & & & 2 & 5 & 4 & 3 & 1 & -6 \\
 & & & & & & 3 & 2 & 1 & -1 & -8 \\
 & & & & & & & -1 & -2 & -4 & -11 \\
 & & & & & & & & -1 & -3 & -10 \\
 & & & & & & & & & -2 & -9 \\
 & & & & & & & & & & -7
\end{array}$$

To compute Kendall's t using this table we simply count the number of positive entries to obtain $n_c=47$ and the number of negative entries giving $n_d=19$. These were obtained in Example 7.2. To obtain n_{cs} we add all positive entries in the matrix. In Exercise 7.3 we ask you to verify that $n_{cs}=1+3+2+4+\ldots+3+2+1=205$. Similarly, adding all negative entries gives $n_{ds}=1+2+1+4+\ldots+10+2+9+7=81$. Substitution in (7.6) gives $r_s=6\times(205-81)/(12\times 143)=0.4336$, in agreement with Example 7.5.

An examination of the triangular matrix helps explain why, in this case, r_s plays down any correlation relative to that indicated by t. A few negative s_{ij} make relatively large contributions to the sum n_{ds}; notably those in the lower half of the final column. This is clearly attributable to the fact that the low rank value $s_{12}=1$ is out of line with the ranks assigned to its near neighbours.

7.1.4 Ties in rank correlation

These occur in measurement data or in ranks when no genuine distinction can be made in the order assigned to two or more items. We adopt the standard procedure for dealing with tied ranks and assign mid-ranks to ties. If we continue to use the formulae (7.4) and (7.5) for t and r_s, little harm is done for only a few ties in fairly large samples although distributional properties of coefficients will not hold exactly. For example, coefficients may not take the value 1 or -1 even when all ranked pairs lie on a straight line and tabulated critical values may no longer be appropriate. Without computer programs to calculate permutation distributions in the presence of ties, significance tests must, in practice, be based on published values for the 'no-tie' case or on asymptotic results. Such approximations may be improved by appropriate modifications to the calculated statistic to allow for ties.

We illustrate these points for Kendall's coefficient with ties, taking an extreme example. Suppose our first pair is ranked 1 in both x and y and that this is followed by rank observations consisting of seven ties in each variable followed by one larger pair ranked 9 for each variable. We allocate a mid-rank of 5 to each of the 7 tied ranks, giving allocated (r_i, s_i):

r	1	5	5	5	5	5	5	5	9
s	1	5	5	5	5	5	5	5	9

It is clear if we plot these rank pairs that they all lie on a straight line of positive slope so it is intuitively reasonable to expect to find $t=1$.

We have yet to consider how to regard ties in determining the number of concordances. Logically, if, for any two rank pairs, there is a tie in either the x- or y-ranks or both this is neither a concordance nor discordance, for the ordering of observations with regard to the tied rank is arbitrary. We ignore any such pair in our count of concordances and discordances, i.e. effectively we score it as zero.

Since in this example the x rankings are ordered, we calculate n_c by counting concordances in the y-rankings that correspond to distinct x-rankings. Clearly $s=1$ gives rise to 8 concordances and *each* of the mid-rank values of 5 gives rise to 1 concordance (with the observation (9, 9), giving a further 7 concordances. Thus $n_c=8+7=15$. Clearly there are no discordances. Since $n=9$, (7.4) gives $t=15/36$, rather than the anticipated $t=1$. The happens because in the presence of ties n_c+n_d no longer equals $\frac{1}{2}n(n-1)$.

We do not prove it, but it can be shown that we get values $+1$ or -1 for tied rank situations that imply a straight-line relationship between ranks or mid-ranks if we replace the denominator in (7.4) by $\sqrt{[(D-U)(D-V)]}$, where

$$D=\tfrac{1}{2}n(n-1),\ U=\tfrac{1}{2}\Sigma u(u-1),\ V=\tfrac{1}{2}\Sigma v(v-1)$$

and u, v refer to the number of consecutive ranks in a tie in the x- and y-rankings respectively, and the summations in U, V are over all sets of tied ranks. We denote this modified statistic, sometimes referred to as ***Kendall's tau-b***, by t_b. In the situation considered here there is one set of consecutive ties in both x- and y- rankings and each is a tie of 7 consecutive ranks. Thus $u=v=7$ and there are no other contributions to U, V, so $U=V=\frac{1}{2}\times 7\times 6=21$.

Since $D=36$ when $n=9$, we replace the denominator in (7.4) by $\sqrt{[(36-21)(36-21)]}=15$. This is equal to the numerator value of n_c-n_d, so the modification gives $t_b=1$.

With no ties the hypothesis test for independence when $n=9$ rejects H_0 at a nominal 5% level in a two-tail test if $t \geq 0.556$ (Table IX). With heavy tying the permutation distribution under H_0 is very different and in this example t_b can take only 5 different values, namely 0, $\pm 7/15$, ± 1. Whereas when $n=9$, the exact permutation distribution gives $\Pr(t=1)=1/(9!)$ $=0.00000276$ with no ties, for the extreme tie situation considered here under the appropriate permutation distribution $\Pr(t_b=1)=(7!)/(9!)=0.0139$.

In practice, with many ties, programs such as STATXACT and DISFREE that provide Monte Carlo estimates of tail probabilities (or that will calculate exact probabilities for very small samples) provide the only satisfactory means of hypothesis testing.

For moderately large samples with only a few ties, use of (7.4) and tables of critical values for the no-tie situation should not be seriously misleading.

Example 7.6

The problem. Life expectancy showed a general tendency to increase during the nineteenth and twentieth centuries as standards of health care and hygiene improved. The extra life expectancy varies between countries, communities and even families. Table 7.4 gives the year of death and age at death for 13 males of a fourth clan, the McDeltas, buried in the Badenscallie burial ground. Is there an indication that life expectancy is increasing for that clan?

Table 7.4 Year of death and ages of 13 McDeltas

Year	1827	1884	1895	1908	1914	1918	1924	1928	1936	1941	1964
Age	13	83	34	1	11	16	68	13	77	74	87
Year	1965	1977									
Age	65	83									

Formulation and assumptions. If there is an increase in life expectancy those dying later will tend to be older than those dying earlier. Years of death, x, are already arranged in ascending order so we may count the number of concordances and discordances by examining all pairs of y and scoring each as appropriate (1 for concordance, -1 for discordance, zero for a tie).

Procedure. There are no ties in the x. The first y entry is 13. In the following entries this is exceeded in 9 cases (concordances); there are two discordances and 1 tie. Similarly for the second y entry (83) there is one concordance, 9 discordances and 1 tie. Proceeding this way we find that

$$n_c=9+1+6+9+8+6+4+5+2+2+0+1=53$$

and

$$n_d=2+9+4+0+0+1+2+0+2+1+2+0=23.$$

Since $n=13$, (7.4) gives $t=(53-23)/(\frac{1}{2}\times 13\times 12)= 0.3846$. We have two ties at $y=13$ and $y=83$. Using the formula for the denominator adjustment we have $D=78$, $U=0$ and $V=\frac{1}{2}\times 2(2-1)+\frac{1}{2}\times 2(2-1)=2$, whence

$$t_b=(53-23)/\sqrt{[78\times(78-2)]}=0.3896.$$

This small difference between t and t_b does not affect our conclusion.

Conclusion. If we ignore ties and accept that a one-tail test is appropriate, using Table IX we reject H_0 : *no increasing trend of life expectancy* at the nominal 5% level.

Comments. 1. If there are only a few ties and we ignore them and use (7.4) our test is conservative, for clearly if we allow for ties by using t_b the denominator is less than that in (7.4).

2. A one-tail test is justified because we are testing whether data for the McDelta clan follow established trends in most developed countries.

2. Care is needed in counting concordances if there are ties in both x and y. For example, for the data set:

x	2	5	5	7	9
y	5	1	3	8	1

the first pair (2, 5) gives one concordance (7, 8) and 3 discordances; the second pair (5, 1) gives 1 concordance (7, 8) and no discordances because of the tied values in x at (5, 3) and in y at (9, 1); the next pair (5, 3) gives one concordance (7, 8) and one discordance (9, 1).

Computational aspects. With more than a few ties a program that gives at least a Monte Carlo estimate of the relevant tail probabilities is useful. The Jonckheere–Terpstra test procedure in STATXACT is one such program.

Ties in the Spearman coefficient may be dealt with relatively easily, bearing in mind that it is effectively the Pearson coefficient for ranks. If we use mid-ranks for ties the appropriate replacement for (7.5) is the calculated Pearson coefficient using ranks and mid-ranks. We may use (7.1) directly, but since the sum of ranks and mid-ranks for n observations is always $\frac{1}{2}n(n+1)$, simplification is possible. If we denote the 'scores' (ranks and mid-ranks) for the x, y by r_i, s_i, respectively, it is not dificult to show that in this case (7.1) reduces to

$$r_s = \frac{\Sigma_i r_i s_i - C}{\sqrt{[(\Sigma_i r_i^2 - C)(\Sigma_i s_i^2 - C)]}} \tag{7.7}$$

where $C = n(n+1)^2/4$.

With only a few ties it is often not misleading to use critical values appropriate to the 'no-tie' case, but with many ties a program giving at least Monte Carlo estimates of tail probabilities is desirable. Thomas (1989) shows that in the presence of ties r_s computed by (7.5) is greater than or equal to the correct value given by (7.7). On the other hand, (7.6) tends to underestimate the magnitude of the correct value.

Example 7.7

The problem. Using the data in Example 7.6 (Table 7.4), compute the Spearman coefficient r_s and compare your conclusions with those for Example 7.6.

Formulation and assumptions. We wish to test $H_0 : \rho_s = 0$ against $H_1 : \rho_s > 0$.

Procedure. We require a table of ranks corresponding to the data in Table 7.4. The reader should verify these are correctly given in Table 7.5.

Table 7.5 Ranks for McDelta data

Year (r)	1	2	3	4	5	6	7	8	9	10	11	12	13
Age (s)	3.5	11.5	6	1	2	5	8	3.5	10	9	13	7	11.5

One might ignore the few ties and use (7.4) with the appropriate 'tied-rank' differences, e.g. $r_1 - s_1 = 1 - 3.5 = 2.5$, but we prefer to use (7.7). In Exercise 7.4 we ask you to verify that this gives $r_s = 0.507$ and also that (7.4) gives $r_s = 0.508$. When $n = 13$, Table X gives 0.484 as the critical value for significance in a one-tail test at the 5% level.

Conclusion. Since $0.507 > 0.484$ we reject the hypothesis of no increase in life expectancy.

Comments. 1. The result is in line with that for Kendall's coefficient. Although the two coefficients led to different conclusions in Examples 7.3 and 7.5, such differences are the exception rather than the rule. However, the two coefficients usually have unequal numerical values.

2. In Exercise 7.5 we ask whether the Cox–Stuart trend test would reject the null hypothesis.

3. One might expect the larger samples for the McAlphas, McBetas and McGammas described in earlier chapters also to give significant results. For each of these groups the calculated r_s and t are positive but just fail to

reach significance at the 5% level in one-tail tests. We may have been lucky to get a significant result with our small McDelta sample. We drew attention in Example 3.6 to to the bimodal form of death statistics due to a tendency for deaths to occur in infancy or else after survival to fairly mature years. In the case of the McDeltas the several youthful deaths occurred early in the time sequence. The significant result would have been negated had there been a death in infancy in the 1980s for that clan.

4. Positive correlation coefficients for each of the four clans (even though only one was significant) suggest a trend towards increasing life expectation. Indeed, if available data are combined for all clans (though there are dangers in combining data in samples from different populations) we should enhance our prospect of a significant result. For the combined sample of 115 deaths (year of death was not available for two individuals) $r_s = 0.2568$, a value significant at the 1% level in a one-tail test when $n = 115$.

5. While the test result formally supports the hypothesis of increasing life expectancy, we must bear in mind that not all members of a clan will be buried near their place of birth; those that are may not be typical of the whole clan. Some will be lost by emigration and this may not be taking place at a steady rate throughout the period. The infamous Highland Clearances had an impact on local death statistics in parts of Scotland for several generations. Time-dependent changes in preference for cremation as opposed to burial and differences in such preferences between age groups may also be relevant. Such 'real-world' points must be taken into account when interpreting a statistically significant result.

Computational aspects. Whether or not there are tied ranks any program for calculating the Pearson coefficient may be used to obtain the Spearman coefficient if we use ranks or mid-ranks as data, but in general one will have to resort to tables or the asymptotic result given in Section 7.1.5 to decide whether a result is significant. STATXACT and DISFREE enable exact permutation distribution tail probabilities or Monte Carlo estimates thereof to be calculated for the Spearman coefficient.

An excellent review of the implication of ties in rank correlation coefficients that considers further aspects is given by Kendall and Gibbons (1990, Chapter 3).

7.1.5 Rank correlation in large samples

For large n, tests may be based on the distribution of functions of t and r_s that have an asymptotic normal distribution. The distributions of $z = 3t\{\sqrt{[n(n-1)]}/[2(2n+5)]\}$ and $z = r_s\sqrt{(n-1)}$ are approximately standard normal under $H_0 : \tau = 0$ and $H_0 : \rho_s = 0$ respectively. The former is a reasonable approximation for somewhat smaller n than is the latter. The latter should not be used for $n<20$, whereas the former is reasonable for $n>15$, some writers even suggesting its use for $n>10$. Using the former for Example 7.6 where $t=0.3846$, we get $z=3\times0.3846\times\sqrt{[(13\times12)/(2\times31)]}=1.83$. This exceeds the critical value of 1.64 for significance in a one-tail test but is less than the value 1.96 required for significance in a two-tail test at the 5% level. Thus our asymptotic result is consistent with our finding in Example 7.6.

While a correlation coefficient significantly different from zero indicates certain types of association, the implications differ between coefficients, and the same degree of association as indicated by a particular data set will lead in general to a different value of each of the coefficients we have discussed. Even the extreme values ±1 have different interpretations for each coefficient depending on the context. While these values of the Pearson coefficient apply linearity of continuous data, for the Kendall or Spearman coefficient they only imply monotonicity of continuous bivariate data if the the ranks are derived from observation (x_i, y_i), although they do imply linearity of ranks. In a situation where the ranks are the primary data, as, for example, when they represent orders of preference of two people for the same set of items, a value $+1$ for t or r_s indicates complete agreement between the people doing the ranking, whereas a value of -1 represents complete disagreement or reversed orders of ranking in 'no-tie' cases.

As increasing magnitude of a correlation coefficient indicates a strengthing of a monotonic relationship, it is sometimes of interest to attach confidence intervals to coefficients. This can only be done easily using asymptotic approximations, and these may have the bizarre effect that one of the limits lies outside the range $(-1, 1)$. A difficulty here – as indeed it is for the Pearson coefficient – is that the distributions of t, r_s are not simple for non-zero values of τ, ρ_s. However, the mechanics of forming a conservative interval using an asymptotic result are straightforward though algebraically complicated. We quote a result here without proof for 95% confidence limits for the population coefficient corresponding to an observed Kendall t. These are τ_1, τ_2, which are given by the roots $\{t\pm1.96\sqrt{[(2/n)(1+7.68/n-t^2)]}\}/(1+7.68/n)$ of a certain quadratic equation. For a full discussion of confidence intervals for τ and ρ_s see Kendall and Gibbons (1990, Chapters 4 and 5).

7.1.6 Some further measures of correlation

Several other nonparametric tests for correlation have been suggested. Pitman (1937b) introduced permutation tests based on the Pearson coefficient for continuous data. The appropriate test is based on the fact that if we fix the order of the x values then all $n!$ permutations of the y values are equally likely under H_0 : *no association*. Permutations giving values of r near ±1 indicate rejection of H_0. This test has similar disadavantages to other permutation tests based on 'raw scores', namely that fresh computation of the permutation distribution, or at least of relevant tail probabilities, is required for each data set; also the test lacks robustness. When normality assumptions hold, the Pitman efficiency of this test is 1. It is easily seen from (7.1) that the permutation distribution of r corresponds to that of $\Sigma_i x_i y_i$ since the other quantities in (7.1) are all invariant under permutation of the y_i.

For rank correlation, Gideon and Hollister (1987) proposed a measure of correlation that is easy to calculate, although this advantage is partly negated for small n by the fact that the coefficient takes only relatively few possible values and the permutation distribution under the null hypothesis is less easily established than that for the Kendall or Spearman coefficients. However, Gideon and Hollister give fairly extensive tables of nominal critical values. They also compare the power of the test for monotonic association using their coefficient to those using other coefficients, with generally favourable results for the former.

We confine our account to the 'no-tie' case. The basic idea they use is that when there is perfect positive agreement between x-ranks and y-ranks for n paired observation, then a shift of origin in a scatter diagram of ranks stepwise from (0, 0) to the points (1, 1), (2, 2), . . . , (n, n), $(n+1, n+1)$ will result in a uniform transfer of the plotted ranks from the first quadrant (x, y both positive) to the third quadrant (x, y both negative). If there is not complete agreement, then at some stage in the shift one or more points will be located in the fourth quadrant (x negative, y positive). We call such a point a **discrepancy**. The number of discrepancies reaches a maximum if there is complete negative correlation (reversal of ranks), in which case at one step in the shift the number of discrepancies will be $\frac{1}{2}n$ if n is even or $\frac{1}{2}(n-1)$ if n is odd. If there is no association the maximum number has expectation $\frac{1}{4}n$. We denote the maximum attained by D_-. A similar pass is now made starting with an axial system with origin at $(0, n+1)$ and moving stepwise to $(1, n)$, $(2, n-1)$, . . . , $(n, 1)$, $(n+1, 0)$. If there is perfect negative association, points will move uniformly from the second quadrant (where x and y are both negative) to the fourth quadrant. If there is complete positive agreement at some stage the number in the third quadrant (discrepancies) now reaches a maximum of $\frac{1}{2}n$ or $\frac{1}{2}(n-1)$,and if there is no association this maximum has expectation $\frac{1}{4}n$. We denote the observed maximum by D_+. The relevant coefficient is $r_g=(D_+-D_-)/([\frac{1}{2}n])$ where $[\frac{1}{2}n]$ denotes the **integral part** of $\frac{1}{2}n$, i.e. the greatest integer less than or equal to $\frac{1}{2}n$. It is not difficult to show that if the x-ranks are arranged in ascending order and the y-rank corresponding to the x-rank i is s_i, then in the first pass if the origin is at (i, i) the number of discrepancies d_i is the number of ranks $s_1, s_2, \ldots, s_i$ that are greater than i. D_- is the maximum of all d_i, $i \leq n$. Similarly, in the second pass, if we write $u_i=n+1-s_i$, then if the origin is at $(i, n+1-\text{i})$ the number of discrepancies d_i^* is the number of $u_1, u_2, \ldots, u_i$ that exceed i. D_+ is the maximum d_i^*. Gideon and Hollister also consider problems with ties.

Example 7.8

The problem. Compute r_g for the ranked data in Table 7.3 and test for evidence of positive association between x- and y- ranks.

Formulation and assumptions. The null hypothesis is that of *no monotonic association.* Computation of r_g involves the steps outlined above and Gideon and Hollister (1987, Table 3) may be used to test for significance.

Procedure. The ranks in Table 7.3 are reproduced below.

r	1	2	3	4	5	6	7	8	9	10	11	12
s	2	3	5	4	6	7	9	12	11	10	8	1

We easily calculate the d_i by inspection. For example, when $i=4$ we see that only one of the $s_j, j \leq i$, namely $s_3=5$, exceeds 4. Thus $d_4=1$. We find $d_1=1, d_2=1$, etc., and proceeding in this way the complete set of d_i is (1, 1, 1, 1, 1, 1, 1, 2, 2, 2, 1, 0). Thus $D_-=2$. To compute the d_i^* we replace the y-ranks s_i by $u_i=n+1-s_i$, giving

r	1	2	3	4	5	6	7	8	9	10	11	12
u	11	10	8	9	7	6	4	1	2	3	5	12

Proceeding as before we find, for example, that $d_5^*=5$, since u_1, u_2, u_3, u_4, u_5 all exceed 5. The complete set of d_i^* is (1, 2, 3, 4, 5, 5, 4, 3, 2, 1, 0 , 0). Thus $D_+=5$, whence, since $n=12$, $r_g=(5-2)/6=0.5$.

Conclusion. Gideon and Hollister's Table 3 indicates significance in a one-tail test but not in a two-tail test at a nominal 5% level if $r_g=0.5$, so we reject the hypothesis of no association in favour of one of positive association.

Comments. This result agrees with our conclusion using Kendall's coefficient.

Computational aspects. Calculation of the coefficient is straightforward. The author is not aware of any program giving explicit tail probabilities for the relevant permutation test although, in theory, it is not difficult to produce an appropriate algorithm. Gideon and Hollister used Monte Carlo simulations to estimate tail probabilites for $n>10$.

Further correlation-like measures of association, some used extensively in the social sciences, are given by Siegel and Castellan (1988, Chapter 9). We consider association in categorical data in Chapters 9 and 10.

7.2 RANKED DATA FOR SEVERAL VARIABLES

If we observe more than two variables for each experimental unit we may want to test for agreement in rankings for all variables.

We indicated in Section 6.3.1 that the Friedman test may be applied to rankings of objects (different varieties of raspberry, placings in a gymnastics contest by different judges, ranking of candidates by different examiners) to test whether there is evidence of consistency between those making the rankings. M.G. Kendall, independently of Friedman, proposed the use of a function of the Friedman statistic which is often referred to as **Kendall's coefficient of concordance** and tabulated some small-sample critical values relevant to testing the hypothesis that the rankings were essentially random against the alternative of evidence of consistency. Kendall regarded his coefficient of concordance as an extension of the concept of correlation to more than two sets of rankings. Whereas in the bivariate case we may use Kendall's correlation coefficient to measure both agreement (positive association) and disagreement (negative association), concordance, whether measured by Kendall's original statistic or the Friedman modification, is one-sided in the sense that rejection of the null hypothesis indicates agreement. For example, if four judges A, B, C, D rank 5 objects in the order given in Table 7.6 we would not reject the hypothesis of no association, for the Friedman statistic (6.5) would take the value zero.

Table 7.6 Rankings of 5 objects by 4 judges

	Judge			
Object	A	B	C	D
I	1	5	1	5
II	2	4	2	4
III	3	3	3	3
IV	4	2	4	2
V	5	1	5	1

There is complete agreement here between judges A and C and between judges B and D: but the latter pair are completely at odds with judges A and C. The Kendall and the Friedman statistics do not detect such patterns.

Partial correlation coefficients for ranks have been devised, but extreme care is needed in interpretation. The reader interested in this topic should refer to the detailed discussion of partial rank correlation given by Kendall and Gibbons (1990, Chapter 8). Partial correlation is also discussed by Conover (1980, Section 5.4) and Siegel and Castellan (1988, Section 9.5).

7.3 FIELDS OF APPLICATION

Political science

Leaders of political parties may be asked to rank issues such as the economy, health, education and current affairs in order of importance. In comparing orderings given by leaders of two parties, rank correlations may be of interest. If leaders of more than two parties are involved the coefficient of concordance may be appropriate, or we may prefer to look at pairwise rank correlations, for leaders of parties at different ends of the political spectrum may tend to reverse, or partly reverse, rankings.

Psychology

A psychologist might show 12 different photographs separately to a twin brother and sister and ask each to rank them in order of the pleasure they give. We might use Blomqvist's, Spearman's, Kendall's or the Gideon–Hollister coefficients to test for consistency in the rankings made by brother and sister.

Business studies

Market research consultants list a number of factors that may stimulate sales such as consistent quality of goods, reasonable guarantees, keen pricing, efficient after-sales service, clear operating instructions, etc. They ask a manufacturers' association and a consumers' association each to rank these characteristics in order of importance. A rank correlation coefficient will indicate the level of agreement between manufacturers' and consumers' views on relative importance.

Personnel management

A personnel officer ranks 15 salesmen on the basis of total sales by each over a 12-month period. His boss suggests he should also rank them on the basis of numbers of customer complaints received about each. A rank correlation could be used to see how well the rankings relate. A Pearson product moment correlation might be of interest if we had figures for sales and precise numbers of customer complaints for each salesman.

Horticulture

Leaf samples may be taken from each of 20 trees and magnesium and calcium content determined by chemical analysis. A Pearson coefficient might be used to see if levels of the two substances are related, but this coefficient can be distorted if one or two trees have levels of these chemicals very different from the others; such influential observations are not uncommon in practice. A rank correlation coefficient may give a better picture of the correlation. If a third chemical, say cobalt, is also of interest, a coefficient of concordance might be appropriate.

7.4 SUMMARY

The most widely used measures of rank correlation are **Kendall's rank corrrelation coefficient (tau)** (Section 7.1.2) and **Spearman's rank correlation coefficient (rho)** (Section 7.1.3). For the former (7.4) and for the latter (7.5) apply for the 'no-tie' case; modifications with ties for either are discussed in Section 7.1.4 and asymptotic results in Section 7.1.5.

Blomqvist's median coefficient (Section 7.1.1) is usually estimated by (7.2) or preferably by (7.3) with appropriate modification for ties.

The **Gideon–Hollister coefficient** (Section 7.1.6) is simple to calculate and robust.

For multivariate ranked data **Kendall's coefficient of concordance** (Section 7.2) is equivalent to Friedman's test statistic for ranked data in randomized blocks given in Section 6.3.1.

EXERCISES

7.1 Compute the probabilities associated with each possible value of Kendall's t statistic under the hypothesis of no association when $n=4$ to verify the results quoted in Example 7.4.

7.2 Compute the probabilities analogous to those sought in Exercise 7.1 for Spearman's r_s statistic.

7.3 Verify the numerical values in the triangular matrix of s_{ij} given on p. 175 for the data in Table 7.3 and also the values of n_{cs} and n_{ds}.

7.4 Verify the value of r_s obtained in Example 7.7.

7.5 In Table 7.4 ages at death are ordered by year of death. Use the Cox–Stuart trend test (Section 2.1.6) to test for a time trend in life spans.

7.6 A china manufacturer is investigating market response to 7 designs of dinner set. The main markets are the British and American. To get some idea of preferences in the two markets a survey of 100 British and 100 American housewives is carried out and each housewife is asked to rank the designs in order of preference from 1 for favourite to 7 for least acceptable. For each country the 100 rank scores for each design is totalled. The design with the lowest total is assigned rank 1, that with the next lowest total rank 2, and so on. Overall rankings for each country are:

Design	A	B	C	D	E	F	G
British rank	1	2	3	4	5	6	7
American rank	3	4	1	5	2	7	6

Calculate the Spearman, Kendall and Gideon–Hollister correlation coefficients. Is there evidence of a positive association between orders of preference in the two countries?

7.7 The manufacturer in Exercise 7.6 later decides to assess preferences in the Canadian and Australian markets by the same methods and the rankings obtained are:

Design	A	B	C	D	E	F	G
Canadian rank	5	3	2	4	1	6	7
Australian rank	3	1	4	2	7	6	5

Calculate the Spearman and Kendall correlation coeficients. Is there evidence of a positive association between orders of preference in the two countries?

7.8 Perform an appropriate analysis of the ranked data for all four countries in Exercises 7.6 and 7.7 to assess the evidence for any overall agreement about ranks. Comment on the practical implications of your result.

7.9 In a pharmacological experiment involving β-blocking agents, Sweeting (1982) recorded for a control group of dogs, cardiac oxygen consumption (MVO) and left ventricular pressure (LVP). Calculate the Kendall and Spearman correlation coefficients. Is there evidence of correlation?

Dog	A	B	C	D	E	F	G
MVO	78	92	116	90	106	78	99
LVP	32	33	45	30	38	24	44

7.10 Bardsley and Chambers (1984) gave numbers of beef cattle and sheep on 19 large farms in a region. Is there evidence of correlation?

Cattle	41	0	42	15	47	0	0	0	56	67	707
Sheep	4716	4605	4951	2745	6592	8934	9165	5917	2618	1105	150
Cattle	368	231	104	132	200	172	146	0			
Sheep	2005	3222	7150	8658	6304	1800	5270	1537			

7.11 Paul (1979) discusses marks awarded by 85 different examiners to each of 10 scripts. The marks awarded by six of these examiners were:

	Script									
Examiner	1	2	3	4	5	6	7	8	9	10
1	22	30	27	30	28	28	28	28	36	29
2	20	28	25	29	28	25	29	34	40	30
3	22	28	29	28	25	29	33	29	33	27
4	24	29	30	28	29	27	30	30	34	30
5	30	41	37	41	34	32	35	29	42	34
6	27	27	32	33	33	23	36	22	42	29

Use rank tests to determine (i) whether the examiners show reasonable agreement on ordering the scripts by merit and (ii) whether some examiners tend to give consistently higher or lower marks than others.

8

Regression

8.1 BIVARIATE LINEAR REGRESSION

Correlation is concerned with qualitative aspects of a relationship like its existence and nature (e.g. whether it is linear, monotonic, linear in ranks, etc.). Regression is concerned with quantitative aspects (e.g. the slope or intercept values for a straight line; the values of the $p+1$ constants in a polynomial of degree p that provides, in some sense, the best fit to data). There is equivalence between some aspects of the two approaches; e.g. in straight-line regression a test of zero slope is equivalent to a test of zero correlation and a value of $+1$ or -1 for the Pearson product moment correlation coefficient for a sample of n paired observations (x_i, y_i) tells us that all the observed points lie on a straight line. Other relationships between correlation and regression emerge later in this chapter.

Least squares is the classic method of fitting a straight line to bivariate data. The method has optimal properties subject to well-known independence and homogeneity assumptions; and with certain normality postulates there are well-established procedures for hypothesis testing and estimation.

Given a set of bivariate observations (x_i, y_i), often the assumptions needed to validate established least squares methods for estimation and hypothesis testing do not hold. This may result in misleading or invalid inferences. In earlier chapters we met situations where a t-test may be insensitive when samples are clearly not from a normal distribution; for regression similar problems arise. The method of least squares is relatively insensitive to some departures from basic assumptions, but may be seriously affected by others. In the past 25 years many diagnostic tests have been proposed to detect observations that might cause difficulty with standard regression methods because these observations are influential in the sense that relatively small changes in them may have a marked effect on estimates. Various robust methods have been developed that are little affected by a few aberrant observations, but which behave almost as well as least squares when the latter is optimal. Such techniques are not confined to regression problems: we say more about one important class of robust methods – using what are called m-estimators – in Chapter 11. In this chapter, after a brief discussion of the role of least squares in a parametric and nonparametric context, we explore a regression procedure closely related to Kendall's correlation coefficient. Classical least squares regression is more closely related to Pearson's correlation coefficient.

8.1.1 Least squares regression

Most experimenters meet bivariate least squares regression as the method that gives, in some sense, a line or curve that best fits some data. The data are paired observations (x_i, y_i), $i=1, 2, \ldots, n$. The x_i are observed values of a mathematical variable, e.g. date or time after application of a treatment, the values being either at the experimenter's choice or else randomly selected. It is also assumed they are measured without error, or at least that any measurement error is negligible. In regression the y_i are always observed values of a random variable. What an experimenter often implies when saying data lie 'almost on a straight line' is that but for some 'nuisance' variation the points would lie exactly on a straight line. In a biological situation this nuisance variation might be genetic variability; in manufacturing processes, machine variability or varying quality of raw material; for a psychologist, vagaries of the human mind. The experimenter does not know *a priori* the slope or position of the idealized straight line. A plot of the observations may produce a graph like that in Figure 8.1.

In classical least squares linear regression we assume that for any given x, say $x=x_i$, our observed y_i is a realization (or a sample value) of a random variable Y_i which has a normal distribution with a mean conditional on x_i of the form $E(Y_i|x=x_i)=\alpha+\beta x_i$ and variance σ^2, this variance being independent of x_i. Here the notation $E(Y_i|x=x_i)$ means the expectation of

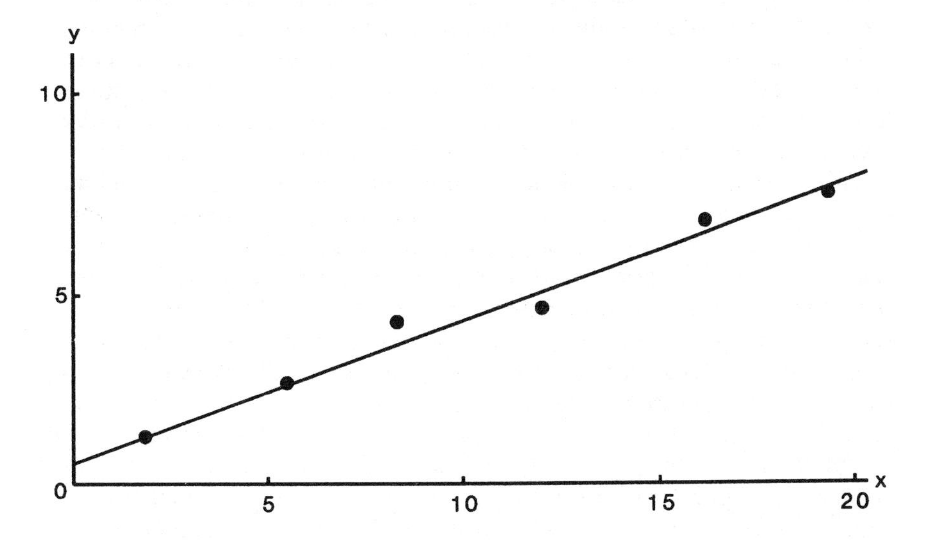

Figure 8.1 Scatter diagram and a least squares regression line.

Y_i **conditional upon** (or when) $x=x_i$; this mean equals $\alpha+\beta x_i$. The regression problem is to estimate the unknown α, β (and sometimes σ^2). Another way of stating this is to say that, for each given (x_i, y_i), we have

$$y_i=\alpha+\beta x_i+\epsilon_i \tag{8.1}$$

where each ϵ_i is an unobservable random variable; the ϵ_i are independently normally distributed with mean 0 and variance σ^2.

We also meet linear regression in another context – that of estimating the conditional mean of Y for a given X where we have observations (x_i, y_i) from a bivariate normal distribution where X and Y are correlated. This estimation problem has the same solution as that for the experimental situation above; however, the straight line is now an estimate of the locus of the conditional mean of Y for specified values of X. The observed points do not necessarily lie close to this straight line; how close depends on the value of the correlation coefficient ρ. We do not pursue this problem here, but return to the more practical research version described above. The assumption of independent normality of the ϵ_i is needed to justify commonly used testing and estimation procedures. If we assume only that the ϵ_i are all independent and are each distributed with mean zero and the same (usually unknown) variance σ^2, there are good reasons why least squares is still an appropriate method of estimation. Without a normality assumption it is often correct to base test and estimation procedures on the relevant permutation distribution theory, details of which are given in Maritz (1981, Chapter 5), who also gives an excellent discussion of other aspects of the theory of nonparametric straight-line regression. As with other permutation tests using continuous (raw) data, least squares give similar results to normal theory tests when these are appropriate but, like other such tests, critical values require fresh computation for each data set and the tests also tend not to be robust against departures from assumptions like that of all ϵ_i having the same variance. For these reasons we do not discuss permutation distribution theory least squares in detail.

With the aid of an example we look at difficulties that may arise with a least squares analysis if certain assumptions are clearly violated. Least squares gives estimates a, b of α, β that minimize $\Sigma_i(y_i-a-bx_i)^2$, i.e. the sum of squares of deviations of the observed points from the fitted line where these deviations are measured in a direction perpendicular to the x-axis (parallel to the y-axis). For data like those plotted in Figure 8.1 the fit is good and is shown on that figure. We now consider a case where a least squares fit gives cause for concern.

Example 8.1

The problem. Use the method of least squares to fit a straight line to the following points:

x	0	1	2	3	4	5	6
y	2.5	3.1	3.4	4.0	4.6	5.1	11.1

Formulation and assumptions. In Section A6 we obtain the estimators a, b as:

$$a=\overline{y}-b\overline{x}$$

and

$$b=[\Sigma_i (x_i y_i)-(\Sigma_i x_i)(\Sigma_i y_i)/n]/[\Sigma_i (x_i^2)-(\Sigma_i x_i)^2/n]=c_{xy}/c_{xx}$$

where $\overline{x}=(\Sigma_i x_i)/n$, $\overline{x}=(\Sigma_i y_i)/n$ and c_{xy}, c_{xx} are as defined for (7.1).

Procedure. Here $n=7$, $\Sigma_i x_i=21$, $\Sigma_i y_i=33.8$, $\Sigma_i x_i^2=91$, $\Sigma_i(x_i y_i)=132.4$, whence $b=1.107$ and $a=1.508$.

Conclusion. The least squares regression line is

$$y=1.508+1.107x. \tag{8.2}$$

Comments. The points are plotted in Figure 8.2 along with the line (8.2). The fit is disturbing. If we substitute the given values of x in (8.2) we get estimates of the means of y conditional upon each of these x. We often denote these estimates by $\hat{y}$. For each data point we may calculate $e_i=y_i-\hat{y}_i$. The e_i are called the **residuals.** For our data the residuals in the order of the data points are 0.992, 0.485, −0.322, −0.829, −1.336, −1.943, 2.950. The magnitude of the residual is the vertical distance from each point to the line; the sign is negative if the observation lies below the line, positive if it is above (see Figure 8.2).

If our fitted line were the true unknown idealized line the residuals would correspond to the ϵ_i in (8.1). Therefore, we might regard the e_i as estimates of the unobservable 'error' random variables ϵ_i. If the ϵ_i are independent of the x – a standard assumption in least squares – it is reasonable to expect the residuals to be **randomly** distributed with mean zero. This looks to be the case in Figure 8.1, but clearly not in Figure 8.2.

The e_i for Example 8.1 are plotted against x in Figure 8.3. The residuals decrease almost linearly as x increases from 0 to 5, changing from positive to negative, then there is a sudden jump when $x=6$ to $e_7=2.950$. Inspection of Figure 8.2 explains why. The first six data points (x_i, y_i) lie close to a straight line but the final point (6, 11.1) is well away from the line suggested by these other points. Either the homogeneous error assumptions implicit in justifying least squares are not appropriate for these data, or there is something peculiar about the last observation or a straight-line fit is inappropriate. The y value for this point is indeed about double the value we would expect on the basis of the other points. This sort of 'data mishap' is not uncommon. If the y values are the average of two measurements did

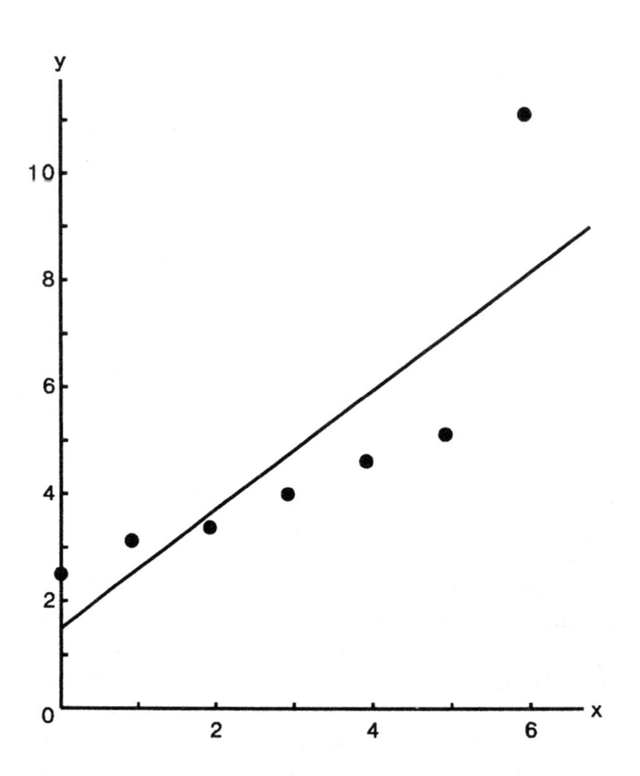

Figure 8.2 An illustration of the influence of an outlier on least squares regression.

someone forget to divide by 2 in this last case? The data themselves cannot tell us, but a data feature like this suggests we should enquire about it and take remedial action if appropriate.

Problems like this stimulated interest in regression diagnostics, especially in methods to detect suspect observations or influential points. These topics are discussed fully by Cook and Weisberg (1982) and by Atkinson (1986). The type of difficulty in Example 8.1 becomes more important in multiple regression (where there are several independent or explanatory or regressor variables rather than just a single x): graphical techniques are then not easily available and an elementary study of residuals may no longer suffice. Here diagnostic techniques are especially important.

An alternative approach to detecting possible difficulties in bivariate (and multiple) regression is to fit by least squares and also by a robust method that reduces the influence of any suspect point. If the two methods agree there is usually no problem. If there is a large discrepancy between results using the two approaches with the same data, one must explore why the difference arises. Good robust methods give results similar to least squares

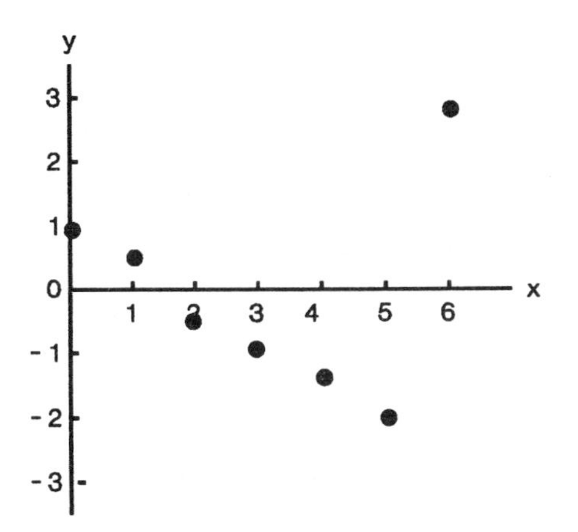

Figure 8.3 Plot of residuals against x for the data in Example 8.1.

when the latter is appropriate, so it has been argued that one should always use robust methods and simply accept the answer. One reason for not accepting this argument is that, for some data, fitting a straight line may be the wrong thing to do: even a robust method may not reveal this without further diagnostic tests. Sometimes we need more data to resolve an anomaly. If we had additional points to those in Example 8.1 with x values between 5 and 6 we might end up with a plot like Figure 8.4. Here clearly some nonlinear curve would fit the data reasonably well. Without such information, and if no reason can be found for the odd behaviour of y when $x=6$, it is sensible to collect more data if this is feasible. If it is only possible to make observations at integral x every effort should be made to get a repeat measurement at $x=6$, and, if possible, additional ones at $x=7$ or $x=8$ to evaluate the continuing trend.

We indicated that there is a relationship between least squares and the Pearson correlation coefficient. It is well known that a test of $H_0 : \beta=0$ in the context of least squares linear regression is equivalent to the test that Pearson's $\rho=0$. A further link with correlation comes from the property of least squares that the estimators a, b are so chosen (see Section A6) that $\Sigma_i\, [x_i(y_i-a-bx_i)]=0$. Since we have defined $e_i=y_i-a-bx_i$ this implies that $\Sigma_i(e_i\, x_i)=0$. It is also well known, and follows from the first estimating equation in Section A6, that $\Sigma_i\, e_i=0$. In Exercise 8.1 we ask you to verify that these results hold for the data in Example 8.1. An important implication is that the sample Pearson product moment correlation coefficient between the e_i and the x_i is always zero. This property provides a basis for hypothesis tests or for estimation procedures about the slope β. The tests and estimation procedures might be based on normal theory if appropriate, or on a permutation distribution basis. These latter tests are

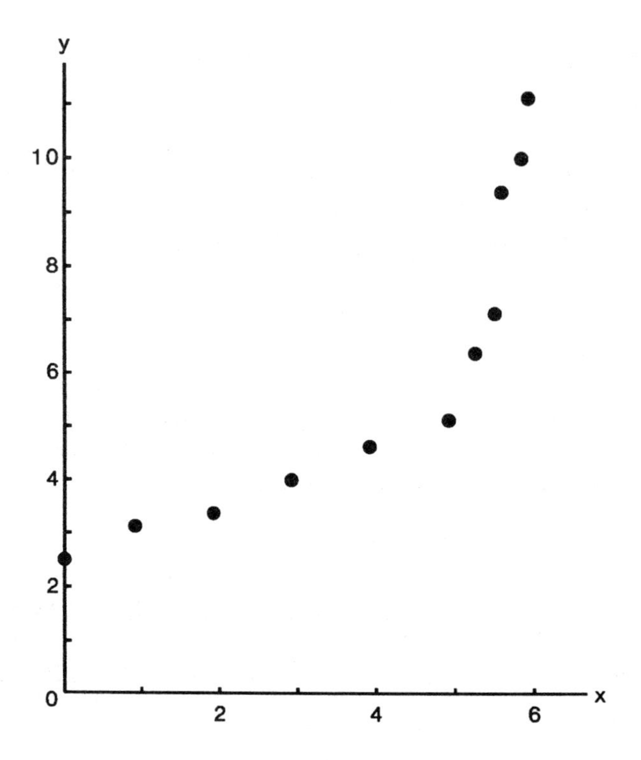

Figure 8.4 A relationship that is clearly nonlinear.

tedious to carry out, but the theory (together with asymptotic approximations) is given by Maritz (1981, Chapter 5),

The fact that $\Sigma_i(e_i x_i)=0$ in Example 8.1 amplifies the point made in Section 7.1 that a correlation coefficient may be zero when there is a nonlinear relationship such as that indicated for x_i, e_i by Figure 8.3.

In view of the lack of robustness for least squares, Brown and Maritz (1982) suggested that we might consider certain generalizations of the least squares estimating equations given in Section A6. As pointed out above, if we write $e_i=y_i-a-bx_i$, the estimating equations may be written

$$\Sigma_i(e_i x_i)=0 \text{ and } \Sigma_i e_i=0$$

Brown and Maritz suggested replacing x_i and e_i in these equations by functions $h(x_i)$, $\phi(e_i)$ that retain the ordering of the x_i and e_i and using estimating equations

$$\Sigma_i[h(x_i)\phi(e_i)]=0 \text{ and } \Sigma_i \phi(e_i)=0$$

In particular, they consider the following choices:

(i) $h(x_i)=\text{sgn}[x_i-\text{med}(x_i)]$, $\phi(e_i)=\text{sgn}[e_i-\text{med}(e_i)]$
(ii) $h(x_i)=\text{rank}(x_i)$, $\phi(e_i)=\text{rank}(e_i)-\frac{1}{2}(n-1)$

where $\text{sgn}(x)=-1$ if $x<0$, $\text{sgn}(x)=0$ if $x=0$, $\text{sgn}(x)=1$ if $x>0$ and 'med' means 'median'.

They obtain tests and confidence intervals for the unknown slope β and extend their method to two regressor variables. The interested reader should refer to their paper for details. Among many papers on the use of ranks in regression, that by Adichie (1967) is basic.

8.1.2 Theil's regression method

Theil (1950) suggested a different approach to distribution-free regression. He proposed estimating the slope of a regression line by the median of the slopes of all lines joining pairs of points with different x values. For the points (x_i, y_i) and (x_j, y_j) the relevant slope is

$$b_{ij}=\frac{y_j-y_i}{x_j-x_i} \tag{8.3}$$

We assume all x_i are different, and for convenience order all observations in ascending order of x. Clearly $b_{ij}=b_{ji}$ for all i, j, so for n observations there are $\frac{1}{2}n(n-1)$ algebraically distinct b_{ij} and it is convenient to write them in an upper triangular matrix:

$$\begin{matrix} b_{12} & b_{13} & b_{14} & . & . & . & b_{1n} \\ & b_{23} & b_{24} & . & . & . & b_{2n} \\ & & b_{34} & . & . & . & b_{3n} \\ & & & . & . & & . \\ & & & & . & & . \\ & & & & & & b_{n-1,n} \end{matrix}$$

The numerical matrix generally has a pattern and even without a computer program the median is usually easy to detect for moderate n without further ordering. The process is similar to that for obtaining the Hodges–Lehmann point estimator in the Wilcoxon signed rank procedure.

Denoting the median estimator of β by b^*, Theil suggested estimating α by a^*, the median of all $a_i=y_i-b^*x_i$; alternatively, we might choose $\hat{a}=\text{med}(y_i)-b^*\text{med}(x_i)$. If we use the latter our fitted line passes through the median of all observations, whereas the least squares line passes through the mean.

Example 8.2

The problem. Estimate the straight-line regression of y on x for the data in Example 8.1 using Theil's method.

Formulation and assumptions. We calculate the b_{ij} and their median and then calculate the median of all a_i as described above.

Procedure. The data are:

x	0	1	2	3	4	5	6
y	2.5	3.1	3.4	4.0	4.6	5.1	11.1

We first calculate

$$b_{12}=(y_2-y_1)/(x_2-x_1)=(3.1-2.5)/(1-0)=0.6$$

and proceeding this way we may write the results in a triangular matrix:

0.600	0.450	0.500	0.525	0.520	1.433
	0.300	0.450	0.500	0.500	1.600
		0.600	0.600	0.567	1.925
			0.600	0.550	2.367
				0.500	3.250
					6.000

The matrix contains 21 b_{ij} so the median is the 11th largest. Inspection shows $b^*=0.567$. To obtain a^* we calculate $a_1=2.5-0.567\times 0=2.5$, and similarly $a_2=2.533$, $a_3=2.266$, $a_4=2.299$, $a_5=2.332$, $a_6=2.265$, $a_7=7.698$. The median of the a_i is $a^*=2.332$.

Conclusion. The fitted line is $y=2.332+0.567x$.

Comments. 1. Figure 8.5 shows this line and the data points. Clearly, compared to least squares, the point (6, 11.1) has a much reduced influence in determining the line. This is because the other pairings all give values of b_{ij} between 0.3 and 0.6: only the six b_{ij} involving the suspected 'rogue' point give widely different values ranging from 1.433 to 6.000.

2. The presence of these rogue b_{ij} lifts the median only slightly, the median for all remaining b_{ij} being 0.520.

3. If we estimate α by $\hat{\alpha}$ =median$(y_i)-b^*$median(x_i) we find $\hat{\alpha}$ =2.299, reasonable close to $a^*=2.332$. The estimator a^* has the appealing property that the number of points above and below the fitted line are approximately equal. They will be exactly equal if no point lies on the fitted line.

4. When there are no rogue observations one feels intuitively that more weight should be given to b_{ij} for pairs of points distant from one another than is given to points close together. This is an argument for taking some sort of weighted median. Jaeckel (1972) proposed weighting each b_{ij} with a weight

$$w_{ij}=(x_j-x_i)\ /\ \Sigma_{i<j}\ (x_j-x_i)$$

and recommended estimating β by the median of the b_{ij} with these assigned weights. The procedure has certain optimal properties when there are no

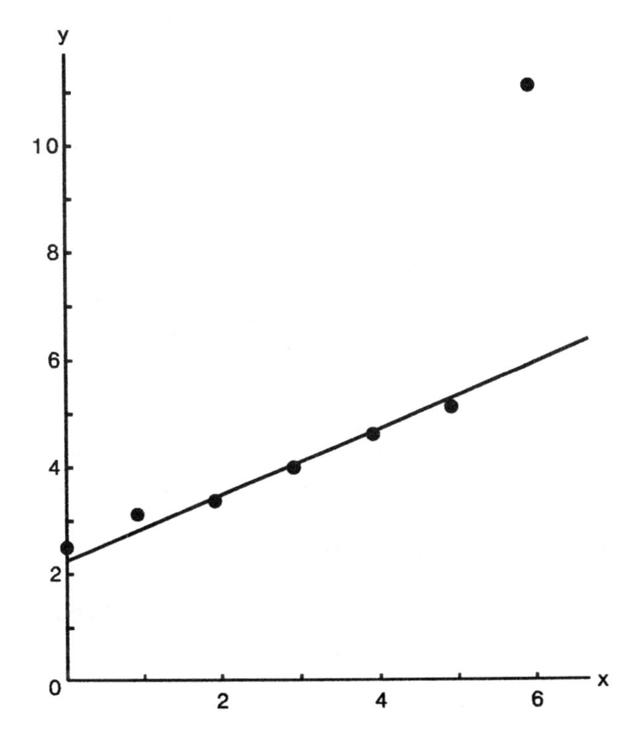

Figure 8.5 The Theil regression line fitted to the data in Example 8.2.

outliers, but this and other weighting schemes may lead to less robust estimators than Theil's method when there are outliers. Alternative weighting schemes are discussed by Scholz (1978), Sievers (1978), Kildea (1978) and Olkin and Yitzhaki (1992). In passing we note that the least squares estimator is equal to the weighted mean of the b_{ij} with weights w_{ij} proportional to $(x_j - x_i)^2$. Median estimates in linear regression are also discussed by Brown (1980).

Computational aspects. For large n, computing problems are formidable with only a pocket calculator. This has lead to modifications to reduce computation. We give one in Section 8.1.3. Although programs for Theil's procedure exist – and would not be too difficult to write – the author is not aware of any in current releases of the major statistical packages, but this situation may well change in the near future. Daniel (1990, Section 10.6) gives general references to some nonparametric regression programs.

The fact that the data in Examples 8.1 and 8.2 lead to markedly different fits using least squares and Theil's method does not tell us which is the more appropriate. Example 8.1 suggests that the least squares fit may not be appropriate since the data evidently do not lie near **any** straight line with identically distributed independent errors. It leaves unresolved the question of whether the point (6, 11.1) is seriously in error or whether the relationship is not truly linear. As indicated in Section 8.1.1, more data might resolve this point: so might a data check (e.g. did someone forget to divide by 2?). If we find sound reason to ignore the point (6, 11.1) we might use a least squares fit to the remaining points, or be content to use that given by Theil's procedure. Similar results could be obtained using other robust methods based on principles outlined in Section 11.1.

In simulation studies introducing what were effectively outliers by replacing errors distributed normally with ones from long-tailed distributions, Hussain and Sprent (1983) found that Theil's method was almost as efficient as least squares when classic normality and homogeneity of variance assumptions were valid and that it showed a marked improvement in efficiency with a long-tailed error distribution, especially with $n<30$. There was an even more marked inmprovement in the estimate of α, although this is commonly of less interest than β.

Hussain and Sprent also found that estimators based on weighted medians performed on the whole no better, and often less well than Theil estimators, in the presence of outliers. Theil's method is further discussed by Maritz (1979).

8.1.3 The abbreviated Theil method

If computational facilities are not available for the complete procedure, Theil suggested an abbreviated method for n distinct x values. If n is even we take $b=\text{med}(b_{i,\, i+\frac{1}{2}n})$ with $i=1, 2, \ldots, \frac{1}{2}n$ and if n is odd we take $b=\text{med}(b_{i,\, i+\frac{1}{2}(n+1)})$ with $i=1, 2, \ldots, \frac{1}{2}(n-1)$. The procedure is not recommended unless $n>12$. If $n=14$, say, we compute only $b_{1,8}$, $b_{2,9}$, $b_{3,10}$, $b_{4,11}$, $b_{5,12}$, $b_{6,13}$ and $b_{7,14}$. For odd n the procedure is very similar and all observations except that corresponding to $x_{\frac{1}{2}(n+1)}$ are used.

The method is called the **abbreviated Theil method.** It is usual to estimate α as for the full Theil method, as this involves only n straightforward calculations if a^* is used, or one simple calculation if $\hat{a}$ is used.

Example 8.3

The problem. We give below modal length (y cm) of samples of Greenland turbot of various ages (x yr) based on data given by Kimura and Chikuni (1987). Fit a straight line to these data using the abbreviated Theil method.

Age (x)	4	5	6	7	8	9	10	11	12	13	14
Length (y)	40	45	51	55	60	67	68	65	71	74	76
Age (x)	15	16	17	18	19	20					
Length (y)	76	78	83	82	85	89					

Formulation and assumptions. For 17 given points we calculate $b_{10,1}$, $b_{11,2}, \ldots, b_{17,8}$ and obtain their median.

Procedure. Since the x_i are equally spaced the denominators in each of the required b_{ij} are all equal to $x_{10}-x_1=13-4=9$, it suffices to calculate the relevant y_j-y_i, obtain their median and divide it by 9 to determine b. Thus $y_{10}-y_1=74-40=34$, $y_{11}-y_2=76-45=31$, etc. The complete set of differences is 34, 31, 25, 23, 23, 15, 17, 24. The median of these is 23.5, so $b=23.5/9=2.61$. To compute $\hat{a}$ we easily establish that $\text{med}(y_i)=71$ and $\text{med}(x_i)=12$, whence $\hat{a}=71-2.61\times 12=39.68$. In Exercise 8.2 we ask the reader to confirm that $a^*=36.80$.

Conclusion. Depending on whether we choose $\hat{a}$ or a^* to estimate α, our fitted regression lines are $y=39.68+2.61x$ or $y=36.80+2.61x$.

Comments. We implicitly accepted that fitting a straight line was reasonable. The data certainly suggest a not unexpected positive relationship between length and age. It is instructive to plot the points. The plot in Figure 8.6 suggests a not quite linear fit.

The two regressions computed in this example are shown by a continuous line (using a^*) and a broken line (using $\hat{a}$). The disposition of points suggests a more or less monotone trend of increasing median length with increasing age but that a straight line is not an entirely adequate summary of this trend. While growth rate is steady until about age 10, it is slower in later years. This situation would be better described by some more complex relationship. One possibility would be to fit a quadratic curve, or alternatively what is sometimes called a two-phase linear regression, consisting of two lines intersecting at $x=10$, that with the greater slope being relevant when $x<10$ and that with the lesser slope when when $x>10$. Another possibility is to fit a monotonic regression of the type we describe in Section 8.2. This possibility is explored in Exercise 8.5.

8.1.4 Hypothesis tests and confidence intervals based on Theil's method

Good estimators a, b of α, β should be such that the residual associated with each observation, denoted by e_i, where $e_i=y_i-a-bx_i$, is equally likely to be positive or negative. This implies an assumption that the e_i are

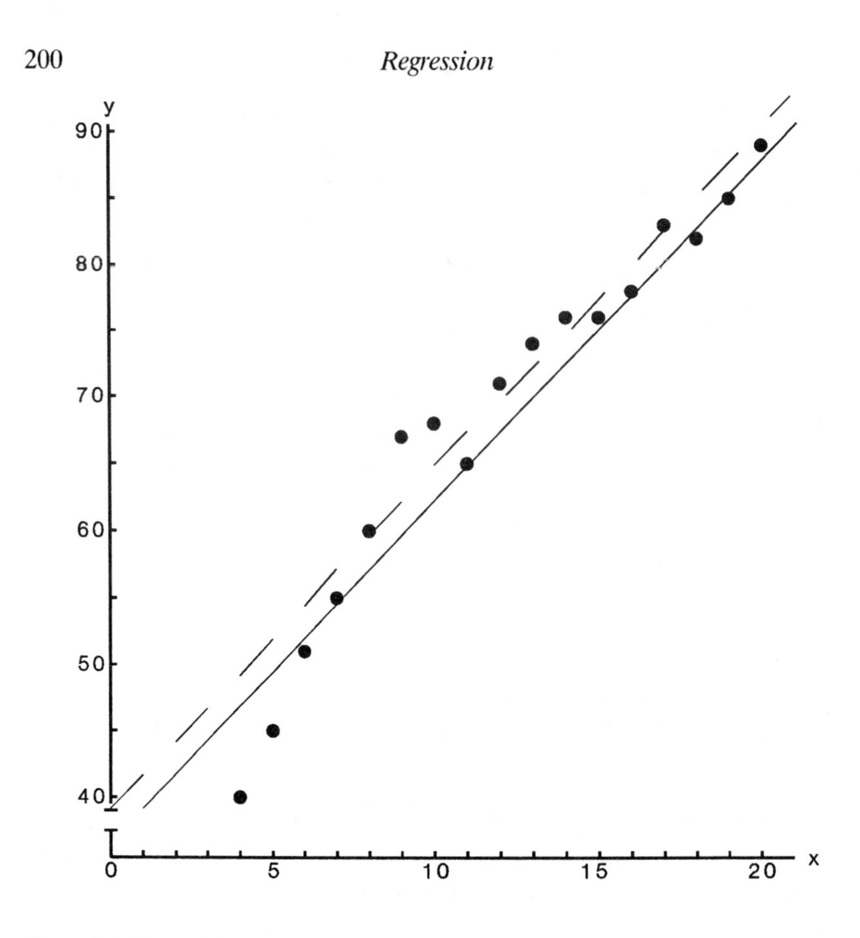

Figure 8.6 Scatter diagram of data in Example 8.3 together with fitted lines based on the abbreviated Theil method.

randomly distributed with median zero and independently of the x_i. Now

$$b_{ij} = \frac{y_j - y_i}{x_j - x_i} = \frac{a + bx_j + e_j - (a + bx_i + e_i)}{x_j - x_i} = b + \frac{e_j - e_i}{x_j - x_i} \tag{8.4}$$

Equation (8.4) implies any b_{ij} will be greater than b if (x_i, e_i) and (x_j, e_j) are concordant and any b_{ij} will be less than b if these are discordant in the sense used for Kendall's correlation coefficient in Section 7.1.2. Choice of $b = \text{med}(b_{ij})$ as estimator of β ensures half the pairs are concordant and half discordant. Fortunately, this result does not depend on our estimate a of the intercept, although as indicated in Example 8.3 only a^* guarantees equal probability that e_i will be positive or negative. Any choice of b for which we would accept the hypothesis $H_0 : \tau = 0$ (i.e. zero correlation) for Kendall's coefficient of correlation between observed x and corresponding residuals, e, is acceptable in the sense that we do not then reject H_0, i.e. it is consistent with zero population correlation between x and the residuals.

In other words, we accept any b which does not give a number of discordances (or concordances) that is too small or too large. Since $n_c+n_d=N=½n(n-1)$ is the total number of b_{ij} generated from n observations with distinct x_i, we reject $H_0 : \tau=0$ in a two-tail test at the 5% level if $|(n_c-n_d)|/[½n(n-1)]$ is greater than the critical value given in Table IX.

This provides a basis for a 95% confidence region for β. Without loss of generality we assume $n_c>n_d$. Then if t_c is the critical value for rejection of H_0 at the nominal 5% level we reject any hypothesized value of β for which $n_c-n_d \geq (n_c+n_d)t_c$, i.e. for which $n_d \leq =n(n-1)(1-t_c)/4$. Denoting the value of the right-hand side of this inequality by r, the lower limit to our confidence interval is the $(r+1)$th smallest b_{ij}, i.e. we reject the r smallest b_{ij}. By symmetry the upper limit is obtained by rejecting the r largest b_{ij}. Due to round-off in published tables, the computed value of r may take a value such as $r=4.9996$. It is appropriate to round this to $r=5$.

Example 8.4

The problem. For the data in Example 8.2 obtain a 95% confidence interval for β.

Formulation and assumptions, We use the Kendall coefficient approach given above.

Procedure. Entering Table IX with $n=7$ we find $t_c=0.7143$. Thus $r=42\times(1-0.7143)/4=2.99985$, so we take $r=3$ and reject the 3 largest and 3 smallest b_{ij}. The next smallest and largest give confidence limits. If the b_{ij} calculated in Example 8.2 are arranged in ascending order we easily establish the nominal 95% confidence limits as 0.500 and 1.900.

Conclusion. A nominal 95% confidence limit for β is (0.500, 1.900).

Comments. 1 The procedure is reminiscent of that for confidence intervals in the Wilcoxon signed rank and Wilcoxon–Mann–Whitney procedures. The point estimator med(b_{ij}) is the Hodges–Lehmann estimator. This relationship is not surprising because we saw in Section 7.1.2 that a test for the Kendall coefficient is a special case of the Jonckheere–Terpstra procedure, which in turn is a generalization of the Wilcoxon–Mann–Whitney procedure.

2. Unlike normal theory least squares confidence intervals for β, the interval is not symmetric about the point estimator med(b_{ij}). Had the point (6, 11.1) not being out of line with the other observations a more symmetric interval would have been obtained. Here lack of symmetry draws attention to the outlier. However, cases arise where one or more outliers call for robust estimation, yet Theil confidence intervals may still be nearly symmetric.

Sen (1968) extends the above arguments for confidence intervals to cases where there are tied x or e values. For the abbreviated Theil method confidence intervals are easily calculated because the b_{ij} are each calculated from different pairs of points and are therefore mutually independent. Thus for an acceptable estimator each relevant $(e_j-e_i)/(x_j-x_i)$ in (8.4) has equal probability of being positive or negative. If we use N independent b_{ij} in our modified Theil procedure we set confidence limits in exactly the way we set such limits for the median of a sample of N independent observations when using the sign test (see Section 2.1.2).

Example 8.5

The problem. Obtain a nominal 90% confidence interval for the abbreviated Theil estimator of slope obtained in Example 8.3.

Formulation and assumptions. We arrange the eight b_{ij} in ascending order and use Table I to determine the interval, treating the ordered b_{ij} as a sample of 8 for which we wish to determine a confidence interval for the median as in the sign test.

Procedure. As in Example 8.3, equal spacing of the x lets us work with the relevant y differences. Ordering the values found in that example, we get 15, 17, 23, 23, 24, 25, 31, 34. From Table I we find we reject a median value at a nominal 10% (exact 7%) significance level if we have 0, 1, 7 or 8 minus signs. This implies the confidence limits are 17 and 31 for the above y differences. We divide these by 9 to obtain the corresponding b.

Conclusions. A nominal 90% (exact 93%) confidence interval for β is (17/9, 31/9), i.e. (1.89, 3.44).

Comments. The full Theil estimation procedure interval based on Kendall's coefficient should be shorter as it is based on more information.

Using either the full or abbreviated Theil method for a test of the form $H_0: \beta=\beta_0$ against $H_1: \beta\neq\beta_0$, we would reject H_0 at the $100\alpha\%$ significance level if β_0 lies outside the $100(1-\alpha)\%$ confidence interval for β.

8.1.5 A test for parallelism

Using the abbreviated Theil method to fit lines to two data sets gives a simple test for parallelism. We compute the appropriate b_{ij} for each line. These are two independent samples so we may test whether they are from populations with the same median (the parallel case) using the Wilcoxon–Mann–Whitney test. An example is given by Marascuilo and McSweeney (1977, Section 11.13). We need 10 or more observations in each sample.

8.2 MONOTONIC REGRESSION

Often a regression of y on x is clearly not linear but has a monotonic property in that as x increases the mean value of y always increases (or at any rate never decreases), or as x increases the mean of y always decreases (or at least never increases).

Conover (1980, p.274) gives an example where x is the amount of sugar added to standard-size jars of grape juice and y the time taken to complete fermentation. Increasing amounts of sugar speed up fermentation on average, but the rate of increase is by no means constant per unit of additional sugar, which would be a requirement for linear regression.

With many learning processes additional tuition or training will decrease the average completion time for a task. Again, the relation between tuition time and average performance time is often monotonic but not linear. This does not mean that observed (x, y) need always exhibit montonicity, since there will be variations from average for individuals.

A moment's reflection shows that if we have a monotonic regression and transform the data to ranks, the rank will show near linearity. In a few fortunate cases the rank pairs will be exactly in order, or inverse order, giving values $+1$ or -1 for a rank correlation coefficient, the rank regressions then being a straight line of slope ± 1. Random variation usually results in a non-exact match of ranks; then we may use least squares or Theil's method to fit a regression of s on r, the ranks of the observed y on the ranks of the observed x. Back-transformation from the estimated $\hat{s}_i$ corresponding to each r_i gives an expected $\hat{y}$ corresponding to an observed x. There are variants on this approach to monotonic regression. We illustrate the basics by an example which slightly modifies the method given by Conover (1980, Section 5.6).

Example 8.6

The problem. Fifteen people are given varying instruction (x hours) on carrying out a machine operation. After completion of instruction they are tested and the time (y minutes) each takes to complete the operation is noted. Fit a monotonic regression to the data.

x	0	0	1	1.6	3	4	4	5	6.5	8	8	10	12	12.6	14
y	18.4	17.4	16.2	16.4	14.4	10.5	11.2	10.8	9.0	8.4	7.0	7.2	6.6	5.0	5.6

Formulation and assumptions. The x are in increasing order. We form paired ranks (r, s) using mid-ranks for ties and compute the least-squares regression of s on r. Writing this equation $s=a+br$, for each given r_i we calculate $\hat{s}=a+br_i$. We obtain estimates $\hat{y}_i$ of the mean y corresponding to each observed x using, if necessary, linear interpolation between adjacent rankings in the manner described below under *procedure*. Straight-line segments between adjacent x values represent the monotonic regression.

Table 8.1 Calculating a monotone regression

x	y	r	s	$\hat{s}$	$\hat{y}$
0	18.4	1.5	15	14.38	17.78
0	17.4	1.5	14	14.38	17.78
1	16.2	3	12	12.91	16.38
1.6	16.4	4	13	11.93	16.07
3	14.4	5	11	10.95	14.24
4	11.2	6.5	10	9.47	10.99
4	10.5	6.5	8	9.47	10.99
5	10.8	8	9	8.00	10.50
6.5	9.0	9	7	7.02	9.03
8	8.4	10.5	6	5.55	7.86
8	7.0	10.5	4	5.55	7.86
10	7.2	12	5	4.07	7.01
12	6.6	13	3	3.09	6.64
12.6	5.0	14	1	2.11	5.71
14	5.6	15	2	1.13	5.08

Procedure. The complete procedure is summarized in Table 8.1. The first two columns are the data and those headed r and s give corresponding ranks used to compute the coefficients a, b for the least squares linear regression of s on r. The regression equation is then used to compute entries in the column headed $\hat{s}$ using the corresponding r_i. Finally, $\hat{y}$ is calculated for each $\hat{s}$ using linear interpolation as described below.

The least squares regression of s on r used to obtain $\hat{s}$ is

$$\hat{s} = 15.856 - 0.982r$$

If using a pocket calculator we save labour by noting that for tied ranks $b=(\Sigma r_i s_i - C)/(\Sigma r_i^2 - C)$, where $C=n(n+1)^2/4$, and that $a=\frac{1}{2}(1-b)(n+1)$. Thus to calculate $\hat{s}_6$ corresponding to $r_6=6.5$, $\hat{s} = 15.856-0.982\times 6.5=9.47$. Other entries in the penultimate column are calculated in a similar way.

Entries in the last column are obtained by linear interpolation between adjacent ranks. For, example $\hat{s}_3=12.91$ lies between the ranks $s=12$ and $s=13$. Now 12.91 represents 91/100 of the interval between 12 and 13, so $\hat{y}$ corresponding to this is 91/100 of the difference between the y value ranked 12 and that ranked 13 added to the y value ranked 12. The relevant y values are 16.2 and 16.4 with difference 0.2, and $0.2\times(91/100)=0.182$. Adding this to 16.2 and rounding to 2 decimal places gives $\hat{y} = 16.38$.

A general formula for linear interpolation between neighbouring assigned y-ranks (which need not be integers if mid-ranks are used for ties) is obtained as follows: let s_i and s_j be the closest 'observation' ranks above and below the relevant $\hat{s}_k$. If y_i, y_j are the observations with ranks s_i, s_j then

$$\hat{y}_k = y_j + \frac{\hat{s}_k - s_j}{s_i - s_j}\,(y_j - y_i)$$

This formula may be used to calculate all $\hat{y}$. For example, $\hat{s}$ =3.09 lies between s_{13}=3 and s_{11}=4; also y_{13}=6.6 and y_{11}=7.0 whence

$$\hat{y}_{13}=6.6+[(3.09-3)/(4-3)]\times(7.0-6.6)=6.636$$

In Table 8.1 this value is rounded to 6.64.

Since we use linear interpolation between corresponding ordered x values it is appropriate to obtain our estimated monotonic regression by plotting each $(x_i, \hat{y}_i)$ and joining points corresponding to adjacent distinct x values by straight-line segments.

Conclusion. The monotonic regression is represented by the line segments in Figure 8.7.

Comments. 1. Sometimes the lowest or highest $\hat{s}$ is less than the least or greater than the greatest s_i. Then one cannot interpolate between successive ranks, so no $\hat{y}_i$ can be calculated. The monotone regression function described here cannot be extrapolated.

2. To determine $\hat{y}_0$ corresponding to x_0 lying between two observed x_i, one may read this from the fitted segmented regression or equivalently estimate it by linear interpolation using a 'rank' r_0 corrresponding to x_0. The

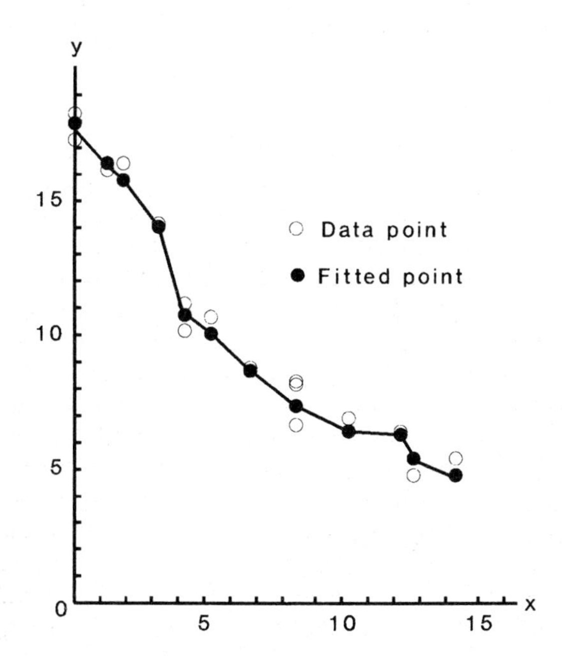

Figure 8.7 A piecewise monotonic regression fitted to the data in Example 8.6.

regression equation is used to calculate the corresponding $\hat{s}_0$, then $\hat{y}_0$ is obtained by interpolation.

3. In situations such as that described a piecewise curve is not entirely satisfactory; logically, a smooth curve with continuous derivatives would be more acceptable. The method given above is an elementary approach largely of interest as a demonstration of how a rank transformation may be used to give a more objective fit than one obtained by eye.

4. If any $\hat{s}$ corresponds to an s_j for an observation, the corresponding $\hat{y}$ is the observed y with rank s_j.

A recent sophisticated development in nonparametric regression involves the concept of splines. Both the theory and practice of these methods require a training both in numerical analysis and somewhat sophisticated statistical practice, or the availability of a sound computer program to carry out the computations. A technical account is given by Silverman (1985) and an application is described by Silverman and Wood (1987). A general review of the use of splines in statistical problems is given by Wegman and Wright (1983). One difficulty with fitting smooth curves is that they may become 'too smooth' in that they pick up local random irregularities and incorporate them into the curve. As an extreme example, a polynomial of degree n may be fitted exactly to pass through any set of $n+1$ points with distinct x values. The problem of over-smoothing is avoided by using what is termed a penalty function, but details are beyond the scope of this book.

8.3 MULTIVARIATE DATA

We often observe not just two variables but three or more for each experimental unit. Multiple regression deals with some problems in this area. Extensions of Theil's method have been considered but these usually lead to heavy computation and may or may not be robust. In their simulation study, Hussain and Sprent (1983) found the most satisfactory of those they examined was one proposed by Agee and Turner (1979). It is mathematically sophisticated, requiring orthogonalization of data and an iterative process using repeated applications of Theil's method; suitable computer software is needed. The interested reader should refer to the paper by Agee and Turner for details. Alternatively, some robust methods outlined in Chapter 11 might be used. Again suitable computer software is needed.

8.4 FIELDS OF APPLICATION

Since regression is a statistical technique used (and sometimes overused) in practically every field of application, we differ our format from that used for this section in other chapters, making only general comments on use of the technique. In bivariate regression where least squares is widely used,

Theil's method provides a robust nonparametric alternative. We briefly mentioned some nonparametric methods relevant to multiple regression in Section 8.3 and other robust methods are mentioned in Sections 11.1.4 and 12.3.2.

In the bivariate case, before fitting a regression line to data it is good practice to plot the points. This indicates whether a straight line will be adequate, whether we are likely to need a robust method of fitting due to outliers, whether a more sophisticated curve (e.g. a quadratic) might be appropriate, or whether we would do better to fit a monotonic regression.

8.5 SUMMARY

Least squares regression, providing errors are independent and have the same variance, may be appropriate even without a normality assumption (Section 8.1.1). However, though inferences then may appropriately be based on the relevant permutation distribution, there is a lack of robustness against certain departures from homogeneity of error.

Theil's method (Section 8.1.2) is robust against many departures from error homogeneity. The point estimator of slope is the median of the slopes of all pairwise joins of data points having different x co-ordinates. An **abbreviated Theil procedure** (Section 8.1.3) only requires computation of a small sub-set of slopes for the joins of pairs of independent points. **Confidence intervals** for Theil's method (Section 8.1.4) depend on Kendall's tau for the complete procedure and on the sign test for the abbreviated procedure.

Monotonic regression (Section 8.2) may be used if the mean of y is an increasing or decreasing nonlinear function of x. It depends on a transformation to ranks, fitting of a linear regression to ranks, then back-transformation of the fitted y-rank estimates to y values.

EXERCISES

8.1 For the data in Example 8.1 verify that the sum of residuals e_i and the sum of products of residuals with their corresponding x_i is zero.

8.2 For the data in Example 8.3 verify that $a^*=36.80$.

8.3 The numbers of rotten oranges (y) in 10 randomly selected boxes from a large consignment are counted after they have been kept in storage for a stated number of days (x). Use Theil's method to fit a straight line to these data.

x	3	5	8	11	15	18	20	25	27	30
y	2	4	7	10	17	23	29	45	59	73

Plot these data and the fitted line. Does the straight line fit seem reasonable?

8.4 Comment critically on and explain the meaning of the following statement. Given n observations (x_i, y_i) where the x_i are equally spaced, the Cox–Stuart test for trend (Section 2.1.6) in the y_i is essentially equivalent to testing whether the abbreviated Theil estimator is or is not consistent with zero slope.

8.5 Fit a monotonic regression to the Greenland turbot data used in Example 8.3.

8.6 Mattingley (1987) quotes data based on the US census of agriculture which gives at approximately 10-year intervals from 1920 to 1980 the percentage of US farms with tractors and farms with horses.

Percentage with tractors	9.2	30.9	51.8	72.7	89.9	88.7	90.2
Percentage with horses	91.8	88.0	80.6	43.6	16.7	14.4	10.5

Fit a monotone regression of horse percentages on tractor percentages. Why would it be pointless, or wrong, to fit a linear regression to these data?

8.7 Gat and Nissenbaum (1976) give ammonia concentrations (y mg l^{-1}) at various depths (x m) in the Dead Sea. Fit a linear regression for concentration on depth using Theil's method and obtain a nominal 95% confidence interval for β.

x	25	50	100	150	155	187	200	237	287	290	300
y	6.13	5.51	6.18	6.70	7.22	7.28	7.22	7.48	7.38	7.38	7.64

8.8 Wahrendorf, Becher and Brown (1987) give data for doses in micrograms of a tumour-inducing compound B(a)P on the percentage of animals developing tumours at each dose. Fit a monotonic regression to these data. Does the fit seem reasonable? Comment critically on the usefulness (if any) of this regression.

Dose	0.05	0.5	5	20	50	100	300
% tumours	1.0	0.0	4.9	44.2	30.0	86.5	56.9

8.9 Katti (1965) gave data for weight of food eaten (x) and gains in weight (y) for 10 pigs given one type of food and for 10 given a second type of food. Use the abbreviated Theil method to fit linear regressions to each and test whether we should accept the hypothesis that the true slopes β_1, β_2 are equal.

First	x	575	585	628	632	637	638	661	674	694	713
	y	130	146	156	164	158	151	159	165	167	170
Second	x	625	646	651	678	710	722	728	754	763	831
	y	142	164	149	160	184	173	193	189	200	201

9

Categorical Data

9.1 CATEGORIES AND COUNTS

This and the next chapter deal with problems that appear very different from those previously considered, yet many are solved using procedures we have already developed. We find relationships between methods that unify much permutation distribution theory.

Data often consist of counts of the number of units (people, institutions, towns, countries, items, etc.) with given attributes. These may be presented in one-, two-, three- or higher-dimensional tables commonly referred to as one-way, two-way, three-way, etc. contingency tables. Each *dimension* or *way* corresponds to a classification into categories representing attributes which may be either **explanatory** (e.g. dose levels of a drug; the names of several different drugs; gender; religious affiliations; age groupings; income levels) or **responses** (e.g. side effects of drugs classified as none, moderate, severe; blood pressure changes after administration of a drug; numbers of candidates giving each of several possible answers to multiple choice questions). The attributes are often qualitative; if there is no natural ordering they are described as **nominal** (e.g. religious affiliations; matrimonial status). Attributes that may be arranged in a natural order are described as **ordinal** (e.g. reactions to a drug classified as slight, moderate, severe; grouping by age under 50 and age 50 and over). Clearly, ordinal attributes may be qualitative or quantitative.

This chapter is mainly concerned with two-way tables consisting of r rows and c columns, referred to as $r \times c$ (in speech 'r by c') tables. Typically; each of the r rows represents either a level of an explanatory attribute or a level of response, and each of the c columns will represent a level of response. The entry at the intersection of the ith row and jth column (referred to as cell (i, j)) represents the number of items exhibiting that combination of row and column attributes. Table 9.1 is an example where $r=2$ and $c=5$. A common question is whether there is evidence that the incidence rate of side-effects differs betwen drugs. The null hypothesis of independence is often expressed as one of **no association** between row and column categories. Precise specification of the null and alternative hypotheses depends to some extent on the nature of these categories.

Table 9.1 Numbers of patients exhibiting side-effects for two drugs

		Side-effect level				
		None	Slight	Mod	Severe	Fatal
Treatment	Drug A	23	8	9	3	2
	Drug B	42	8	4	0	0

In Table 9.1 the row categories are explanatory and nominal. However, had each row referred to different dose levels of the same drug they would have been ordinal explanatory categories. The column categories refer to ordinal responses arranged in increasing severity.

We have already met situations where we formed $r \times c$ tables deduced from measurement data by counting the numbers of measurements falling into defined categories. For instance, Tables 5.1 and 6.1 involve counts from $r=2$ and $r=6$ samples, respectively, of numbers in each of $c=2$ categories (above and below an overall sample median M). The column categories are ordinal responses ('$<M$' and '$>M$'). The row categories (samples) are explanatory. Whether they are ordered depends upon the nature of the samples. If successive samples correspond to increasing dose levels of a drug they are ordinal; if the samples represent religious affiliations they are nominal.

Many tests for lack of association appropriate to counts with nominal categories in both rows and columns may still be applied when either or both of the row and column categories are ordered, but generally more powerful procedures exist that take account of ordering.

In Table 9.2 both rows and columns represent response categories. Here the same stimulant is given to all subjects and its effect noted on changes in *appetite* (rows) and *attitude* (columns), it being understood that all subjects can be classified into one and only one combination of row and column categories, e.g. no individual becomes both more talkative and also sexually aroused, nor shows a desire for both more food and more liquid; if such combinations occurred we would need additional categories for such dual responses. Table 9.2 includes row and column totals and the total number of counts, 31, in the bottom right. Showing these totals is common practice for contingency tables as they play a prominent part in analyses. In Table 9.2 both row and column categories are nominal, although we might argue that there is a partial ordering for rows in that no effect/more food and no effect/more liquid are both ordered, but more food/more liquid is not ordered. The hypothesis of no association for this, or any $r \times c$ table, implies that expected numbers in each row are proportional to column totals and that expected numbers in each column are proportional to row totals. In Table 9.2 association appears likely, for sexual arousal is a relatively more common attitude response among those who want more food or liquid than it is among those whose appetite is unaffected.

Table 9.2 Categorical responses to a stimulant

Appetite	*Attitude* More talkative	Desire for fresh air	Sexually aroused	Total
No effect	7	4	1	12
Want more food	0	2	9	11
Want more liquid	1	1	6	8
Total	8	7	16	31

9.1.1 Some models for counts in $r \times c$ tables

Categorical data analysis in parametric and nonparametric contexts has recently received considerable attention. Books on the subject include Bishop, Fienberg and Holland (1975), Breslow and Day (1980), Plackett (1981), Agresti (1984, 1990), Christensen (1990) and Everitt (1992).

It is convenient to represent a general $r \times c$ table in the form of Table 9.3. Here n_{ij} is the count in the categories specified by row i and column j, i.e. in cell (i, j) and n_{i+} is the total count for row i, i.e. $n_{i+}=\Sigma_j n_{ij}$. Similarly, n_{+j} is the total count for column j, and $N=\Sigma_i n_{i+}=\Sigma_j n_{+j}=\Sigma_{i,j} n_{ij}$ is the total count over all cells.

The methods we develop are nonparametric in the sense that, although several parametric models may give rise to an observed distribution of counts in a contingency table, the inference procedures we use often do not depend upon which of these models we choose or upon the precise values of certain parameters, since these are not specified in our hypotheses. Technically, we avoid the problem of specifying parameter values by a device known as conditional inference, which makes use of properties of

Table 9.3 A general $r \times c$ contingency table

		Column categories 1	2	3	.	.	.	c	Totals
	1	n_{11}	n_{12}	n_{13}	.	.	.	n_{1c}	n_{1+}
	2	n_{21}	n_{22}	n_{23}	.	.	.	n_{2c}	n_{2+}
Row	.	.	.	.	.	.	.	.	.
categories	.	.	.	.	.	.	.	.	.
	.	.	.	.	.	.	.	.	.
	r	n_{r1}	n_{r2}	n_{r3}	.	.	.	n_{rc}	n_{r+}
Totals		n_{+1}	n_{+2}	n_{+3}	.	.	.	n_{+c}	N

sufficient and ancillary statistics, a topic beyond the scope of this introductory text. Those familiar with the basic theory of maximum likelihood will find an excellent account of inference theory for categorical data in Agresti (1990, Chapter 3).

In making inferences we nearly always regard row and column totals as fixed; this is the *conditional* feature. It is justified by the concept of conditional inference, but is **not** in general a consequence of our experimental set-up. Sometimes neither row nor column totals are fixed *a priori* by the nature of the experiment, in other cases one of these sets is fixed but not the other and only exceptionally are both fixed. For example, if we are comparing two drugs, A and B, that will result in either an improvement (a success, denoted by S) or no improvement (a failure, denoted by F) we may have 19 patients and allocate 9 patients at random to receive drug A, the remaining 10 to receive drug B, and observe the following results.

	S	F	Total
Drug A	7	2	9
Drug B	5	5	10
Total	12	7	19

(9.1)

Here row totals are fixed by our choice of numbers to be allocated to each drug. If we regard the 19 patients as the entire population of interest then under the null hypothesis of *independence or no association* (i.e. that the drugs are equally effective) we are justified in regarding the total number of successes, 12, as fixed irrespective of how we allocate patients to the groups of 9 and 10 to the two drugs. Just as in the situation in Example 1.4, it is extreme outcomes under randomization like none (or 9) of the 12 successes being among the 9 patients allocated to drug A that make us suspect the null hypothesis is not true. As in other permutation tests, we may here compute the probability of each possible outcome under H_0 when the margins are fixed. The situation is slightly different when we sample our patients as two groups of 9 and 10 from a larger population, for clearly, even under H_0, if we took another sample of 9 and 10 (the fixed row totals) it is unlikely that the combined number of successes for these different patients would again be 12. We might get an outcome like that in (9.2).

	S	F	Total
Drug A	6	3	9
Drug B	3	7	10
Total	9	10	19

(9.2)

However, the hypothesis of independence does not involve the **acutal** probability of success with each drug, but stipulates only that the probability of success, p, say, is the same for each drug. In practice, we do not know the value of p. However, if we assume there is such a common value for both drugs, we can obtain an estimate based of our total number of successes, 12 from 19, in our sample of 19 in our observed situation, i.e. that in (9.1). We can use this information to compute the expected number of successes in sub-samples of 9 and 10 from the total of 19 under the assumption that H_0 is true. Again, if we observe big discrepancies between the observed and expected numbers of successes computed in this way for the two drugs, this is an indication of association, i.e. that the probabilities of success are different for drug A and for drug B. In effect, what we are doing is looking at a permutation distribution once more, but only in possible samples which have fixed column totals (in our case 12, 7 respectively) as well as fixed row totals. Such conditional inference is perfectly valid, and indeed it is the entries in the body of any such table with fixed marginal totals that provide the evidence for or against independence. The marginal totals tell us something about the likely value of p, the probability of success if indeed H_0 is true, but they contain very little information as to whether there is a common probability, p, of success for each drug. It is the individual cell values that contain virtually all the information on that point. Similar arguments apply for regarding all marginal totals in an $r \times c$ table as fixed for nearly all the models we consider in this and the next chapter. Methods for inference that do not require all marginal totals to be fixed have been proposed, but there is near-unanimity among statisticians that conditional inference with fixed marginal totals is the most appropriate way to test for independence (although independence is not the only thing we may want to test in contingency tables). Conditional inference is convenient, valid and overcomes technical problems with significance tests in this context. For enlightening discussions of this approach, see Yates (1984), Upton (1992) and a wide-ranging paper by Agresti (1992). We show in this and the next chapter that many permutation distribution tests already met in this book can be reformulated as tests involving contingency tables with fixed row and column totals.

Experimentally, we might have arrived at the outcome in (9.1) without pre-fixing even the row totals. For example, the data might have been obtained from what is called a retrospective study, by simply going through hospital records of all patients treated with the two drugs during the last 12 months and noting the responses of each. Our method of testing for no association using conditional inference would be the same, i.e. conditional upon the observed row and column totals.

For completeness, we describe briefly several models used to describe counts, indicating where each might be appropriate, and show that they lead to the same conditional inference procedures. We confine this discussion largely to 2×2 tables. The reader who is not famliar with the multinomial and Poisson distributions and the concept of conditional and marginal distributions may wish to omit the remainder of this section and move directly to Section 9.2, where we describe applications.

A typical model for the situation exemplified by (9.1), where the row classifications are explanatory and the column classifications are responses, is to assume that the two levels of the explanatory categories each correspond to a binomial distribution with probabilities of success p_A and p_B respectively, the other parameters n_A, n_B being the row totals. In (9.1) $n_A=9$ and $n_B=10$. The probabilities associated with each cell in (9.1) are thus:

$$p_{11}=\Pr(n_{11}=7)={}^9C_7\, p_A{}^7(1-p_A)^2 \qquad p_{12}=1-p_{11}$$
$$p_{21}=\Pr(n_{21}=5)={}^{10}C_5\, p_B{}^5(1-p_B)^5 \qquad p_{22}=1-p_{21}$$

As already indicated, the hypothesis of no association in this case is equivalent to $H_0 : p_A=p_B$, but the common value p, say, is unspecified. Independence of response from patient to patient and the additive property of binomial variables together mean that under H_0 the probability of observing a total of 12 successes in 19 observations (the column total line) is

$$p_{+1}={}^{19}C_{12}\, p^{12}(1-p)^7$$

Since the classifications of individuals are independent, it follows from standard definitions of conditional and independent probabilities that the probability of observing 7 successes $(=n_{11})$ in 9 $(=n_{1+})$ trials, conditional upon having observed a total of 12 successes in 19 trials, is

$$P=p_{11}p_{21}/p_{+1}.$$

Thus, when $p=p_A=p_B$,

$$P=[{}^9C_7\, p^7(1-p)^2\times{}^{10}C_5\, p^5(1-p)^5]/[{}^{19}C_{12}\, p^{12}(1-p)^7]$$
$$=[{}^9C_7\times{}^{10}C_5]/[{}^{19}C_{12}]=[9!10!12!7!]/[19!7!2!5!5!]$$

As indicated in Section 5.1.1, for a general 2×2 table with cell entries

$$\begin{matrix} a & b \\ c & d \end{matrix}$$

this generalizes to

$$P=\frac{(a+b)!(c+d)!(a+c)!(b+d)!}{(a+b+c+d)!a!b!c!d!} \qquad (9.3)$$

Equation (9.3) represents the probability mass function for the **hypergeometric distribution**. Again, as indicated in Section 5.1.1, when we fix marginal totals, P depends essentially only on a, since, given a, then b, c and

d are automatically fixed to give the correct marginal totals. The fact that fixing a fixes all cell entries is summarized by saying that a 2×2 contingency table has one degree of freedom. The value of P given by (9.3) is relevant to testing independence (no association). The test based on the hypergeometric distribution is **Fisher's exact test.** We illustrated its application to a 2×2 table in Example 5.1. By calculating P for all values of a we may set up critical regions in the tail of the distribution for rejection of H_0 in either one- or two-tail tests, as we did in Example 5.1. We sometimes require estimates of the expected values of each n_{ij} under H_0. If we knew p we could obtain exact expected values, but when p is unknown we base our estimates on the marginal totals, using these to estimate p. Since n_{+1} has expected value Np, we estimate p by $\hat{p}=(n_{+1})/N$. We estimate the expected value of n_{11} as $m_{11}=n_{1+}\hat{p}=(n_{1+}n_{+1})/N$. Similarly, we establish the general result $m_{ij}=(n_{i+}n_{+j})/N$. These expected values sum to the marginal totals.

More generally, if there is association in a 2×2 table, then clearly, in (9.1) for example, if the binomial probabilities were known to be p_A, p_B the expected values in (9.1) would be

$$\begin{matrix} 9p_A & 9(1-p_A) \\ 10p_B & 10(1-p_B) \end{matrix} \qquad (9.4)$$

The quotients, $p_A/(1-p_A)$ and $p_B/(1-p_B)$, represent the odds on the outcome success. A quantity of interest in determining association is the odds ratio, defined as

$$\theta=\frac{p_A/(1-p_A)}{p_B/(1-p_B)}=\frac{p_A(1-p_B)}{p_B(1-p_A)}$$

From (9.4) the odds ratio is also the ratio of cross products of the cell expected frequencies m_{ij}, i.e. $\theta=(m_{11}m_{22})/(m_{12}m_{21})$. Clearly, if $p_A=p_B$ then $\theta=1$. This is also true for the estimated expected values calculated under H_0, i.e. $m_{ij}=(n_{i+}n_{+j})/N$. When we do not know p_A, p_B we may estimate θ by the sample odds ratio, defined as $\theta^*=(n_{11}n_{22})/(n_{12}n_{21})$. In Chapter 10 we consider tests for association, one of which tests whether θ^* is consistent with $H_0 : \theta=1$.

If $p_A=¾$ and $p_B=½$ it is intuitively obvious that we expect more successes with drug A than with drug B and from (9.4) we find $m_{11}=6.75$ and $m_{21}=5$. These expected values no longer add to the column total as do the expected values under H_0; this should cause no surprise as we are dealing with a situation where the probabilities of success are known and different for each drug. For these p_A, p_B values $\theta=(¾)(½)/[(½)(¼)]=3$. If $p_A=½$ and $p_B=¾$ we find $\theta=⅓$. Intuitively, we feel that both these cases reflect the same degree of association, but in opposite directions. This is reflected if we use $\phi=\log\theta$ as our measure of association, for now

$$\log 3=-\log(⅓)$$

and generally

$$\log\theta = -\log(1/\theta)$$

Possible values of ϕ range from $-\infty$ to ∞ and when there is no association $\phi=0$. While the above relationships are true for logarithms to any base, most formulae for testing and estimation are based on natural logarithms using the base 'e'. Such logarithms are sometimes denoted by 'ln' rather than 'log', but we use the latter.

The above binomial model is not appropriate if our data are for two response classifications, like those in the tableau (9.5) giving numbers of patients responding in specified ways to a drug.

		Blood cholesterol levels			
		Increased	Unchanged	Total	
Blood	Higher	14	7	21	(9.5)
pressure	Lower	11	12	23	
	Total	25	19	44	

Here all four cells are on a par in the sense that all we can say is that with an unknown probability p_{ij} any of the $N=44$ subjects may be allocated to cell (i, j). The multinomial distribution is an extension of the binomial distribution relevant to independent allocation of n items to $k>2$ categories with specified probabilities associated with each category. Here we have 4 categories, one corresponding to each cell. The probabilities are subject to the constraint $p_{11}+p_{12}+p_{21}+p_{22}=1$, since each person is allocated to one and only one cell. We specify by p_{1+}, p_{2+} the marginal probabilities that items will fall in row 1 or row 2, and likewise we specify marginal probabilities p_{+1}, p_{+2} for columns. For independence between rows and columns, standard theory gives the relationship $p_{ij}=p_{i+}p_{+j}$ for $i, j=1, 2$.

Even if there is independence we still do not known the value of the p_{ij}, a problem we again overcome by conditional inference. Assuming no association (i.e. independence), we base our estimates on marginal totals, estimating p_{i+} by n_{i+}/N and p_{+j} by n_{+j}/N. Since $p_{ij}=p_{i+}p_{+j}$ we estimate p_{ij} by $\hat{p}_{ij}=n_{i+}n_{+j}/N^2$. It follows from properties of the multinomial distribution that we may then obtain expected estimated counts as $m_{ij}=N\hat{p}_{ij}=n_{i+}n_{+j}/N$, exactly as in the two-sample binomial model, and proceed to tests appropriate to that model. It is easily verified that under independence the odds ratio, here defined as $\theta=(p_{11}p_{22})/(p_{12}p_{21})$, is unity. (Hint: put $p_{11}=p_{1+}p_{+1}$, etc.) If the p_{ij} are estimated under H_0 in the way indicated above, using these estimates or the estimated expected frequencies m_{ij} gives $\theta=1$. The sample odds ratio is $\theta^*=(n_{11}n_{22})/(n_{12}n_{21})$.

Another model often relevant is that in which counts in each cell have independent Poisson distributions with a mean (Poisson parameter) m_{ij} for cell (i, j). If $\theta=(m_{11}m_{22})/(m_{12}m_{21})=1$, there is no association between rows and columns. In general, we do not know the m_{ij}, but it can be shown that, again under H_0, these may be estimated from marginal totals in the way we do for the row-wise binomial and the overall multinomial models.

Extension of the three models above to $r\times c$ tables is basically straightforward. For $c>2$, binomial sampling within rows generalizes to multinomial sampling within rows. For this, or the overall multinomial sample or the Poisson count situation, under the null hypothesis of no association the expected cell frequencies are estimated in the notation of Table 9.3 by $m_{ij}=n_{i+}n_{+j}/N$, $i=1, 2, \ldots, r$ and $j=1, 2, \ldots, c$. An odds ratio may be defined for any two rows i, k and any two columns j, l as

$$\theta_{ijkl}=(m_{ij}m_{kl})/(m_{il}m_{kj})$$

and $\theta_{ijkl}=1$ for all i, j, k, l if there is *no association*. The sample odds ratios are $\theta^*_{ijkl}=(n_{ij}n_{kl})/(n_{il}n_{kj})$.

Fisher's exact test extends to $r\times c$ tables in a way described in Section 9.2.1.

9.2 INDEPENDENCE WITH NOMINAL ATTRIBUTE CATEGORIES

Tests developed in this section take no account of order within row or column categories. They may (and often are) used for tests of no association with ordered attributes, but we see in Section 9.3 that more powerful tests are then available.

We describe three commonly used test procedures. Confusion exists in the literature because each often (but not invariably) leads to essentially equivalent permutation distributions for 2×2 tables, but to different asymptotic test statistics even in this case: for general $r\times c$ tables both the exact permutation distribution of the test statistics and the asymptotic test statistics differ. However, one should not exaggerate the importance of these differences, for unless the choice of a precise $100\alpha\%$ significance level is critical it is seldom that the three possible exact tests lead to markedly different conclusions. However, there may be considerable differences between exact and asymptotic tests in some circumstances. Although the numerical values of the three test statistics often differ considerably, even for moderately large samples, each statistic has asymptotically a χ^2 distribution with $(r-1)(c-1)$ degrees of freedom under the null hypothesis of independence (no association between row and column classifications).

The three tests are (i) the **Fisher exact test** (also known as the Fisher–Irwin test or the Fisher–Freeman–Halton test); (ii) the **Pearson chi-squared test**; and (iii) the **likelihood ratio test.**

Practice has often been to use (i) when asymptotic theory is clearly inappropriate and either (ii) or (iii) when asymptotic results are thought to be appropriate. Despite differences in numerical value of the relevant test statistics and the fact that exact permutation distribution theory differs in each case, there seems no compelling reason to depart from this practice.

Tests based on exact permutation theory, except in the case of 2×2 tables (where Fisher's exact test is commonly used) are not practical without relevant computer software. STATXACT allows exact permutation theory tests based on (i), (ii) and (iii) above and some facilitics arc also included in TESTIMATE and DISFREE, so there is now less reason to use asymptotic theory when it is of doubtful validity. This is particularly relevant for 'sparse' contingency tables in which there are many low cell counts even though N may be large. Such situations are not uncommon in medical applications where responses may be recorded on a limited number of diseased patients because resources are in short supply or responses in certain categories are uncommon (see Exercise 9.23).

9.2.1 Fisher's exact test

Fisher's exact test for 2×2 tables based on the hypergeometric distribution specified in (9.3) was extended to $r \times c$ tables by Freeman and Halton (1951). In the notation of Table 9.3, the generalized hypergeometric distribution relevant to an $r \times c$ table gives the probability of an observed configuration under H_0, conditional upon fixed marginal totals as

$$P = \frac{\Pi_i(n_{i+}!)\ \Pi_j(n_{+j}!)}{n!\ \Pi_{i,j}(n_{ij}!)} \tag{9.6}$$

The probabilities for all configurations with the given marginal totals having the same or lower probability than that observed are summed. We reject H_0 : *no association* at significance level $100\alpha\%$ if the sum of these probabilities does not exceed α.

Example 9.1

The problem. A rare infection gives rise to severe irritation of one of the nose, throat or ears. In each of 6 regions A to F with large populations, the site of irritation for victims and the numbers in each category are:

	Region					
	A	B	C	D	E	F
Nose	1	1	0	1	8	0
Throat	0	1	1	1	0	1
Ears	1	0	0	0	7	1

Is there evidence of association between regions and sites of infection?

Formulation and assumptions. We use Fisher's exact test for independence. This requires suitable computer software. We use a program described in Chapter 5 of the STATXACT manual.

Procedure. STATXACT gives the probability of observing this configuration under H_0 : *no association* as 0.00005769 and the probability of this or a less probable configuration as 0.0261. It also gives a probability of 0.0039 of obtaining a configuration with the same probability as that observed, indicating that more than one configuration has this probability.

Conclusion. We reject the hypothesis of no association at an actual 2.61% significance level.

Comments. 1. It is evident that the main cause of association is the high incidence of nose and ear (but not throat) irritation at site E. Clearly, this region has incidences out of line with the general trend, whereas the other regions produce reasonably consistent results for all three sites of irritation.

2. The finding of dependence is not surprising, so it is relevant to note that asymptotic theory here does not indicate significnce at the 5% level. We discuss the more commonly used asymptotic statistics in Sections 9.2.2 and 9.2.3, but see also the comments below under *Computational aspects*.

3. While it is impractical to compute manually all relevant permutation distribution probabilities, it is not too difficult to verify that for the given configuration (9.6) shows P=0.00005769, as indicated by the STATXACT output. (See Exercise 9.1).

4. We are not testing whether incidence levels differ between regions, but whether the proportions at irritation sites differ between regions (or equivalently, whether the proportions for regions differ between irritation sites). Had there been 8 cases of throat irritation at E we would have accepted the hypothesis of *no association*, although clearly the numbers of cases at all three irritation sites would then be higher for region E.

5. The test of association is esentially two-tail, the situation being broadly analagous to that in the Kruskal–Wallis test where we test for differences in location which may be in any direction, or in different directions in populations corresponding to different samples.

Computational aspects. A program such as that in STATXACT is invaluable in situations where asymptotic results may be misleading. Although seldom used in practice, a statistic is given in Chapter 5 of the STATXACT manual which is a monotonic function of P given by (9.6) and which has an asymptotic χ^2 distribution with $(r-1)(c-1)$ degrees of freedom. This statistic, denoted by F^2, takes the value 14.38 for this example, and $\Pr(\chi^2 \geq 14.38)=0.1597$. This is in sharp contrast to the result of the exact test, for we do not establish significance at any conventional level with the asymptotic test. In practice, the asymptotic tests given in Sections 9.2.2 and 9.2.3 are more widely used.

Many general packages include the Fisher exact test for 2×2 tables only.

Example 9.1 is not atypical; the STATXACT manual gives very similar data for location of oral lesions in a survey of three regions in India. The asymptotic test has $(r-1)(c-1)$ degrees of freedom in an $r \times c$ table because, if we allocate arbitrarily cell values in, say, the first $r-1$ rows and $c-1$ columns then the last row and column values are fixed to ensure consistency with the row and column totals.

9.2.2 The Pearson chi-squared test

An alternative statistic for testing independence of row and column categories is the Pearson chi-squared statistic. Its appeal is (i) ease of computation; (ii) fairly rapid convergence to the chi-squared distribution as count size increases; and (iii) intuitive reasonableness.

An exact test may be based on the permutation distribution of this statistic over all sample configurations having the fixed marginal totals. Computation of probabilities is no easier than it is for the Fisher exact test and only for 2×2 tables are the exact tests often equivalent.

Denoting as before the expected number in cell (i, j) by $m_{ij}=n_{i+}n_{+j}/N$, then $(m_{ij}-n_{ij})^2$ is a measure of departure from expectation under H_0. It is reasonable to weight this measure inversely by the expected frequency, as a departure of 5 from an expected frequency of 100 is less unexpected than a departure of 5 from an expected frequency of 6 for any of the sampling models given in Section 9.1. Pearson (1900) proposed the statistic

$$X^2=\Sigma_{i,j}[(m_{ij}-n_{ij})^2/m_{ij}] \tag{9.7}$$

for a test of association.

For computational purposes it is easier to use the equivalent form

$$X^2=\Sigma_{i,j}(n_{ij}^2/m_{ij})-N \tag{9.8}$$

Example 9.2

The problem. For the data in Example 9.1 calculate X^2. Test for significance using the asymptotic result that X^2 has a χ^2 distribution with $(r-1)(c-1)=10$ degrees of freedom.

Formulation and assumptions. We calculate each $m_{ij}=n_{i+}n_{+j}/N$, then use formula (9.8) for X^2.

Procedure. By computing row and column totals for the data in Example 9.1 we easily establish $m_{11}=2\times 11/24=11/12$. Proceeding in this way, we obtain the following table of expected frequencies:

	A	B	C	D	E	F
Nose	11/12	11/12	11/24	11/12	55/8	11/12
Throat	1/3	1/3	1/6	1/3	5/2	1/3
Ears	3/4	3/4	3/8	3/4	45/8	3/4

whence, from (9.8),

$$X^2=[1^2/(11/12)+1^2/(11/12)+\ \ldots\ +7^2/(45/8)+1^2/(3/4)]-24=14.96$$

With 10 degrees of freedom, χ^2 tables show that for significance at the 5% level we require $X^2 \geq 18.31$. $\Pr(\chi_{10}^2 \geq 14.96)=0.1335$.

Conclusion. We do not reject H_0 at a nominal 5% significance level.

Comments. An asymptotic result may be unsatisfactory with many sparse cells, but here that using X^2 behaves marginally better than that based on F^2 (Example 9.1). See also Exercise 9.23.

Computational aspects. Most standard statistical packages include a program for tests based on X^2 using the asymptotic result. STATXACT also computes $\Pr(X^2 \geq 14.96)=0.1094$ for the given marginal totals using the exact permutation distribution. This not only highlights the fact that the exact Fisher test and the exact Pearson chi-squared statistic tests are not equivalent, but indeed suggests that the tests may differ in power.

For 2×2 tables the exact Fisher test and the exact tests based on X^2 are often equivalent, though the asymptotic test statistics generally differ. It is customary to use the latter in asymptotic tests, and a better approximation is obtained by using a continuity correction commonly known in this context as **Yates' correction.** With this correction we compute

$$X^2=\Sigma_{i,j}\ [(|n_{ij}-m_{ij}|-\tfrac{1}{2})^2/m_{ij}]$$

i.e. we subtract ½ from the magnitude of each difference in the numerator before squaring. An alternative form useful for computation is

$$X^2=n[(|n_{11}n_{22}-n_{12}n_{21})|-\tfrac{1}{2}N]^2/(n_{1+}n_{+1}n_{2+}n_{+2}) \qquad (9.9)$$

Omitting the term $\tfrac{1}{2}N$ gives the X^2 value without continuity correction. The correction is used only in 2×2 tables.

Example 9.3

The problem. A new and expensive drug (A) is only available in preliminary trials for 6 patients; the patients to receive the drug are allocated at random from a total of 50 patients; the remaining 44 receive the standard drug (B). Of those receiving drug A only one shows a side-ffect, while 38 of those receiving drug B exhibit side-effects. Test the hypothesis of no association between side-effects and drug administered, using (9.9) as a test statistic in an asymptotic test.

Formulation. From the information above we deduce the following 2×2 table:

	Side-effect	None	Total
Drug A	1	5	6
Drug B	38	6	44
Total	39	11	50

We use (9.9) to compute X^2 and test using χ^2 with 1 degree of freedom.

Procedure. Substituting in (9.9) we find

$$X^2=50[|(1\times6-5\times38)|-25]^2/(6\times39\times44\times11)=11.16.$$

Tables of the chi-squared distribution confirm this indicates significance at the nominal 0.1% level $[\Pr(\chi^2\geq11.16)=0.0008)]$.

Conclusion. We reject H_0 : *no association* at an actual 0.08% significance level.

Comment. The permutation X^2 test indicates significance at the 0.11% level. Without Yates' correction $X^2=14.95$, overestimating significance.

Computational aspects. Most general statistical packages include a program for the Pearson's chi-squared test for $r\times c$ tables, often using Yates' correction (or providing it as an option) for 2×2 tables.

9.2.3 The likelihood ratio test

An alternative statistic for association is the likelihood ratio

$$G^2=2\Sigma_{i,j}n_{ij}\log(n_{ij}/m_{ij})$$

where 'log' is the natural logarithm, or logarithm to the base 'e'. Asymptotically G^2 again has a χ^2 distribution with $(r-1)(c-1)$ degrees of freedom. For Example 9.1, $G^2=17.33$. Although not significant at the 5% level, this is a better approximation than that given by X^2. The test has some optimal properties for the overall multinomial sampling model. Convergence to the χ^2 distribution under H_0 is often faster than that of X^2.

An exact permutation theory test may be based on G^2 and, except for 2×2 tables, the results will generally differ from those for exact tests based on F^2 or X^2. For the data in Example 9.1 the exact test based on G^2 indicates significance at an actual 4.84% level.

In each of the 3 tests for the data in Example 9.1 the asymptotic test is an unsatisfactory approximation. Although the three exact permutation distribution tests do not agree, two give a clear indication of significance at conventional levels and all give lower tail probabilities than those for asymptotic theory.

The sample has a preponderance of sparse cell entries and although such situations are realistic, in many contingency tables very few cells have low expected values under H_0. There is a variety of conflicting advice in the literature about when it is safe to use asymptotic results with G^2 or X^2. Generally speaking, the larger r and c are, the less we need worry about a small proportion of sparse cell counts. Programs such as STATXACT, and to a certain extent TESTIMATE and DISFREE, that provide both exact and asymptotic probability levels so that these may be compared, are invaluable if one frequently meets contingency tables with sparse data.

Since the tests in Section 9.2 apply to nominal categories, rows or columns may be reordered without affecting the value of the test statistic. As already indicated, for ordered categories there are more appropriate tests that depend on the order of rows and columns.

9.3 ORDERED CATEGORICAL DATA

We consider first nominal row explanatory categories and ordered column response categories; then ordered row explanatory and ordered column response classifications or ordered row and column response categories.

9.3.1 Nominal row and ordered column categories

Table 9.4 repeats Table 9.1 with the addition of row and column totals.

Table 9.4 Numbers of patients exhibiting side-effects for two drugs

		Side-effect level					
		None	Slight	Mod	Severe	Fatal	Total
Treatment	Drug A	23	8	9	3	2	45
	Drug B	42	8	4	0	0	54
	Total	65	16	13	3	2	99

Row explanatory categories, type of drug, are clearly nominal but column reponses are ordinal. Here the Wilcoxon–Mann–Whitney (WMW) test is appropriate, but the data are heavily tied, so, e.g. the 65 reponses 'tied' as *none* are allocated the mid-rank 33, the 16 responses tied as *slight* are given the mid-rank of ranks 66 to 81, i.e. 73.5, and so on. Use of the Mann–Whitney formulation obviates the need to specify mid-ranks, for we simply count the numbers of ranks in drug B equal to or exceeding each in drug A, scoring an equality as ½ and an excess as 1. Thus the 23 responses *none* for drug A each give rise to 42 ties, scoring 21 and 8+4 higher ranks, contributing a score $23(½\times42+8+4)=759$ to the Mann–Whitney U_m.

Example 9.4

The problem. For the data in Table 9.4 compute the Mann–Whitney statistic and test for a significant difference in levels of side-effects.

Formulation and assumptions. We compute either U_m or U_n in the way described above, and which was also illustrated in Example 5.6. As the sample is large we might use (5.9), although in the light of ties a modification of (5.7) is more appropriate. If a suitable computer program is available, with so many ties an exact test may be preferred.

Procedure. We calculate $U_m=23(½\times42+8+4)+8(½\times8+4)+9(½\times4)=841$. Using (5.9) gives $Z=2.63$, whereas the appropriate modification of (5.7) gives $Z=3.118$. Using STATXACT, the exact permutation test gives a two-tail test significance level of 0.19%

Conclusion. With sample sizes 45 and 54, having allowed for ties, asymptotic results should be reasonable, so we may reject H_0 at a nominal 1% significance level in a two-tail test. The exact test level is 0.19%.

Comments. If we ignore ordering and use the the Fisher exact test the actual significance level is 1.27%; an asymptotic Pearson chi-squared test gives a level of 1.93%. These findings are in line with our general remark that we increase the power of tests by taking account of ordering. However, for some configurations we do observe marginally higher significance levels when we ignore ordering.

Computational aspects. Since asymptotic results appear reasonable here, any statistical package for the WMW test that uses mid-ranks and gives the significance level based on asymptotic theory that takes account of ties should be satisfactory. A program such as STATXACT does, however, give additional confidence in that it compares not only exact and asymptotic results for the WMW test, but also allows us to perform Fisher's exact test or a Pearson chi-squared test (exact or asymptotic) and compare results.

In Section 5.1.1 we recommended Fisher's exact test when using a median test to compare treatments. Since we have ordered categories, why did we not recommend the WMW test if we have a program for exact permutation distribution tests? It does not matter which we use for 2×2 tables, for the exact tests are then equivalent, although the asymptotic forms of the two are not quite the same. The position is more complicated for the median test in Section 6.1.1, where we have more than two samples. We discuss this in Section 9.3.2.

Nearly all location tests in earlier chapters can be formulated as tests involving contingency tables with fixed marginal totals. This is a key to efficient computation of exact permutation distributions and is how STATXACT handles data for these tests. For instance, in Example 1.4 we developed a permutation test for what we now recognize as a specific example of the WMW test. In that example we had sets of 4 and 5 patients ranked 1 to 9. In contingency table format we regard the drugs as the row classification and the ordered ranks 1 to 9 as column (response) classifications. We count the number of items allocated to each row and column classification; e.g. this count will be 0 in row 1, column 4 if no patient given drug A was ranked 4. It will be 1 in row 2 column 5 if a person given drug B is ranked 5. There were no ties in Example 1.4, so if we observed the ranks 1, 2, 3, 6 for the four patients given drug A (implying those receiving drug B were ranked 4, 5, 7, 8, 9) we may construct the contingency table:

				Rank					
	1	2	3	4	5	6	7	8	9
Drug A	1	1	1	0	0	1	0	0	0
Drug B	0	0	0	1	1	0	1	1	1

In this table the row totals are 4, 5 (the sample sizes) and in this no-tie situation each column total is 1. In the Wilcoxon formulation of the WMW text the test statistic is $S_i=\Sigma_j jn_{ij}$, where either $i=1$ or $i=2$. In this context j is a **column score** which here is the rank. The permutation distribution obtained in Example 1.4 is obtained by permuting all possible cell entries consistent with the fixed row and column totals and calculating S_i for each permutation. Rank is only one possible column score. Replacing the score j in column j by any ordered scores s_j we obtain, by appropriate choice of s_j, various other location test statistics of the form $S_i=\Sigma_j s_j n_{ij}$, where $i=1$ or $i=2$, discussed in Chapter 5. For example, the s_j may be the van der Waerden scores or ordered data, the latter giving rise to the raw scores permutation distribution test. Choices of column scores in 2×c tables for many well-known tests are discussed by Graubard and Korn (1987). In the case of ties in the WMW test, cell entries correspond to the number of ties in the relevant sample (row) at any given mid-rank and the column scores are the appropriate mid-rank values. Each column total is the total number of ties assigned that mid-rank score. Permutation is over all configurations giving the fixed row and column totals.

In a 'no-tie' situation, under H_0 we expect a good mix of ones and zeros over cells of the $2 \times c$ table. A concentration of ones at opposite corners of the table indicates association and possible rejection of H_0; e.g. an example with the marginal totals above that indicates rejection is

				Rank					
	1	2	3	4	5	6	7	8	9
Drug A	0	0	0	0	0	1	1	1	1
Drug B	1	1	1	1	1	0	0	0	0

If we have more than two nominal explanatory categories for rows and the column responses are ordinal the obvious extension from the WMW test is to the Kruskal–Wallis test (with ties if appropriate). Unless some samples are very small or there is a concentration of many entries in one or two columns (heavy tying) asymptotic results are usually satisfactory. Except for very small samples, even STATXACT cannot cope with the exact permutation distribution for the Kruskal–Wallis test, but a Monte Carlo approximation is available and works well in practice.

Example 9.5

The problem. The following information on times to failure of car tyres due to various faults is abstracted from a more detailed and larger data set made available by Timothy. P. Davis. The classified causes of failure are:

A Open joint on inner lining
B Rubber chunking on shoulder
C Loose casing low on sidewall
D Cracking on the sidewall
E Other causes

One cause of failure, cracking of the tread wall, is omitted from this illustrative example, for it clearly gave a very different pattern of times to failure than that for other faults. Depending on one's viewpoint, the row categories A, B, C, D, E might be regarded as nominal explanatory *or* as nominal response categories. In the sense that they are different types of failure they represent different responses; on the other hand, each tyre failing for a particular reason might be regarded as one of a sample of tyres doomed to fail in that way and if we are interested in times to failure at which different faults occur we might regard the different faults as explanatory variables. For illustrative purposes in this example we take the latter stance, although the analysis would still be appropriate if we regarded these as distinct reponse categories. The column classification is time to failure recorded to the nearest hour, grouped into intervals <100, 100–199, 200–299, 300+ hours.

	Times to failure (hrs)			
Cause	<100	100–199	200–299	300+
A	2	6	9	0
B	0	1	7	0
C	4	10	6	0
D	1	6	5	0
E	0	8	10	2

Formulation. We perform a Kruskal–Wallis test with 5 samples. Mid-ranks are calculated regarding all units in the same column as ties. Thus, the seven ties in the column labelled <100 are all given the mid-rank 4, and so on. The Kruskal–Wallis statistic is calculated as in Section 6.2.2 and the statistic has asymptotically a chi-squared distribution with $r-1=4$ degrees of freedom. Computation is tedious without a computer program.

Procedure. The computation procedure outlined above gives the value 10.44 for the Kruskal–Wallis statistic. This is significant at the nominal 5% level [$\Pr(\chi^2 \geq 10.44 = 0.0336$].

Conclusion. We reject the hypothesis of no difference in 'sample' response distribution of failure times, i.e. we conclude there is association between rows and columns.

Comments. Despite heavy tying, in the light of our finding in earlier chapters, one hopes the asymptotic result is reasonable here. That this is so was confirmed by the STATXACT Monte Carlo procedure for estimating the exact test significance level. This gave the size of the critical region as $\alpha=0.0293$, the 99% confidence interval for α being (0.0256, 0.0330). This suggests the asymptotic result is conservative, falling just above the upper confidence limit for the permutation distribution result.

Computational aspects. Many packages compute the Kruskal–Wallis statistic, so the asymptotic test is readily available. STATXACT may compute the exact distribution siginificance level with no more than 5 rows or columns, but computation may fail on some PCs or prove very time consuming for examples with a moderate number of ties. The Monte Carlo method is relatively quick. Our estimate, $\alpha=0.0293$, was based on 14 000 simulations. Increasing the number of simulations shortens the confidence interval for α.

In a no-tie situation with r samples and a total of N observations the Kruskal–Wallis test may be formulated in contingency table terms as an $r \times N$ table with rows corresponding to samples and all cell entries 1 or 0. The columns are ordered by ranks and the observation of rank j is recorded as 1 in the cell in the jth column in the row corresponding to the sample in which it occurs. All other entries in that column are zero in the no-tie case. The row totals are the sample sizes n_i and each column total is 1. Ranks are the scores used to generate the exact permutation distribution and the statistic is a function of the rank sums for each row. Permutation is over all cell entries in the $r \times N$ table consistent with row and column totals.

9.3.2 Ordered row and column categories

We consider first the case where the row categories are ordinal explanatory categories. One approriate test of no association is then the Jonckheere–Terpstra test with rows corresponding to ordered samples. Asymptotic results may be reasonable if allowance is made for ties. We illustrate the need to allow for ties in a situation not uncommon in drug testing, where side-effects may not occur very often, but may be serious if they do.

Example 9.6

The problem. Side-effects (if any) exhibited by patients at increasing dose levels of a drug are classified as *none*, *mild*, *moderate* or *severe*. Do the following data indicate that side-effects increase with dose level?

	Side-effects			
Dose	None	Mild	Moderate	Severe
100 mg	50	0	1	0
200 mg	60	1	0	0
300 mg	40	1	1	0
400 mg	30	1	1	2

Formulation. The Jonckheere–Terpstra statistic is relatively easy to compute as an extension of the procedure used for the Mann–Whitney formulation of the WMW test used in Example 9.4, scoring the relevant ties in any column as ½. Those in the first column make a substantial contribution to the total.

Procedure. Without a computer program it is tedious but not difficult to compute the relevant statistic, denoted by U in Example 6.4. It is

$$U=50[\tfrac{1}{2}(60+40+30)+7]+60[\tfrac{1}{2}(40+30)+6)]+40(\tfrac{1}{2}\times30+4)+5+3.5+3+2.5=6834.$$

The asymptotic test (6.4), with no correction for ties, gives $\Pr(U>6834)=0.2288$. The two-tail probability is 0.4576 so we would not reject H_0 : *side-effect levels are independent of dose* in either a one- or two-tail test.

Conclusion. We do not reject the null hypothesis that side-effects are independent of dose even in a one-tail test, which might be justified on the general grounds that increasing dose levels of a drug are more likely to produce side-effects if indeed there are any.

Comments. A clinician is unlikely to be happy with this finding. Side-effects are not common, but there is to a clinician (and even to the layman) an indication that these are more likely and more severe at the highest dose level. Either the evidence is too slender to reach a firm conclusion or our test is inappropriate; in particular, can we justify an asymptotic result which does not allow for ties with such a high incidence of these?

Computational aspects. Some general statistical packages include the asymptotic Jonckheere–Terpstra test without allowance for ties. We show below that correcting for ties may have a dramatic effect. STATXACT (version 2.03 or later), includes an asymptotic test that allows for ties.

Example 9.7

The problem. Use the exact Jonckheere–Terpstra test on the data in Example 9.6 to determine if there is acceptable evidence that side-effects become more common as dose increases.

Formulation. The permutation test requires computation of probabilities for the Jonckheere–Terpstra statistic for all permutations of the contingency table in Example 9.6 with the same marginal totals and with the same or a lesser probability than that observed.

Procedure. STATXACT provides a program for these computations which indicates $\Pr(U>6834)=0.0168$.

Conclusion. We reject H_0 in a one-tail test at an actual 1.68% significance level.

Comments. This is a dramatic example of a situation where the exact permutation theory leads to a markedly different conclusion to a commonly used asymptotic test despite a large sample size. See also Exercise 9.23.

We give an amended formula for Var(U) in (6.4) which allows for ties. The formula is adapted from one given by Kendall (1970) and quoted by Lehmann (1975, p. 235). The notation used matches that in Table 9.3.

We calculate

$$\mathrm{Var}(U)=U_1/d_1+U_2/d_2+U_3/d_3 \tag{9.10}$$

where

$$U_1=N(N-1)(2N+5)-\Sigma_i n_{i+}(n_{i+}-1)(2n_{i+}+5)-\Sigma_j n_{+j}(n_{+j}-1)(2n_{+j}+5)$$

$$U_2=[\Sigma_i n_{i+}(n_{i+}-1)(n_{i+}-2)][\Sigma_j n_{+j}(n_{+j}-1)(n_{+j}-2)]$$

$$U_3=[\Sigma_i n_{i+}(n_{i+}-1)][\Sigma_j n_{+j}(n_{+j}-1)]$$

$$d_1=72,\ d_2=36N(N-1)(N-2),\ d_3=8N(N-1)$$

When there are no ties, all $n_{+j}=1$ and $U_2=U_3=0$, and Var(U) reduces after algebraic manipulation to that in (6.4). In Example 9.6 the relevant values are $N=188$, $n_{1+}=51$, $n_{2+}=61$, $n_{3+}=42$, $n_{4+}=34$, $n_{+1}=180$, $n_{+2}=n_{+3}=3$, $n_{+4}=2$. Substituting in (9.10) reduces the denominator in (6.4), $\sqrt{[\mathrm{Var}(U)]}$, from 415.36 (ignoring ties) to 145.18 (Exercise 9.2). The corresponding $Z=2.125$, and standard normal distribution tables give $\Pr(Z\geq 2.125)=0.0168$, agreeing exactly with the result obtained in Example 9.7.

In clinical trials situations like that discussed in Examples 9.6 and 9.7 are common. They apply not only when considering relatively rare side-effects of drugs but also in other situations where certain responses are rare, e.g. incidence of lung cancer in populations exposed to adverse environmental factors. We may observe many thousand people in each explanatory category, only a few of whom develop lung cancer.

In Section 6.1.1 we recommended the Fisher exact test for the median test for $r>2$ samples. This ignores the fact that the responses $<M$ and $\geq M$ are ordered. If the sample classifications are nominal an alternative is the Kruskal–Wallis test. If the samples are ordered so that the natural alternative hypothesis is some systematic increase or decrease in the median, we may do better using the Jonckheere–Terpstra procedure. With only two columns (ordered response categories) there will often be little difference between results using the Fisher test or Kruskal–Wallis or (if appropriate) the Jonckheere–Terpstra procedure. Certainly, if only asymptotic results are used in this last test a correction for ties is important. In the special case of only two ordered response categories we show below that the Jonckheere–Terpstra test is equivalent to the WMW test.

Example 9.8

The problem. In Exercise 6.21 we gave data supplied by Chris Theobald for blood platelet counts and spleen sizes for 40 patients. One aspect of interest to the investigators was whether the proportion of patients with an abnormally low platelet count increased with increasing spleen size. If we define an abnormal count as one below 120, we deduce the contingency table below from the data in Exercise 6.21. Use the Jonckheere–Terpstra test to determine if there is evidence of a direct association between spleen size and platelet count.

	Blood platelet count	
	Abnormal	Normal
Spleen size		
0	0	13
1	2	10
2	5	6
3	3	1

Formulation and assumptions. Exercise 6.21 indicated that platelet counts tend to decrease as spleen size increases. There ties were minimal, but here we must take them into account in an asymptotic test.

Procedure. Proceeding as in Examples 9.6 and 9.7, we find $U=183$ and $E(U)=287.5$. Using (9.10) we find the standard deviation of U is $\sqrt{[\text{Var}(U)]}=30.67$, whence, from (6.4), $Z=-3.407$. From tables of the standard normal distribution we find $\Pr(Z\leq-3.407)=0.0003$. We ask for confirmation of these results in Exercise 9.3.

Conclusion. There is strong evidence that the proportion of individuals with abnormal counts increases as spleen size increases.

Comments. 1. Computing Z by (6.4), ignoring ties, gives $Z=-2.551$, indicating significance in a one-tail test at the 0.54% level (compared with the level 0.03% obtained above). An exact permutation test using STATXACT with these data indicates significance at a 0.02% level in a one-tail test, in close agreement with the asymptotic result taking ties into account.

2. In general, ties have a greater effect on Var(U) when they occur between samples (in columns) than if they occur largely within one sample (in a row).

3. Regarding counts less than 120 as 'abnormal' is arbitrary. Using an extended data set, Bassendine *et al.* (1985) considered counts below 100 as 'abnormal'. With such arbitrary decisions about categories, one choice may lead to a significant result, while another does not. In Section 9.4.2 we discuss a similar problem arising with grouped continuous data.

4. Since we have only two columns, they remain 'ordered' if we interchange them. We might look upon 'abnormal' and 'normal' as designating samples from two populations and spleen sizes as rankings for

those populations, and apply the WMW test as in Example 9.4. Doing so, we get exactly the same results as we do with Jonckheere–Terpstra.

The expression for Var(U) in (9.10) is unaltered if we interchange rows and columns in an $r\times c$ table. However, the statistic U and the mean E(U) are altered by this change. It is not difficult to show that if we write U_1 for the Jonkheeree–Terpstra statistic for the original $r\times c$ table and U_2 for the corresponding statistic when rows and columns are interchanged then $U_2=U_1+\frac{1}{4}(\Sigma_i n_{i+}-\Sigma_j n_{+j})$ and $E(U_2)=E(U_1)+\frac{1}{4}(\Sigma_i n_{i+}-\Sigma_j n_{+j})$. Clearly, the asymptotic tests using U_1 and U_2 are identical. Although U_1 and U_2 differ by the constant value given above, their permutation distributions are otherwise identical, i.e., for all k, $\Pr(U_1=k)=\Pr(U_2=k+\frac{1}{4}(\Sigma_i n_{i+}-\Sigma_j n_{+j})$.

This equivalence may seem surprising as the Jonckheere–Terpstra test was introduced as a test for a monotone trend between samples. However, if there are no ties in p samples with $n_1, n_2, \ldots, n_p$ observations respectively the problem may be stated in a contingency table format with p rows and $N=\Sigma_i n_i$ columns. In each column all cells except one have zero entries and the remaining cell has an entry unity. The entries in row i are constrained so that the sum of the cell entries that are unity (all others being zero) is n_i. If we interchange rows and columns the equivalent test problem is one in which we have N samples each of one observation and these take only p different values of which there are n_1 tied at the lowest value, n_2 tied at the next lowest, and so on. An extreme case is that in which we have N samples, each of one observation and none of these are tied. We may represent this situation by an $N\times N$ table with all column totals and all row totals equal to 1. This implies that in any row or column there is one cell value unity and all others are zero. In this case clearly $U_2=U_1$. As pointed out in Section 7.1.2, the Jonckheere–Terpstra test procedure is here equivalent to that for Kendall's tau. U_1 or U_2 are the numbers of concordances in this 'no-tie' situation, and this is sufficient for determination of Kendall's t.

Any situation with ties, and where the concept of linear rank association implicit in Kendall's tau is relevant, may be looked upon as generating an $r\times c$ matrix which 'shrinks' an $N\times N$ matrix of zeros and ones by superimposing rows with identical 'x' values and superimposing columns with identical 'y'. In particular, if the ith sample has n_i observations we shrink n_i adjacent rows to a single row by adding all column entries in those n_i adjacent rows. The x variable values in this situation are effectively labels that distinguish the (ordered) samples. We have pointed out that in the 'no-tie' case the Jonckheere–Terpstra statistic is equivalent to the number of concordances in Kendall's tau. With ties the two statistics use slightly different counting systems. The Kendall statistic counts concordances as unity, ties as zero and discordances as -1; for the Jonckheere–Terpstra statistic, concordances count as 1, ties as ½ and discordances as zero.

If rows and columns are both response categories we may be interested in whether there is a patterned association (akin to a correlation) between row and column classifications in that high responses in one classification tend to be associated with high responses in the other, or high reponses in one are associated with low responses in the other. Again the Jonckheere–Tepstra test is relevant.

Several statistics are used as measures of association but none is completely satisfactory in all circumstances. For some of those suggested, an exact Jonckheere–Terpstra test is appropriate to test for significant association in the form of a monotonic trend. However, for non-monotonic associations such tests may fail to detect patterns of association. For example, in an ordered-categories contingency table with cell entries

$$\begin{array}{ccc} 0 & 17 & 0 \\ 5 & 0 & 5 \\ 3 & 0 & 3 \end{array} \qquad (9.11)$$

there is a clear, but obviously not monotonic, association. Here it is easily verified (Exercise 9.4) that the asymptotic Jonckheere–Terpstra statistic Z takes the value 0, giving $\alpha=0.50$ for a one-tail test! Apart from a discontinuity effect the exact test confirms this probability.

A desirable property of a measure of monotonic association, akin to that of a correlation coefficient, is that it takes the value 1 when there is complete positive association, the value zero if there is no rank association, and the value -1 if there is inverse association, e.g. high category row classifications are associated with low category column classifications and vice-versa. An appropriate test for linear rank association using such statistics is often the Jonckheere–Terpstra test, but to measure degree of association there is a case for having a statistic that behaves like a correlation coefficient. We discuss one such coefficient here. This and other measures of association are described in considerable detail by Kendall and Gibbons (1990, Chapter 3) and by Siegel and Castellan (1988, Chapter 9). These and other authors give guidance on relevant asymptotic tests which take account of the tied nature of the data.

The **Goodman–Kruskal gamma statistic** is usually denoted by G. We count the number of concordances and discordances between row and column classifications as for Kendall's tau, but make no allowance for ties. When both row and column classifications are ordered, for each count in cell (i, j) there is concordance between that count and the count in any cell below and to the right. Thus, if we denote by N_{ij}^+ the sum of all elements below and to the right of cell (i, j), which itself has count n_{ij}, the total number of concordances is

$$C=\Sigma_{i,j} n_{ij} N_{ij}^+, \quad 1\le i\le r-1, \quad 1\le j\le c-1$$

This is like the count for the Jonckheere–Terpstra test, except that for counts in cell (i, j) we ignore ties in column j (each counted as ½ in Jonckheere–Terpstra). Similarly, denoting the sum of all elements to the left of and below the cell (i, j) by N_{ij}^{-}, the total number of discordances is

$$D=\Sigma_{i,j}\, n_{ij}N_{ij}^{-}, \quad 1\leq i\leq r-1, \quad 2\leq j\leq c.$$

The Goodman–Kruskal G is defined as $G=(C-D)/(C+D)$. Clearly, if $D=0$, $G=1$. If $r\geq c$, $G=1$ if all entries in any column except the last are zero, unless $i=j$; in the last column they must be zero if $i<c$. If $c\geq r$, $G=1$ if all entries in any row except the last are zero, unless $i=j$; in the last row they must be zero if $j<r$. If $r=c$, $G=1$ if all entries for which $i\neq j$ are zero. In Exercise 9.5 we ask for confirmation that $G=1$ in each of the tables

3	0	0	0
0	4	2	7

7	0	0
0	1	0
0	0	6
0	0	7
0	0	5

It is easy (Exercise 9.6) to obtain analogous conditions for $G=-1$. When there is a good scatter of counts over all cells, C and D will not differ greatly from one another and G will have a near-zero value. Near-zero values of G are also possible if association is not monotonic. It is easily verified (Exercise 9.7) that $C=D=0$, hence $G=0$, in the highly patterned contingency table in (9.11).

Like Kendall's tau, G takes no account of ties in both x and y, yet unlike Kendall's tau it may equal 1 even when there is heavy tying because the denominator is $C+D$. We omit detail, but it is not difficult to see that in the case of ties, this falls well short of the denominator in Kendall's tau for the no-tie case. In Section 7.1.4 we indicated that it may be better to use a modification of Kendall's tau (often denoted by t_b) where an allowance is made in the denominator for ties. This is a compromise between ignoring them (as in the Kendall t) and deleting their effect if in the same row or column as we do in G. Calculation of t_b for contingency tables is straightforward and follows the method outlined in Section 7.1.4. Other measures of association and some appropriate tests are given by Siegel and Castellan (1988, Chapter 9); we concur with their comment that 'all of them should be useful when appropriately applied.' Other approaches to exploring association in contingency tables are given in Chapter 10.

9.4 GOODNESS-OF-FIT TESTS FOR DISCRETE DATA

The Pearson chi-squared test and the likelihood ratio test (Sections 9.2.2 and 9.2.3) may be looked upon as tests of goodness of fit of data in a contingency table to a model of independence (no association) between row and column classifications. The asymptotic chi-squared test is widely used as a goodness-of-fit test to data from any discrete distribution.

Table 9.5 lists the number of times each digit 0 to 9 occurs as the penultimate digit in one column of the 1986 telephone directory for Tayside and North Fife, Scotland; e.g. the penultimate digit in the number 32759 is 5. These entries might be accepted as an effectively (though not truly) random sample of penultimate digits from all entries in that telephone directory and they form a 1×10 contingency table. We may want to test some hypothesis about the probabilities that these penultimate digits take particular values. An obvious hypothesis is that any digit value is equally likely, i.e. that the digits are uniformly distributed.

Table 9.5 Frequency of occurrences of digits 0–9 in penultimate position in telephone numbers

Digit	0	1	2	3	4	5	6	7	8	9
Freq.	13	12	10	13	11	16	12	13	17	13

This is a test about a parameter, p, that specifies the expected frequency of any digit. However the **goodness-of-fit test** we develop is distribution-free in that it may be used to test hypotheses about any discrete distribution.

We might want to perform other nonparametric tests with data like those in Table 9.5. Are odd digits more common than even digits? Does the frequency of digits increase as the digit values increase from 0 to 9? For the latter we might use the Cox–Stuart test for trend, although the sequence is necessarily short and could only provide a meaningful critical region for a one-tail test; a nonparametric test based on correlation or regression might be more appropriate.

Tests of goodness of fit to any discrete distribution (often appropriate for counts) may be based on the chi-squared statistic X^2 in (9.7). Expectations are calculated for a hypothesized distribution which may be binomial, negative binomial, Poisson, uniform or any discrete distribution. Sometimes one or more parameters must be estimated from the data. If we have r counts or cells the test will have $r-1$ degrees of freedom if no parameters are estimated; if a parameter (e.g. p for the binomial or λ for the Poisson distribution) is estimated from data, one further degree of freedom is lost for each parameter estimated. While tests of the Kolmogorov–Smirnov type are sometimes also applied to discrete data, generally speaking the test criteria are no longer exact and the tests are often inefficient.

Example 9.9

The problem. It is often suggested that recorded ages at death are influenced by two factors. The first is the psychological wish to achieve ages recorded by 'decade', e.g. 70, 80, 90, and that a person nearing the end of a previous age decade who knows his or her days may be numbered will struggle to attain such an age before dying. If so, ages at death with final digits 0, 1 should be more frequent than higher digits. A second factor is that elderly people may be unsure about their age, tending to round it; e.g. if they are in the mid-seventies they tend to say they are 75 if anywhere between about 73 and 77. Similarly a stated age of 80 may correspond to a true age a year or two above or below. If these factors operate, final digits in recorded ages at death would not be uniformly distributed. Table 9.6 gives final digits at age of death for the 117 males recorded at the Badenscallie burial ground (see Section A7). Is the hypothesis H_0 : *any digit is equally likely* acceptable?

Table 9.6 Recorded last digit of age at death, Badenscallie males

Digit	0	1	2	3	4	5	6	7	8	9
Freq.	7	11	17	19	9	13	9	11	13	8

Formulation and assumptions. For 117 deaths the expected number of occurrences of each of the 10 digits, if all are equally likely, is 11.7. Since the denominator in X^2 is always the expected number, $m=11.7$, we calculate X^2 by taking the differences between the observed numbers n_i and 11.7, squaring these, adding and finally dividing the total by 11.7.

Procedure. The differences $n_i - m$ are respectively −4.7, −0.7, 5.3, 7.3, −2.7, 1.3, −2.7, −0.7, 1.3, −3.7. These differences sum to zero – a property of deviations from the mean. The sum of squares of differences is 136.1. Division by the expected number $m=11.7$ gives $X^2=136.1/11.7=$ 11.63. The minimum value of χ^2 for significance at the 5% level with $10-1=9$ degrees of freedom is 19.02.

Conclusion. We do not reject H_0 : *the digits may be random.*

Comments. If our suppositions in posing the problem hold we expect a build-up of digits around 0, 1 and perhaps near 5. Variation in digit frequency is not as we predicted. Any build-up is around 2 and 3. As the overall result is not significant we must not read too much into this, but it may represent some tendency, especially for the more elderly when aged about 82 or 83 or about 92 or 93, to feel they are unlikely to outlive that decade and so make no special effort to prolong life; but this is speculation. There is certainly no evidence in these data that age may be erroneously reported to the nearest half decade, e.g. as 75, 80, 85.

9.4.1 Goodness of fit with a parameter estimate

We often have data that might belong to a specific distribution such as the Poisson distribution with unknown mean. We may use the statistic X^2 to test this (although for the Poisson and some other distributions alternative parametric tests are more powerful against a wide range of alternatives).

For the asymptotic test using X^2 a difficulty arises if expected numbers in some 'cells' are small. Traditional advice is to group such cells to give a group expected number close to 5 (but this is only a rough guide). There is a corresponding reduction in degrees of freedom.

Example 9.10

The problem. A factory employs 220 people. The numbers experiencing 0, 1, 2, 3, . . . accidents in a given year are recorded.

No. of accidents	0	1	2	3	4	5	≥6
No. of people	181	9	4	10	7	4	5

Are these data consistent with a Poisson distribution?

Formulation and assumptions. The maximum likelihood estimate, $\hat{\lambda}$, of the Poisson parameter λ, is the mean number of accidents per person. The expected numbers having r accidents are $E(X=r)=220\hat{\lambda}^r \exp(-\hat{\lambda})/r!$; this follows because $\Pr(X=r)=\lambda^r \exp(-\lambda)/r!$ (see Section A1.1).

Procedure. To obtain the mean number of accidents per person we multiply each number of accidents by the number of people having that number of accidents, add these products and divide the sum by 220, giving

$$\hat{\lambda}=(0\times181+1\times9+2\times4+3\times10+4\times7+5\times4+6\times5)/220=125/220=0.568$$

This is approximate because we treat 'six or more' as 'exactly 6'. In practice, this has little influence on our result, but it is a limitation we should recognize. A pocket calcuator with an e^x key aids calculation. An iterative algorithm may be used, noting that $E(X=r+1)=E(X=r)\times\hat{\lambda}/(r+1)$.

The expected numbers are

No. accidents	0	1	2	3	4	5	6 or more
Expected nos	124.7	70.8	20.1	3.8	0.5	0.1	0

Grouping results for 3 or more accidents gives an associated total expected number 4.4. We calculate X^2 using (9.8) as

$$X^2=(181^2/124.7)+(9^2/70.8)+(4^2/20.1)+(26^2/4.4)-220=198.3$$

The degrees of freedom are 2 since we have four cells in the final test and we have estimated one parameter, λ. The critical value of χ^2 for significance at the 0.1% significance level is 13.81.

Conclusion. There is strong evidence that the data are not consistent with a Poisson distribution.

Comments. Accident data seldom follow a Poisson distribution; they would if accidents occurred entirely at random and each person had the same probability of experiencing an accident, and having one accident did not alter the probability of that person having another. In practice, a better model is one that allows people a differing degree of accident proneness and also allows for the occurrence of one accident to change the probability victims having a further accident, perhaps reducing it because they become more careful, or increasing it because their concentration or confidence is lowered, making them more accident prone. These factors may act differently from person to person, or for individuals exposed to different risks. Multiple accidents, in which a number of people are involved in one incident, also affect the distribution of numbers of accidents per person.

Computational aspects. Nearly all standard statistical packages include a program for chi-squared goodness-of-fit tests.

9.4.2 Goodness of fit for grouped data

The chi-squared test is sometimes applied to test goodness of fit of grouped data from a specified continuous distribution, commonly a normal distribution with specified mean and variance, or with these estimated from the data. This is not recommended unless the grouped data are all that are available and grouping is on some natural basis. One arbitrary grouping may result in rejection of a hypothesis, whereas another grouping may not.

A situation where a test for normality based on Pearson's X^2 might be used is that for sales of clothing of various sizes. A large retailer might note numbers of sales of ready-made trousers with nominal leg lengths (cm) 76, 78, 80, ... The implication is that customers requiring a leg length between 75 and 77 cm will purchase trousers with leg length 76, those requiring leg lengths between 77 and 79 will purchase trousers of length 78, and so on. The sizes are the centre value for each group. To test whether sales at nominal lengths are consistent with a normal distribution, mean and variance unspecified, these parameters must be estimated from the grouped data and 2 degrees of feeedom deducted to allow for this.

Given complete sample data from a continuous distribution it is better to use goodness-of-fit tests of the Kolmogorov/Lilliefors type as appropriate, or other tests relevant to particular continuous distributions. If grouping is used, and tests are to be based on X^2, parameter estimates should be based on the grouped data rather than the original data even if the latter are available. Anomalies may arise if this procedure is not adopted; these are discussed by Kimber (1987).

9.5 EXTENSION OF McNEMAR'S TEST

We introduced McNemar's test in Section 4.2. There is a chi-squared approximation equivalent to the normal approximation to the binomial. For the data in Table 4.4, for example, we effectively test whether 9 succcesses and 14 failures (or vice versa) are consistent with a binomial distribution with $p=½$ and $n=9+14=23$. We perform a goodness-of-fit test. The expected numbers of successses or failures is 11.5 and the test statistic is $X^2=(2.5)^2/11.5+(2.5)^2/11.5=1.09$. This is compared with the relevant critical value for chi-squared with 1 degree of freedom. Recalling that the relevant cells in a 2×2 table for the McNemar test are the off-diagonal cells n_{12}, n_{21}, it is easily verified that if we have these numbers of successes or failures ($n_{12}=9$ and $n_{21}=14$ in Table 4.4) then

$$X^2=(n_{12}-n_{21})^2/(n_{12}+n_{21})$$

In Section 6.4.4 we introduced Cochran's test for comparing several treatments with binary (0,1) responses. When $t=2$ it is not difficult to show (Exercise 9.17) that Cochran's Q is equivalent to the form given above for X^2. In this sense the Cochran test is a generalization of the McNemar test.

Another generalization of the McNemar test is from a 2×2 contingency table to an $n\times n$ table in which we test for off-diagonal symmetry. One relevant test was proposed by Bowker (1948). A situation in which the test may be used is that where a new formulation of a drug which it is hoped will reduce side-effects is under test. If we have a number of patients who have been treated with the old formulation and records are available of any side-effects we might now treat each patient with the new formulation and note incidence of side-effects. Table 9.7 shows a possible outcome for such an experiment.

Table 9.7 Side-effects with old and new formulations of a drug

		Side-effect levels – new formulation			
		None	Slight	Severe	Total
Side-effect levels – old formulation	None	83	4	3	90
	Slight	17	22	5	44
	Severe	4	9	11	24
	Total	104	35	19	158

Inspecting the table, we note that each off-diagonal count below the diagonal exceeds the count in the symmetrically placed cell above the diagonal, e.g. $n_{31}=4>3=n_{13}$, etc. This gives the impression that the trend is towards less severe side-effects with the new formulation, although a few who suffered no side-effects with the old formulation do show some with the new formulation. Bowker proposed as a test statistic to determine whether at least one pair of probabilities associated with the symmetrically placed-off diagonal cells differ in an $n \times n$ table, a generalization of the X^2 statistic used in the McNemar test that takes the form

$$X^2 = \Sigma(n_{ij} - n_{ji})^2/(n_{ij} + n_{ji}) \quad (9.12)$$

where the summation is over all i from 1 to $n-1$ and $j>i$. Under the null hypothesis of symmetry the asymptotic distribution of X^2 in (9.12) is chi-squared with $\frac{1}{2}n(n-1)$ degrees of freedom.

Example 9.11

The problem. Do the data in Table 9.7 provide evidence that side-effects are less severe with the new formulation of the drug?

Formulation and assumptions. The Bowker statistic given by (9.12) is appropriate. Here $n=3$, $n_{12}=4$, $n_{13}=3$, etc. The null hypothesis is that for all off-diagonal counts in the table the associated probabilities are such that all $p_{ij}=p_{ji}$. The alternative is that for at least one such pair $p_{ij} \neq p_{ji}$.

Procedure. Substitution of the relevant values in (9.12) gives

$$X^2 = (4-17)^2/21 + (3-4)^2/7 + (5-9)^2/14 = 9.33$$

Under the null hypothesis, X^2 has a chi-squared distribution with 3 degrees of freedom, establishing significance at a nominal 5% (actual 2.6%) level.

Conclusion. There is evidence of a differing incidence rate for side-effects under the two formulations. From Table 9.7 it is clear that this difference is towards less severe side-effects under the new formulation.

Comments. Marascuilo and McSweeney (1977, Section 7.6) discuss an alternative test based on marginal totals due to Stuart (1955; 1957). Agresti (1990, Chapter 10) discusses other generalizations of the McNemar test.

9.6 FIELDS OF APPLICATION

Tests of independence for categorical data occur widely. Here are a few situations.

Rail transport

A rail operator may wonder if its image is different for standard class and for first class passengers. It may ask samples of each to grade the service as excellent, good, fair or poor, setting up a 2×4 table of response numbers.

Television viewing

A public service broadcasting channel competes with a commercial service. Samples of men and women are asked which they prefer. The results may be expressed in a 2×2 table, any differences between the sexes in preference ratings being of interest.

Drug addiction

A team of doctors compares 3 treatments allocated at random to subjects for curing drug addiction. For each subject, withdrawal symptoms are classed as severe, moderate, mild or negligible. The resulting 3×4 table of counts is tested to see if there is evidence of an association between treatments and severity of withdrawal symptoms. If the treatments can be ordered (e.g. by degree of physiological changes induced) a test for association of the Jonckheere–Terpstra type might be used. In the absence of such ordering a Kruskal–Wallis test would be appropriate.

Sociology

A social worker may be interested in whether blonde or brunette teenage girls who frequent public houses prefer to drink spirits, wine, beer or soft drinks; the resulting 2×4 table may be used in a test for lack of association between hair colour and drink preference.

Public health

After a contaminated food episode on a jumbo jet some passengers show mild cholera symptoms. The airline wants to know if those previously inoculated show a higher degree of immunity. They find out which passengers have and have not been inoculated; records also show how many in each category exhibit symptoms, so they can test for association.

Rain making

In a low-rainfall area rain-bearing clouds are 'seeded' to induce precipitation. Randomly selected clouds are either seeded or not seeded and we observe whether or not local rainfall occurs within the next hour.

Educational research

Children are to be exposed to an ordeal which some may fear but which holds no terror for others. An explanation of the ordeal might be given before it is faced. This may placate a child who was initially frightened but implant fear in one who was initially fearless. If numbers who were influenced by a prior explanation can be obtained, McNemar's test may be used to indicate whether an explanation does more harm than good.

Medicine

To test whether a drug is effective it and a placebo are given to patients in random order; it is noted which, if either, they claim to be effective. Some claim both are, some neither, some one but not the other. McNemar's test could be used to see if results favour the placebo, the new drug, or neither.

Here are examples of goodness-of-fit tests.

Genetics

Genetic theory may specify the proportions of plants in a cross that are expected to produce blue or white flowers, round or crinkled seeds, etc. Given a sample in which we know the numbers in each colour/seed-shape combination, we may use a chi-squared goodness-of-fit test to see if these are consistent with theoretical proportions.

Sport

It is often claimed that the starting positions in horse-racing, athletics or rowing events may influence the chance of winning. If we know the starting positions and winners for a series of rowing events in each of which there are six starters we might test the hypothesis that the numbers of wins from each starting position is consistent with a uniform distribution.

Horticulture

The positions at which leaf buds or flowers form on a plant stem are called nodes. Some theories suggest a negative binomial distribution for the node number (counted from the bottom of the stem) at which the first flower forms. Given data, a chi-squared goodness-of-fit test is appropriate.

Commerce

A car salesman may doubt the value of advertising. He might compare sales over many weeks in which he has advertised in either 0, 1, 2, 3 or 4 newspapers. He would expect a uniform distribution if advertising is worthless. In this example it might be necessary to adjust for trends with time in sales; the effect of this might be minimized if weeks were chosen at random for each number of advertisements over a sufficiently long period.

Queues

Numbers of people entering a bank, post office or superstore during one-minute intervals are recorded over a long period. If the process were completely random with constant mean, the numbers of such intervals in which 0, 1, 2, 3, . . . people enter should follow a Poisson distribution. Would you be surprised if a test rejected this hypothesis? If so, why?

9.7 SUMMARY

Tests for independence between row and column classifications in contingency tables are usually based on conditional inference, regarding the marginal totals as fixed (Section 9.1).

For $r \times c$ tables with nominal categories exact tests for association are most commonly based on **Fisher's exact test** (Section 9.2.1), while asymptotic tests are usually based on **Pearson's chi-squared statistic** X^2 (Section 9.2.2) or the **likelihood ratio statistic** G^2 (Section 9.2.3). Exact permutation tests may also be based on X^2 and G^2; while the permutation distributions differ from that for Fisher's exact test (except in 2×2 tables) conclusions for all three approaches are often broadly in line.

For categorical tables with ordered rows and columns, tests equivalent to the **Jonckheere–Terpstra test** may be used (Section 9.3.2). Adjustments for ties are important if asymptotic theory is used. The **Goodman–Kruskal gamma** statistic, G, (Section 9.3.2) is closely related to Kendall's tau as a measure of association. An adjusted form may be appropriate with ties.

The **chi-squared goodness-of-fit test** (Section 9.4) is appropriate for testing goodness of fit to discrete distributions. It uses the Pearson X^2 statistic, the degrees of freedom depending on the number of parameters that are estimated from the data.

McNemar's test for off-diagonal symmetry in 2×2 tables may be tested using a statistic (Section 9.5) which has an asymptotic chi-squared distribution. This may be extended to **Bowker's test** for off-diagonal symmetry in $n \times n$ tables.

EXERCISES

9.1 Use (9.6) to verify that the probability of the configuration in Example 9.1 given by the Fisher exact test is $P=0.00005769$.

9.2 Using the modified form of Var(U) for the Jonckheere–Terpstra test given in (9.10) for the data in Examples 9.6 and 9.7, show that the standard deviation of U is 145.18, whereas it is 415.36 if ties are ignored.

9.3 Confirm the results quoted in Example 9.8 for association between blood platelet counts and spleen size.

9.4 Verify for the contingency table (9.11) that the asymptotic Jonckheere–Terpstra statistic Z takes the value zero. [Note it is not necessary to calculate Var(U) to confirm this.]

9.5 Confirm that the Goodman–Kruskal statistic G takes the value $G=1$ for each of the contingency tables on p. 234.

9.6 Determine conditions similar to those given for the case $G=1$ to ensure that the Goodman–Kruskal statistic takes the value $G=-1$.

9.7 Verify that $C=D$ and $G=0$ for the Goodman–Kruskal statistic for the table in (9.11).

9.8 In a psychological test for pilot applicants, each candidate is classed as extrovert or introvert and is subjected to a test for flying aptitude which he or she may pass or fail. Do the results suggest an association between aptitude and personality type?

	Introvert	Extrovert
Pass	14	34
Fail	31	41

9.9 A manufacturer of washing machines issues instructions for their use in English for the UK and US markets, French for the French market, German for the German market and Portuguese for the Portuguese and Brazilian markets. The manufacturer conducts a survey of randomly selected customers in each of these markets and asks them to classify the instructions (in the language appropriate to that country) as excellent, reasonable, or poor. Do the reponses set out below indicate the instructions are more acceptable in some countries than in others?

It is sensible to consider UK/USA separately – also Portugal/Brazil – because of differences in idiom that may affect understanding of the instructions: e.g. British houses have 'taps', while American houses have 'faucets'.

	Excellent	Reasonable	Poor
UK	42	30	28
USA	20	41	19
France	19	29	12
Germany	26	22	12
Portugal	18	31	21
Brazil	31	42	7

9.10 In Palpiteria all who visit a doctor must pay. A political party claims that poorer people are thus inhibited from seeking medical aid. Data in the table below are for a random sample of wage earners, and incomes are stated in palpiliras (P), the country's unit of currency. Do they substantiate the claim that the poor make proportionately less use of the services? In selecting an appropriate test take into account the fact that the row and column categories are both ordered.

	Income		
Time since last visit to doctor	Over 10 000P	5000–10 000P	Under 5000P
Under 6 months	17	24	42
6–12 months	15	32	45
Over 12 months	27	142	271
Never been	1	12	127

9.11 Would your conclusions in Exercise 9.10 have been different if the data had been in only two income groupings: (i) under 5000P and (ii) over 5000P?

9.12 Prior to an England *v*. Scotland football match 80 English, 75 Scottish and 45 Welsh supporters are asked who they think will win. Do the numbers responding each way indicate that the proportions expecting each side to win are influenced by nationality?

	English	Scottish	Welsh
English win	55	38	26
Scottish win	25	37	19

9.13 A machine part is regarded as satisfactory if it operates for 90 days without failure. If it fails in less than 90 days it is unsatisfactory and this results in a costly replacement operation. A supplier claims that for each part supplied the probability of a satisfactory life is 0.95. Each machine requires 4 of these parts and all must be functional for satisfactory operation. In a 90-day test run it may be assumed no part will fail more than once. To test the supplier's claim a buyer runs each of 100 machines initially fitted with 4 new parts for a 90-day test period. The numbers of parts (0–4) surviving 90 days are recorded for each machine as follows:

No. surviving	0	1	2	3	4
No, of machines	2	2	3	24	69

Do these results substantiate the supplier's claim?

9.14 It is claimed that a typesetter makes random errors at an average rate of 3 per 1000 words set, giving rise to a Poisson process. 100 randomly chosen sets of 1000 words from his output are examined and the mistakes in each counted. Are the results below consistent with the above claim?

No. of errors	0	1	2	3	4	5	6	7
No. of samples	6	11	26	33	12	6	4	2

9.15 Responses to highly emotive questions may be influenced by factors such as the age, race, sex and social background of the person asking the question. A random sample of 500 women aged between 30 and 40 are further divided into 5 groups of 100, and each group is allocated to one of the following interviewers:

A: A 25-year-old white female with secretarial qualifications
B: A middle-aged clergyman
C: A retired army colonel
D: A 30-year-old Pakistani lady
E: A non-white male university student

Each interviewer asks each of the 100 people allocated to him or her: 'Do you consider marriages between couples of different ethnic groups socially desirable?' The numbers answering 'yes' in each group are given below. Assess the evidence that response may be influenced by the type of person conducting the interview.

Interviewer	A	B	C	D	E
No. of 'yes'	32	41	18	57	36

9.16 To measure abrasive resistance of cloth, 100 samples of a fabric are each subjected to a 10-minute test under a series of 5 scourers, each of which may or may not produce a hole. The number of holes (0 to 5) is recorded for each sample. Are the data consistent with a binomial distribution with $n=5$ and p estimated from the data? (Hint: Determine the mean number of holes per sample. If this is x then an appropriate estimate of p is $x/5$.)

No. of holes	0	1	2	3	4	5
No. of samples	42	36	14	3	4	1

9.17 The McNemar test data in Table 4.4 on climbs can be reformulated in a way that makes the Cochran test given in Section 6.4.4 appropriate with $t=2$. Denoting a success by 1 and a failure by 0, we may classify each of the 108 climbers' outcomes for the first and second climb as either 0 or 1 in a 2×108 table. Show that the Cochran Q statistic is in this case identical with the McNemar X^2 statistic given in Section 9.5.

9.18 Aitchison and Heal (1987) give numbers of OECD countries using significant amounts of only 1, 2, 3 or 4 fuels in each of the years 1960, 1973, 1983. Are the proportions in the different categories of use changing significantly with time?
Think carefully about an appropriate test and the interpretation of your findings. The data are in no sense a random sample from any population of countries. The categories are ordered, but do you consider that the linear trends in association implicit in the conditions to validate, say, a Jonckheere–Terpstra test are likely to be relevant for these data?

	Year		
No of fuels	1960	1973	1983
1	7	10	1
2	13	11	13
3	5	4	9
4	0	0	2

9.19 Marascuilo and Serlin (1979) report a survey in which a number of women were asked whether they considered the statement 'The most important qualities of a husband are determination and ambition' to be true or false. The respondents were asked the same question at a later date. Numbers making the possible responses were as follows:

First response	Second response	Numbers
True	True	523
True	False	345
False	True	230
False	False	554

Is there evidence that experience significantly alters attitudes of women towards the truth of the statement?

9.20 Jarrett (1979) gives the following data for numbers of coal mine disasters involving 10 or more deaths between 1851 and 1962.

Day of week	Sun	Mon	Tue	Wed	Thu	Fri	Sat
Number	5	19	34	33	36	35	29

Month	Jan	Feb	Mar	Apr	May	Jun	Jul	Aug
Number	14	20	20	13	14	10	18	15
Month	Sep	Oct	Nov	Dec				
Number	11	16	16	24				

Test whether accidents appear to be uniformly spread over days of the week and over months of the year. What are the implications? Do they surprise you?

9.21 Dansie (1986) gives data for a survey in which 800 people were asked to rank 4 makes of car A, B, C, D in order of preference. The number of times each rank combination was specified is given below in brackets after each order. Do the data indicate that preference may be entirely random? Is there a significant preference for any car as first choice?
ABCD(41), ABDC(44), ACBD(37), ACDB(36), ADBC(49), ADCB(41), BACD(38), BADC(38), BCAD(25), BCDA(22), BDAC(33), BDCA(25), CABD(31), CADB(26), CBAD(40), CBDA(33), CDAB(33), CDBA(35), DABC(23), DACB(39), DBAC(30), DBCA(21), DCAB(26), DCBA(34).

9.22 Noether (1987b) asked students to select by a mental process what they regarded as random pairs from the digits 1, 2, 3, repeating that process four times. Noether recorded frequency of occurrences of the last digit pair written down by each of 450 students. The results were:

		First digit		
		1	2	3
	1	31	72	60
Second digit	2	57	27	63
	3	53	58	29

What would be the expected numbers in each cell of the above table if digits were truly random? Test whether one should reject the hypothesis that the students are choosing digits at random. How do you interpret your finding?

9.23 For the following 2×12 table calculate the Pearson X^2 and the likelihood ratio G^2 statistics. Explain the basic cause of any difference between their values. If suitable computer software is available, determine the exact tail probabilities corresponding to each test statistic.

0	0	0	0	0	0	0	0	0	0	1	1
3	4	17	2	5	1	8	6	4	11	3	0

10

Association in categorical data

10.1 THE ANALYSIS OF ASSOCIATION

If we reject the hypothesis of no association in a contingency table, analysis of the nature of the association may be important. In Section 9.3.2 we considered one aspect of association in tables with ordinal row and column categories, recommending the Jonckheere–Terpstra test for monotonic association. In this chapter we consider how to assess several aspects of association. The subject is a vast one and our treatment is indicative and not comprehensive. References are given to more detailed accounts. Approaches used to assess association range from parametric modelling to distribution-free methods. Parametric approaches depend specifically on the mechanism generating the counts. One such parametric model is the logistic regression model for binary (e.g. success or failure) reponses. General descriptions of this model are given by Agresti (1990, Chapter 4) and by Everitt (1992, Chapter 6) and a detailed treatment covering many applications is presented in Cox and Snell (1989). A simple situation where the logistic model is particularly appropriate is that of an $r \times 2$ table where the rows correspond to r explanatory categories and the columns represent binary responses with binomial distributions. The model being parametric, we do not consider it further in this book.

In Chapter 9 we considered two-way $r \times c$ tables. Three-way tables are common; they usually require detailed, often subtle, analysis to elucidate patterns of association. Techniques involved range from extensions of some of the nonparametric methods outlined for a few special cases in this chapter to analyses based on **generalized linear models.** A detailed account of the latter will be found in McCullagh and Nelder (1989) and a more elementary treatment with practical applications is given by Dobson (1990).

Table 10.1 shows a common way of presenting data from a three-way classification using two-way cross-sectional tables. The counts in Table 10.1 cover 197 patients in a study of coronary heart disease where the first categorizatation is into presence (CHD) or absence (no CHD) of coronary heart disease. The second classification is at three levels (A, B, C) of cholesterol and the third classification is at five blood pressure levels. We show two cross-sections of the second and third classifications, one corresponding to each of the two levels of the first classification. Marginal totals relevant to each cross-section and a set of combined totals are given. The data may be presented in other ways (Exercise 10.2).

Table 10.1 Categorization of 197 patients in a study of coronary heart disease

		Blood pressure group					Total
		I	II	III	IV	V	
	Cholesterol						
	A	2	3	1	0	4	10
CHD	B	2	1	5	3	0	11
	C	4	7	8	6	2	27
	Total (1)	8	11	14	9	6	48
	A	16	14	11	8	6	55
No CHD	B	22	18	5	3	2	50
	C	15	13	10	5	1	44
	Total (2)	53	45	26	16	9	149
	Total (1+2)	61	56	40	25	15	197

Tests for complete independence extend from two- to multi-way tables, and tests of independence may be applied to two-way cross-sectional tables, but such analyses are usually insufficient (and often inefficient) because we are more interested in exploring the nature of association.

A particular type of three-way classification that has received considerable attention is the $k \times 2 \times 2$ table; the most appropriate method of analysis depends on the questions of interest and on the nature of the categories in each classification, e.g. whether they are explanatory or response categories, and whether they are nominal or ordered. It is often appropriate to consider the problem as one involving k tables, each 2×2. For instance, if two drugs are being compared and we are interested in the binary response *side-effects* or *no side-effects* we may suspect that the responses are affected by age and obtain data for patients in each of $k=5$ age groups, e.g. 20–29, 30–39, 40–49, 50–59, over 60. We may find evidence of association between drugs and side-effects in each age group, and a natural question to ask is whether there is evidence of a change in the level of this association between age groups, or we may want some measure of the levels of association. For 2×2 tables an approriate measure of association is the odds ratio, or (better in many circumstances) the logarithm of the odds ratio. A classic example of data of this type which is essentially a $2 \times 2 \times 2$ table was first discussed by Bartlett (1935). The data are survival numbers of plant (plum rootstock) cuttings in each of four batches of 240 cuttings, and the numbers surviving in each batch are given in Table 10.2.

Table 10.2 Survival numbers for early and late planted short and long cuttings

	Planted early		*Planted late*	
Length of cutting	Alive	Dead	Alive	Dead
Long	156	84	84	156
Short	107	133	31	209

The first two data columns and last two data columns in Table 10.2 represent the $k=2$ cross-sectional tables corresponding to the two planting times. We might equally well have formed cross-sectional tables each corresponding to different lengths of cuttings, as in Table 10.3.

Table 10.3 Survival numbers for short and long cuttings planted early and late

	Long cuttings		*Short cuttings*	
Planting time	Alive	Dead	Alive	Dead
Early	156	84	107	133
Late	84	156	31	209

In Tables 10.2 and 10.3 planting times and lengths of cutting are explanatory categories, and survival and non-survival are responses. We discuss methods of analysing these data for association in Sections 10.2.1 and 10.3, but first we turn our attention to partitioning the test statistic G^2 introduced in Section 9.2.3 to examine association in $r \times c$ tables.

10.1.1 Partitioning of G^2 in studies of association

We noted in Sections 9.2.2 and 9.2.3 that for an $r \times c$ table both the Pearson X^2 and the likelihood ratio G^2 statistic have asymptotically a χ^2 distribution with $(r-1)(c-1)$ degrees of freedom. It is well known that a variate having a χ^2 distribution with υ degrees of freedom may be expressed as the sum of υ independent variates each having a χ^2 distribution with 1 degree of freedom. The G^2 statistic may be partitioned similarly into components each of which has asymptotically a χ^2 distribution with 1 degree of freedom. Rules for partitioning are complicated for the general $r \times c$ case and usually there is more than one possible partitioning. Necessary conditions for the single-degree-of-freedom components to be additive are given by Agresti (1990, p. 53). We do not consider the general partitioning problem, but in Example 10.1 we illustrate a possible partitioning and discuss interpretation of results for a straightforward example using broadly realistic data, simplified only to make the arithmetic easy for demonstration purposes.

Example 10.1

The problem. Two drugs used in chemotherapy are tested on 100 patients and numbers treated with each drug who exhibit the presence or absence of two specific side-effects, hair loss (HL) and visual impairment (VI) are as follows:

	Side-effect status			
	HL	VI	HL + VI	None
Drug A	9	4	16	31
Drug B	3	16	2	19

Partition G^2 to examine association between single side-effects, between single and double side-effects and between overall side-effect status for the two drugs.

Formulation and assumptions. We assume (and verify numerically) that a suitable partitioning of the overall G^2 with 3 degrees of freedom allows us to compare columns 1 and 2, then columns 1+2 with column 3, then columns 1+2+3 with column 4, these components representing appropriate tests for association (i) between single side-effects for each drug, (ii) between a single-side effect (either H or VI) and a double side-effect (both H and VI), and (iii) some side effects and no side effects.

Procedure. We calculate expected frequencies under the hypothesis of independence for each count in the usual way using marginal totals, e.g. $m_{12}=(20\times60)/100=12$. If using a pocket calculator it is convenient to record the data together with marginal totals and expected frequencies in brackets beside each observed cell entry in the following manner:

	H	VI	H and VI	None	Total
Drug A	9 (7.2)	4 (12)	16 (10.8)	31 (30)	60
Drug B	3 (4.8)	16 (8)	2 (7.2)	19 (20)	40
Total	12	20	18	50	100

Calculating $G^2=\Sigma n_{ij}\log(n_{ij}/m_{ij})$ with 3 degrees of freedom, remembering 'log' is the natural logarithm, gives $G^2=22.13$ (Exercise 10.1), indicating significance at a nominal 0.1% level. Consider now columns 1 and 2 only. The expected frequencies calculated afresh for these two columns are:

	H	VI	Total
Drug A	9 (4.875)	4 (8.125)	13
Drug B	3 (7.125)	16 (11.875)	19
Total	12	20	32

whence we find (Exercise 10.1) G_1^2=9.72 with 1 degree of freedom, a value significant at a nominal 0.1% level.

Combining columns 1 and 2 and and comparing with column 3, we get the following table of counts and expected values:

	H or VI	H and VI	Total
Drug A	13 (18.56)	16 (10.44)	29
Drug B	19 (13.44)	2 (7.56)	21
Total	32	18	50

whence (Exercise 10.1) G_2^2=12.24 with 1 degree of freedom, a value significant at the 0.1% level.

Combining columns 1, 2 and 3 and comparing with column 4 we get:

	Side effects	None	Total
Drug A	29 (30)	31 (30)	60
Drug B	21 (20)	19 (20)	40
Total	50	50	100

whence (Exercise 10.1) G_3^2=0.17 with 1 degree of freedom, a result that is clearly not significant. We easily verify that

$$\Sigma G_i^2=9.72+12.24+0.17=22.13=G^2$$

Conclusion. The highly significant value for G^2 indicates association between row and column classifications. The high value of G_1^2 arises because, clearly, when there is only one side-effect it is more likely to be hair loss with drug A and visual impairment with drug B. On the other hand, the high value of G_2^2 arises because drug A is much more likely than drug B to give rise to both hair loss and visual impairment, if indeed there is any side-effect. Non-significance of G_3 implies that the drugs are equally likely to give rise to some side-effects.

Comments. 1. What action should follow from these findings depends on the importance of each side-effect. If visual impairment were slight and temporary it might be rated less serious than hair loss and this would indicate a clear preference for drug B. If visual impairment were severe and permanent it would be more serious than hair loss, but again two side-effects may be regarded as more serious than only one.

2. The above partitioning is not unique; we might reorder the columns, e.g. place column 4 first and combine the new columns as above. The order we used leads to more logical interpretations of the components of G^2.

3. We have used a particular case of a rule that always gives a partitioning of G^2 into $c-1$ additive components in a $2\times c$ table. The rule is to test first in columns 1 and 2, next in columns 1+2 and column 3, next in columns 1+2+3 and column 4, and so on, the final component involving comparison of columns $1+2+3+\ldots+(c-1)$ and column c.

4. Although the Pearson X^2 statistic and G^2 have the same asymptotic distribution, if X^2 is partitioned as above the component values will not in general sum to X^2.

10.2 SOME MODELS FOR CONTINGENCY TABLES

10.2.1 The log-linear model

In any $r\times c$ table with independence the expected frequency in cell (i, j) is $m_{ij}=n_{i+}n_{+j}/N$, where n_{i+}, n_{+j} are the ith row and jth column totals and N is the total number of counts in the table. It follows that

$$\log m_{ij}=\log n_{i+}+\log n_{+j}-\log N \qquad (10.1)$$

i.e. under independence, the logarithms of the expectations are linear functions of the logarithms of the row, column and grand totals.

Readers familiar with the normal theory linear model for a randomized block design, basic to the analysis of variance, will recognize (10.1) as an analogue of the additive linear model expressing a response (e.g. an expected yield) as the sum of an overall experimental mean plus a block effect plus a treatment effect. With that model the observed yield for any unit differs from the expected yield by an additive amount usually regarded as a random error or departure from expectation. By analogy, even when there is no association (independence) between row and column classifications in an $r\times c$ contingency table, in general m_{ij} will not equal the observed count n_{ij}. For the model based on (10.1) we may look upon the difference $\log n_{ij}-\log m_{ij}$ as an additive 'error' or 'departure' term when the model of no association is adequate. If these 'departure' terms are too large to be ascribed to sampling variation, or have some clear pattern, this is evidence against the hypothesis of no association.

Association in contingency tables has analogies with interaction in factorial treatment structures in the analysis of variance and these analogies are exhibited in extensions of the **log-linear** model (10.1) that allow for association. A detailed treatment is beyond the scope of this text but we illustrate some basic ideas for 2×2 and $2\times 2\times 2$ contingency tables and indicate some extensions to simple cases involving $r\times c$ tables. The reader unfamiliar with the analysis of designed experiments with factorial treatment structures may find this section difficult; if so, it may be wise at first reading to skim through it briefly to grasp basic ideas rather than to try to understand detail.

We digress to comment on fundamentals concerning additive effects and interactions in a simple 2×2 factorial treatment structure where observations are measured variables assumed to have some continuous distribution. The simplest factorial treatment arrangement is the 2×2 structure with two factors each at two levels. For instance, we may allow a chemical reaction to proceed for two periods (first factor), e.g. two hours or three hours, and at each of two temperatures (second factor), e.g. 75°C or 80°C. The 'response' is the amount of some chemical produced by a given amount of input material using each factor combination. Such an experiment is repeated (or replicated) and the average (or total) output of the same number of replicates for each factor combination is recorded. Apart from random variation in the form of an additive error, which in the analysis of variance model is assumed to be distributed normally with mean zero and variance σ^2, the output x_{ij} for the first factor at level i and the second at level j $(i, j = 1, 2)$ is specified by a linear additive model. If the effects of each factor are purely additive we speak of a **no-interaction** model.

Ignoring for the moment random variation, suppose for instance, we found that operating for two hours at 75°C gave an output X and that this was increased to $X+3$ if we left time unaltered but increased temperature to 80°C and to $X+8$ if we left the temperature at 75°C but increased the time to three hours. We say we have a no-interaction model if the result of increasing both temperature from 75 to 80°C **and** time from 2 to 3 hours is to increase yield from X to $X+3+8$. These results for **expected** yield are summarized in Table 10.4.

Table 10.4 Expected yields in a no-interaction model

		Temperature (°C) 75	80
Process time (hr)	2	X	$X+3$
	3	$X+8$	$X+3+8$

More generally, for the combination of level i of the row factor and level j of the column factor, we denote the **actual** yields by y_{ij} and **expected** yield (value of output apart from random variation) by x_{ij} . In Table 10.4, $x_{11}=X$, $x_{12}=X+3$, $x_{21}=X+8$, $x_{22}=X+3+8=X+11$; thus $x_{11}+x_{22}=2X+11=x_{12}+x_{21}$, i.e. the diagonal or cross-sums in Table 10.4 are equal. This is a characteristic of a no-interaction model. In a real-life situation we may find the effect of increasing both time and temperature is to boost or diminish the effects of changing only one factor. This would occur if, for example, x_{22} in Table 10.4 had the value $X+17$ or $X+2$ rather than $X+11$. The cross-sums are then no longer equal. For the 2×2 factorial model with no interaction the key requirement is

$$x_{11}+x_{22}-x_{12}-x_{21}=0 \qquad (10.2)$$

This equality will not hold for the observed mean yields y_{ij}, or corresponding total yields, because of random variation. The analysis of variance test for no interaction in this model is essentially one of whether the observed value $y_{11}+y_{22}-y_{12}+y_{21}$ differs significantly from zero. Generally speaking, an interaction implies accepting a hypothesis

$$H_1 : x_{11}+x_{22}-x_{12}-x_{21}=I, \text{ where } I\neq 0 \tag{10.3}$$

In Section 9.1.1 we showed that for a 2×2 contingency table a condition for no association (independence) between row and column categories was that the odds ratio $\theta=(m_{11}m_{22})/(m_{12}m_{21})=1$, or equivalently $m_{11}m_{22}=m_{12}m_{21}$. Taking logarithms, we get the analogue of (10.2) for the log-linear model, that is

$$\log m_{11}+\log m_{22}-\log m_{12}-\log m_{21}=0 \tag{10.4}$$

as the criterion for no association.

For the observed counts n_{ij} the empirical odds ratio $(n_{11}n_{22})/(n_{12}n_{21})$ does not in general equal 1. A test of association (or interaction) in a contingency table is one to determine whether this ratio differs significantly from 1, or equivalently, whether $\log n_{11}+\log n_{22}-\log n_{12}-\log n_{21}$ differs significantly from zero, i.e. whether to reject $H_0 : I=0$ and accept

$$H_1 : \log m_{11}+\log m_{22}-\log m_{12}-\log m_{21}=I, \text{ where } I\neq 0, \tag{10.5}$$

and (10.5) is an analogue of (10.3).

The models extend to 2×2×2 factorial experiments and 2×2×2 contingency tables. First we outline the extension of the linear model to three factors at each of two levels in an analysis of variance context. The interaction measure introduced in (10.3) specifies a **first-order** (or a **two-factor**) interaction. This is the only kind of interaction in a 2×2 factorial treatment structure.

With 3 factors, each at two levels, we represent expected yields at level i of the first factor, level j of the second factor and level k of the third factor by x_{ijk}, where i, j, k take the values 1 or 2. The observed yields, y_{ijk}, differ from these by additive random errors or departures. If we consider the first two factors at level $k=1$ of the third factor we have a first-order interaction between factors 1 and 2 at this level of factor 3 if $x_{111}+x_{221}-x_{121}-x_{211}=I$, $I\neq 0$. A first-order interaction between factors 1 and 2 at level $k=2$ of factor 3 implies $x_{112}+x_{222}-x_{122}-x_{212}=J$, $J\neq 0$. If $I=J$ there is no second-order interaction between the three factors. If $I\neq J$ there is a **second-order** or **three-factor** interaction. If $I=J=0$ we have no interaction. The observed y_{ijk} will in general not satisfy the equalities $I=J$ or $I=J=0$ and the appropriate tests are essentially tests of whether departures from the relevant expectation equalities are significant.

In the context of the log-linear model for 2×2×2 contingency tables we replace the x_{ijk} in the above conditions by $\log m_{ijk}$ where m_{ijk} are the relevant expected counts based on marginal totals in a way described below in Example 10.2. For 2×2×2 tables the no association (no interaction) model corresponds to that of independence between classifications, and in terms of odds ratios may be written

$$(m_{111}m_{221})/(m_{121}m_{211})=(m_{112}m_{222})/(m_{122}m_{212})=1$$

Dependence or association may be first- or second-order (first- or second-order interaction in the log-linear sense). For the first-order interaction model

$$(m_{111}m_{221})/(m_{121}m_{211})=(m_{112}m_{222})/(m_{122}m_{212})=k,\ k\neq 1$$

For the second-order interaction model

$$(m_{111}m_{221})/(m_{121}m_{211})\neq(m_{112}m_{222})/(m_{122}m_{212})$$

Using the log-linear model we may test whether data like those for the plant cuttings given in Section 9.1 can be adequately described by a particular model. It is easily verified that the data do not satisfy an independence model by applying a Pearson X^2 or a G^2 test to either of the 2×2 tables corresponding to long or short cuttings, respectively, in Table 10.3 (see Exercise 10.3). We may ask whether a first-order interaction model fits the data. We must first estimate the m_{ijk}. We must then select test criteria. The reader should consult a specialist text such as Bishop, Fienberg and Holland (1975), Fienberg (1980), Plackett (1981), Agresti (1984; 1990) or Everitt (1992) for details of more general situations.

For our illustrative example we summarize the steps for testing whether a first-order interaction model is relevant. For a 2×2×2 contingency table under the hypothesis of first-order interaction, the maximum likelihood esitmators of m_{ijk}, which we denote by $\hat{m}_{ijk}$, satisfy the condition

$$(\hat{m}_{111}\hat{m}_{221})/(\hat{m}_{121}\hat{m}_{211})=(\hat{m}_{112}\hat{m}_{222})/(\hat{m}_{122}\hat{m}_{212}) \qquad (10.6)$$

with the constraints that they sum to the observed marginal totals over any suffix. In general log-linear estimation, the maximum likelihood estimators are obtained by iterative methods (one such method is called the **iterative scaling procedure**); however, for a 2×2×2 table with a first-order interaction model direct calculation is possible. Once the $\hat{m}_{ijk}$ have been calculated the statistics X^2 or G^2 are used for significance tests. Under H_0 each has 1 degree of freedom in this case. In more general contexts where the test statistic has more than 1 degree of freedom, G^2 is preferred because it may be partitioned into single-degree-of-freedom components in much the way this is done for orthogonal sums of squares in the analysis of variance.

Example 10.2

The problem. Determine whether the data in Table 10.3 on survival of long and short cuttings are consistent with a first-order interaction model.

Formulation and assumptions. We postulate a first-order interaction model. We specify the first classification as early or late, the second as alive or dead and the third as long or short. We calculate expected values to give maximum likelihood estimates to satisfy (10.6) and base our test on the G^2 statistic.

Procedure. Since expectations must all add to the relevant marginal totals we can express all $\hat{m}_{ijk}$ in terms of $\hat{m}_{111}$. It is easily verified by calculating the relevant marginals in the following layout (from Table 10.3)

	Long cutting			*Short cutting*	
Planted	Alive	Dead		Alive	Dead
Early	156	84	Early	107	133
Late	84	156	Late	31	209

that if we denote $\hat{m}_{111}$ by x then $\hat{m}_{121}=\hat{m}_{211}=240-x$, $\hat{m}_{221}=x$, $\hat{m}_{112}=263-x$, $\hat{m}_{122}=x-23$, $\hat{m}_{212}=x-125$, $\hat{m}_{222}=365-x$, whence (10.6) implies

$$\frac{x^2}{(240-x)^2}=\frac{(263-x)(365-x)}{(x-23)(x-125)}$$

This cubic equation in x may be solved numerically with an appropriate computer algorithm. The relevant solution is $x=\hat{m}_{111}=161.1$. The remaining expectations may be calculated using the relations given above and are summarized as follows:

	Long cutting			*Short cutting*	
Planted	Alive	Dead		Alive	Dead
Early	161.1	78.9	Early	101.9	138.1
Late	78.9	161.1	Late	36.1	203.9

Calculating G^2 (where summation is over all 8 cells) we find (Exercise 10.4) that $G^2=2.30$, clearly less than the value 3.84 required for significance at the 5% level with 1 degree of freedom for the asymptotic χ^2 distribution.

Conclusion. Since G^2 is below the value required for significance we accept that the data are consistent with a first-order interaction model.

Comments. 1. Acceptance of a first-order interaction model for a 2×2×2 table is equivalent to accepting the hypothesis that the odds ratios for the cross-sectional tables we have considered do not differ significantly.

2. We give an alternative analysis for these data in Example 10.6.

Computational aspects. SPSS and other standard packages have programs covering many aspects of log-linear models including fitting by maximum likelihood.

In the general $r \times c$ table, if there is association an extra term is needed for each $\log m_{ij}$ given in (10.1). If a separate additive term is included in the model for each cell this may lead to what is called a saturated model, since by suitable choice of these terms the expected count in each cell may be made equal to the observed count. The situation with a saturated model is analogous to that in the analysis of variance of randomized block designs where, with only one replicate of each treament in each block, separate assessment of *block*×*treatment* interaction is not possible. In the standard analysis the *block*×*treatment* mean square is used as the error mean square, assuming there is no interaction between blocks and treatments.

We are often interested in some compromise between the independent model and the saturated model for interaction. One such model is described in Section 10.2.2.

In practice, constraints must be placed on parameters in a log-linear model to ensure uniqueness. These are analogues of constraints used in the linear models for the analysis of variance, e.g. the sums of block and treatment parameters are both zero (although other constraints are possible).

Important theoretical points have been glossed over in this short introduction to log-linear models. The models are applicable to many problems where independence clearly does not hold. A classic example on fingerprint patterns is discussed by, among others, Plackett (1981) and also in the entry on contingency tables in Kotz and Johnson (1983, vol. 3, pp. 161–71). A log-linear model proves illuminating for these data. A comprehensive account of log-linear models is given by Agresti (1990, Chapters 5–9) and applications are discussed by Everitt (1992, Chapters 5 and 7); McCullagh and Nelder (1989, Chapter 6) and Dobson (1990, Chapter 9) deal with log-linear models in the context of generalized linear models.

10.2.2 The linear-by-linear association model

The linear-by-linear association model is essentially a log-linear model in which departures from independence can be modelled with the use of only one additional parameter β in the additive 'interaction' terms. The model may be written

$$\log m_{ij} = \text{independent model parameters} + \beta(u_i - m_u)(v_j - m_v) \quad (10.7)$$

where $u_1 < u_2 < \ldots < u_r$ and $v_1 < v_2 < \ldots < v_c$ are preassigned row and column **scores** with means m_u, m_v respectively. One reason why this model is important is that a number of standard nonparamteric test statistics have exact permutation distributions based on it. The model is relevant to $r \times c$

tables of counts in ordered categories, when, if classifications are not independent, they show a monotonic trend across row and column categories. In Chapter 7 we considered the use of transformation to ranks to produce linearity in situations where we had such monotonicity, or a trend that suggested an underlying monotonicity. The linear-by-linear association model extends these ideas.

The parameter β may be estimated to reflect the extent of any relevant departure from independence. We do not discuss estimation of β, but indicate what we mean by 'relevant' departures in Example 10.3. There cells (i, j) for which $|\beta(u_i-m_u)(v_i-m_v)|$ is large show the greatest departures from independence. Our interpretation of such association depends in part on our choice of u_i and v_j. The simplest choice is $u_i=i$, $v_j=j$, which assigns the ranks of the ordered row and column categories as scores.

Example 10.3

The problem, For the case $r=5$, $c=7$, using scores $u_i=i$, $v_j=j$, determine for each cell in the $r\times c$ table the contribution to $\log m_{ij}$ of the interaction term $\beta(u_i-m_u)(v_j-m_v)$.

Formulation and assumptions. In this example $m_u=3$ and $m_v=4$ and the interaction term in (10.7) for cell (i, j) is $\beta(i-3)(j-4)$.

Procedure. For any β relevant interaction terms in each cell are given in Table 10.5.

Table 10.5 Interaction term in $\log m_{ij}$ using rank scores in a 5×7 contingency table

6β	4β	2β	0	-2β	-4β	-6β
3β	2β	β	0	$-\beta$	-2β	-3β
0	0	0	0	0	0	0
-3β	-2β	$-\beta$	0	β	2β	3β
-6β	-4β	-2β	0	2β	4β	6β

Conclusion. If $\beta>0$, $\log m_{ij}$ is greater than its value under independence for cells at the top left and bottom right of Table 10.5 and less for cells at the top right or bottom left. If $\beta<0$ this situation is reversed. If $\beta=0$ there is no association.

Comments. For more general scores u_i, v_j and for other values of r, c there will be broadly similar patterns for linear-by-linear association.

The term linear-by-linear association reflects the characteristic that within any row, say i, for any chosen u_i, v_j the interaction contribution across columns is a linear function of the v_j with slope $\beta(u_i - m_u)$ and that within any column j the interaction contribution across rows is a linear function of u_i with slope $\beta(v_j - m_v)$.

The relevance of this model to permutation distribution theory is that many permutation tests depend upon a statistic of the form

$$S = \Sigma_{i,j}\, u_i v_j n_{ij} \tag{10.8}$$

with appropriate choices of the u_i, v_j. Independence ($\beta=0$) corresponds to the null hypothesis; we reject H_0 if our observed outcome expressed in the appropriate contingency table format indicates a general pattern of the form exemplified for a particular case in Table 10.5 when $\beta \neq 0$. This corresponds to extreme values of S given by (10.8), whereas intermediate values indicate lack of association.

The Spearman rank correlation coefficient provides an illustration. If there are no ties and n observations the rank outcome can be put in an $n \times n$ contingency table with rows and columns representing x-ranks and y-ranks respectively. If the x-rank i is paired with y-rank j we enter 1 in cell (i, j). All other entries in that row and column are zero. The row and column totals are all 1. We illustrate the pattern for 3 cases when $n=5$.

Case A	x-ranks	1	2	3	4	5
	y-ranks	4	2	5	3	1
Case B	x-ranks	1	2	3	4	5
	y-ranks	2	1	3	5	4
Case C	x-ranks	1	2	3	4	5
	y-ranks	1	2	3	4	5

These are represented by the 5×5 tables (10.9), (10.10), (10.11).

Case A

y-ranks	1	2	3	4	5
x-ranks					
1	0	0	0	1	0
2	0	1	0	0	0
3	0	0	0	0	1
4	0	0	1	0	0
5	1	0	0	0	0

(10.9)

Case B

y-ranks	1	2	3	4	5
x-ranks					
1	0	1	0	0	0
2	1	0	0	0	0
3	0	0	1	0	0
4	0	0	0	0	1
5	0	0	0	1	0

(10.10)

Case C

y-ranks	1	2	3	4	5
x-ranks					
1	1	0	0	0	0
2	0	1	0	0	0
3	0	0	1	0	0
4	0	0	0	1	0
5	0	0	0	0	1

(10.11)

In case A the 1s are well scattered over rows and columns. In case B there is a discernable tendency for the 1s to drift towards top left and bottom right, hinting at some association between x- and y-ranks. Case C clearly represents the closest possible positive association and corresponds to $r_s=1$. The resemblance of these tables to scatter diagrams (with the conventional axes interchanged and the direction of x reversed) is apparent if we imagine the zeros to be deleted and the 1s replaced by dots..

Determination of the exact permutation distribution of the Spearman coefficient requires computation of r_s (or some strictly monotonic function thereof) for all possible arrangements of tables like (10.9) to (10.11) with all marginal totals unity. It is easy to verify that when $n=5$ there are 5! distinct tables. It is not difficult to show that, if r_i, s_i are observed x- and y-ranks, $\Sigma r_i s_i$ may be used as the test statistic in an exact permutation distribution test. This follows from (7.7) since, under permutation of the r_i, s_i all terms in (7.7) except $\Sigma r_i s_i$ remain constant. It immediately follows that if we take the ordered x-ranks and y-ranks as our scores in contingency tables of the form (10.9) to (10.11) then S given by (10.8) is equivalent to $\Sigma r_i s_i$ in (7.7).

For n paired observations (x_i, y_i) and no ties the permutation distribution of the Pearson coefficient (Section 7.1.6) is equivalent to that of $\Sigma x_i y_i$. This is equivalent to (10.8) with scores $u_i=x_{(i)}$, $v_i=y_{(i)}$. Permutation is again over all $n \times n$ tables with row and column sums all unity.

The models extend easily to tied situations. For example, using mid-ranks for the Spearman coefficient with ties, consider the case:

x-ranks	1.5	1.5	3	4	5
y-ranks	1	4	4	2	4

The relevant contingency table with mid-rank scores is now:

y-rank / *x-rank*	1	2	4	Total
1.5	1	0	1	2
3	0	0	1	1
4	0	1	0	1
5	0	0	1	1
Totals	1	1	3	5

To generate the relevant permutation distribution we calculate S using mid-rank scores for all permutations of this table with the given marginal totals. We could equally well use centralized scores of the form $u_i - m_u$, $v_j - m_v$.

STATXACT includes a program giving relevant probabilities under permutation of such tables subject to fixed marginal totals for arbitrarily assigned scores. This is a powerful tool, but one must choose sensible scores. We illustrate such choices in a reanalysis of the data in Example 9.6. The tests are essentially applications of the permutation test for a Pearson product moment correlation coefficient with ties and scores (x, y values) chosen to reflect what may be regarded as appropriate measures of distances between categories.

Example 10.4

The problem. Given the data on side-effects of drugs in Example 9.6, viz.

		Side effects		
Dose	None	Slight	Moderate	Severe
100 mg	50	0	1	0
200 mg	60	1	0	0
300 mg	40	1	1	0
400 mg	30	1	1	2

allocate appropriate scores and perform a linear-by-linear association test for evidence of association between dose and incidence of side-effects.

Formulation and assumptions. One reasonable choice of score would be row and column ranks. For rows this corresponds to dose levels in 100 mg units. For columns it represents an ordering of side-effects consistent with limited information. Our second choice amends column scores in a way a clinician might interpret the data. He may accept 1 and 2 as reasonable scores for no and slight side-effects but regard moderate side effects as ten times as serious as slight ones and severe side-effects as 100 times as serious as slight ones, giving logical column scores $v_1=1$, $v_2=2$, $v_3=20$, $v_4=200$. The STATXACT linear-by-linear test program will perform the relevant exact test using either scoring system. If no suitable program is available we might use the asymptotic result given below.

Procedure. We may use the STATXACT or an equivalent linear-by-linear association program with each choice of scores. For rank scores in rows and columns the exact one-tail probability is 0.0112, and doubling this gives the two-tail probability 0.0224. With the alternative column scores 1, 2, 20, 200 the corresponding tail probabilities are 0.0071 and 0.0142.

Conclusion. A one-tail test is justified as it is logical for side-effects to increase (rather than decrease) with increasing concentration of a drug. Thus for the first model we reject independence at an exact 1.12% significance level and with the second model at an exact 0.71% significance level.

Comments. In Example 9.7 using the Jonckheere–Terpstra test we rejected independence at the 1.68% significance level. As pointed out in Section 9.3.2, the Jonckheere–Terpstra test is equivalent to a test using Kendall's tau. Using ranks for row and column scores is superficially like a test using Spearman's coefficient but we make no allowance for ties. Other row and column scores give a test using the Pearson coefficient with these scores playing the role of x, y values.

An asymptotic test that allows for ties will be adequate in many cases. S given by (10.8) has mean

$$\mathrm{E}(S)=[(\Sigma_i u_i n_{i+})(\Sigma_j v_j n_{+j})]/N$$

and variance

$$\mathrm{Var}(S)=[\Sigma_i u_i^2 n_{i+}-(\Sigma_i u_i n_{i+})^2)/N][\Sigma_j v_j^2 n_{+j}-(\Sigma_j v_j n_{+j})^2)/N]/(N-1)$$

Asymptotically $Z=[S-\mathrm{E}(S)]/\surd[\mathrm{Var}(S)]$ has a standard normal distribution.

Example 10.5

The problem. Apply the asymptotic test to the data in Example 10.4 using rank scores for rows and columns.

Formulation and assumptions. E(S) and Var(S) are computed by the formulae given above and the value of Z is calculated.

Procedure. The rank scores are $u_i=i$, $v_j=j$ and we easily verify that

$$n_{1+}=51,\ n_{2+}=61,\ n_{3+}=42,\ n_{4+}=34$$

$$n_{+1}=180,\ n_{+2}=3,\ n_{+3}=3,\ n_{+4}=2,\ N=188.$$

For the given data simple calculations give $S=484$, $\mathrm{E}(S)=469.7$ and $\mathrm{Var}(S)=35.797$, whence $Z=2.39$, and $\Pr(Z>2.39)=0.0084$.

Conclusion. We reject the hypothesis of independence at the 0.84% siginificance level.

Comments. This result is broadly in line with that for the exact test. In Exercise 10.5 we establish that with the second choice of scores in Example 10.4 the asymptotic significance level is 1.01%. Despite many ties in this example asymptotic results are not seriously misleading.

Many other tests may be formulated as a linear-by-linear association test. Indeed, the Wilcoxon rank sum test is such a test applied to a $2\times c$ table with each row corresponding to a sample and columns ordered by the c observed values. Here the row scores are respectively 0, 1 and the column scores are the ranks (mid-ranks in the case of ties) of the observations. In this test the row categories need not be ordinal because we may calculate the relevant statistic using the ranks in either sample. Replacing the column scores by van der Waerden scores or by original data values will give a normal scores test or the raw scores permutation test. Gehan scores for censored data may also be used or indeed any scores that can be justified logically.

10.3 ANALYSIS OF k 2×2 TABLES

Binomial sampling with two levels of an explanatory variable gives rise to a 2×2 table. Typical situations include comparison of two drugs in which we record for each the numbers of cures or nil responses, or numbers showing side effects or no side-effects, etc.; in a production process two competing types of machine may be compared and in runs from each the numbers of satisfactory items and of flawed items requiring modification might be recorded. Interest centres on the degree of association between explanatory variables and responses. We indicated in Section 10.2.1 that for a 2×2 table a first-order interaction is the only type of association and that this is measured by the odds ratio θ. We have noted in several places that, conditional upon fixed marginal totals, once we know n_{11} the remaining cell counts n_{12}, n_{21}, n_{22} are fixed. It follows that, given the marginal totals and n_{11}, this determines the observed odds ratio. We make free use of the association between n_{11} and the observed odds ratio in many of the test and estimation procedures developed in this section.

In situations of the type described in the preceding paragraph we may introduce further explanatory variables, sometimes called **covariates.** For instance, in comparing binary responses to two drugs it might be suspected that levels of association differ with age of patient, or between hospitals if other aspects of treatment differ between hospitals. In the comparison of output from two machines it might be thought that any association might be affected by different sources of raw material. If there are k levels (sometimes called strata) of a covariate we form k 2×2 tables, one for each of the k levels of this additional explanatory variable. In effect, we then have a three-way k×2×2 table. In the drug example we might subdivide patients into those aged 20−29, 30−39, 40−49, and over 50, recording results for a side-effect as in Table 10.6. Informal inspection indicates that side-effect incidence shows little tendency to be age dependent for drug A, but there is a hint of higher incidence for older patients for drug B. Otherwise there looks to be little difference in side-effect incidence between drugs.

Table 10.6 Side-effects of two drugs grouped by ages

Age group	*Drug*	*Side-effect status*	
		No side-effect	Some side-effect
20−29	A	8	1
	B	11	1
30−39	A	14	2
	B	18	0
40−49	A	25	3
	B	42	2
over 50	A	39	3
	B	22	6

Table 10.7 gives data for numbers of satisfactory and faulty items produced by two types of machine using raw material from three sources.

Table 10.7 Production for two machines with three raw material sources

Material source	*Machine*	*Output status*	
		Satisfactory	Faulty
A	Type I	42	2
	Type II	33	9
B	Type I	23	2
	Type II	18	7
C	Type I	41	4
	Type II	29	12

The key feature of the data in Table 10.7 is that clearly type II machines produce a higher proportion of faulty items. Variations in this difference between raw materials appear to be slight.

STATXACT provides programs which, in each of these cases, enable us to test whether the data support a hypothesis of a common odds ratio for each of the k categories. If we accept this hypothesis a further program provides estimates of this common ratio together with a confidence interval. We return to these matters in Section 10.3.1.

In Example 10.2 we used a log-linear model to test the hypothesis that the odds ratios for survival responses for cuttings planted early or late were the same for long and short cuttings. For any 2×2 table the odds ratio or some monotonic function of it contains all information on association. We indicated in Section 9.1.1 that if we denote the empirical or observed odds ratio by $\theta^*=(n_{11}n_{22})/(n_{12}n_{21})$ we feel intuitively that the odds ratio $1/\theta^*$ obtained by interchange of rows or of columns represents the same degree of association but in the opposite direction. STATXACT in fact defines the odds ratio as this reciprocal of our θ^*. If θ is the expected value of the odds ratio in a 2×2 table with any of the models considered in Section 9.1.1 (e.g. binomial sampling within rows, multinomial sampling overall or Poisson counts) then, if the association model is correct, θ^* is asymptotically normally distributed with mean θ. Under the model of no association between row and column responses, $\theta=1$. Since θ is necessarily non-negative but may take any positive value the distribution of θ^* under the hypothesis of independence is highly skew, so convergence to normality is slow. The statistic $\phi^*=\log\theta^*$ is symmetric about a mean of zero and converges more rapidly to normality. Taking the natural logarithm (to base 'e') the asymptotic variance of ϕ^* (Agresti, 1990, Section 3.4.1) is

$$\mathrm{Var}(\phi^*)=1/n_{11}+1/n_{12}+1/n_{21}+1/n_{22}$$

Difficulties arise if any n_{ij} is zero, resulting in infinite values of ϕ^* and var(ϕ^*). These may be overcome if θ^* is replaced by

$$\theta^+=[(n_{11}+0.5)(n_{22}+0.5)]/[(n_{12}+0.5)(n_{21}+0.5)]$$

Differences in inferences based on θ^* and θ^+ are not marked for moderate to large values of n_{ij}.

For 2×2×2 tables, using the asymptotic distribution of ϕ^* provides an alternative test of the hypothesis $\theta_1=\theta_2$ where θ_1 and θ_2 are the expected odds ratios, e.g. for long and short cuttings in Example 10.3. This obviates the need for computing the expected counts m_{ijk} required for that example.

The hypothesis $\theta_1=\theta_2$ implies $\theta_2/\theta_1=1$ or $\phi_2-\phi_1=0$. Asymptotically ϕ_1^*, ϕ_2^* are independently normally distributed with means ϕ_1, ϕ_2 and variances $\mathrm{Var}(\phi_k^*)=(1/n_{11k}+1/n_{12k}+1/n_{21k}+1/n_{22k})$, $k=1, 2$.

Under $H_0 : \phi_2-\phi_1=0$ it follows that

$$Z = \frac{\phi_2^* - \phi_1^*}{\sqrt{[\text{Var}(\phi_1^*) + \text{Var}(\phi_2^*)]}} \quad (10.12)$$

has a standard normal distribution.

Example 10.6

The problem. For the data in Table 10.3 establish acceptability of the hypothesis $H_0 : \theta_1 = \theta_2$ (implying equal association).

Formulation and assumptions. We evaluate $\phi_k^* = \log\theta_k^*$ and $\text{Var}(\phi_k^*)$ for $k=1$, 2 and substitute in (10.12).

Procedure. From Table 10.3 we find

$$\phi_1^* = \log[(156\times156)/(84\times84)] = 1.2381$$

$$\phi_2^* = \log[(107\times209)/(31\times133)] = 1.6908$$

$$\text{Var}(\phi_1^*) + \text{Var}(\phi_2^*) = 2/156 + 2/84 + 1/107 + 1/209 + 1/31 + 1/133 = 0.0905$$

whence

$$Z = (1.6908 - 1.2381)/\sqrt{0.0905} = 1.51$$

Conclusion. Since $Z < 1.64$ required for significance, even in a one-tail test (not appropriate here) at the 5% level, we do not reject H_0.

Comment. The result is in line with that in Example 10.3 where we found $G^2=2.30$. Under H_0, Z^2 has a chi-squared distribution with 1 degree of freedom. The value $Z^2=1.51^2=2.27$ is close to that of G^2.

Similar approaches can be used in a relatively straightforward manner to find confidence intervals for ϕ_1, ϕ_2 or $\phi_2-\phi_1$, but a disadvantage of the method is that it does not immediately generalize to $k>2$. We discuss possible approaches for that case in the next section.

Example 10.7

The problem. Obtain a 95% confidence interval for the difference $\phi_2-\phi_1$ for the data in Example 10.6.

Formulation and assumptions. Following the usual methods of forming a 95% confidence interval for a normally distributed variate, the requisite interval is of the form

$$(\phi_2^* - \phi_1^*) \pm 1.96\sqrt{[\text{Var}(\phi_1^*) + \text{Var}(\phi_2^*)]}$$

Procedure. Substituting appropriate values calculated in Example 10.6 the relevant interval is clearly:

$$(1.6908-1.2381)\pm 1.96\sqrt{0.0905}=0.4527\pm 0.5896$$

Conclusion, The requisite 95% confidence interval is $(-0.1369, 1.0423)$.

Comment. The interval contains zero, consistent with our decision in Example 10.6 not to reject the hypothesis of no difference in association. In Exercise 10.6 we ask you to show that we reject the hypothesis of no association in either of the corresponding 2×2 tables, using an asymptotic test based on $\phi=\log\theta$.

10.3.1 Asymptotic tests and estimation of odds ratios in k 2×2 tables

In any situation with $k\geq 2$, three questions commonly arising are:

1. Is there evidence of association (i.e. $\theta_s\neq 1$), $s=1, 2, \ldots, k$?
2. If there is association is θ_s the same for all s?
3. If θ_s is the same for all s how do we estimate the common value θ?

A number of asymptotic tests help answer these questions; also exact permutation tests are made feasible by packages such as STATXACT. We outline some basic ideas behind these procedures, quoting relevant formulae without derivation. Some asymptotic tests are provided in general statistical packages and may even be conducted manually in simple cases. Exact permutation tests require specialized packages such as STATXACT.

Many inference procedures, in this and the next section rely heavily on the fact that once we have an observed or expected count for cell (1,1) in any 2×2 table this fixes the observed or expected odds ratio.

A widely used test for overall independence, i.e. H_0 : all $\theta_s=1$, $s=1, 2, \ldots, k$ in situations like those epitomized by Tables 10.6 and 10.7 was proposed by Mantel and Haenszel (1959) and bears their names. In its basic form we should first establish acceptability of a hypothesis of equality of all θ_s, for the test may accept H_0 even when there is clear evidence that some individual $\theta_s\neq 1$. We pursue this point further after Example 10.9.

For the moment we assume acceptance of the hypothesis that all θ_s are equal and we only want to test whether the common value is unity (indicating independence).

In this and the next section we denote the sth table (sth stratum) entries and marginal totals by

			Total
	n_{11s}	n_{12s}	n_{1+s}
	n_{21s}	n_{22s}	n_{2+s}
Total	n_{+1s}	n_{+2s}	N_s

The expected count in cell (1,1) of stratum s under the hypothesis of independence is $m_{11s}=n_{1+s}n_{+1s}/N_s$. The test is based on the fact that if there is no association the n_{11s} will fluctuate more or less randomly about the m_{11s} so that we then expect $\Sigma_s m_{11s}=m_{11+}$ and $\Sigma_s n_{11s}=n_{11+}$ not to differ greatly. Just how different they need be to indicate significance will depend on the size of the marginal totals in each table. It can be shown that, under independence with fixed marginals, $E(n_{11s})=m_{11s}$ and

$$\text{Var}(n_{11s})=n_{1+s}n_{+1s}n_{2+s}n_{+2s}/N_s^2(N_s-1) \qquad (10.13)$$

The Mantel–Haenszel test statistic is

$$M^2=\frac{(m_{11+}-n_{11+})^2}{\Sigma_s \text{Var}(n_{11s})} \qquad (10.14)$$

Asymptotically M^2 has a χ^2 distribution with 1 degree of freedom.

Example 10.8

The problem. For the four 2×2 tables in Table 10.6 compute M^2, and, assuming all k tables have the same expected odds ratios θ, test the hypothesis $H_0 : \theta=1$.

Formulation and assumptions. Assuming all θ_s are equal we compute M^2 by (10.14) and compare this with the critical value of χ^2 with 1 degree of freedom.

Procedure. We find $m_{111}=9\times19/21=8.14$, $m_{112}=15.06$, $m_{113}=26.06$, $m_{114}=36.60$; also $\text{Var}(n_{111})=(19\times9\times12\times2)/(21^2\times20)=0.4653$, $\text{Var}(m_{112})=0.4832$, $\text{Var}(m_{113})=1.1213$, $\text{Var}(m_{114})=1.9096$, whence we obtain $m_{11+}=8.14+15.06+26.06+36.60=85.86$ and $n_{11+}=8+14+25+39=86$ and $\Sigma_s\text{Var}(n_{11s})=0.4653+0.4832+1.1213+1.9096=3.9794$, so that

$$M^2=(0.14)^2/3.9794=0.0049.$$

Conclusion. Clearly $M^2=0.0049$ is not significant at the 5% level using the asymptotic χ^2 test, so we accept H_0.

Comment. 1. Some writers recommend use of a continuity correction, subtracting ½ from $|m_{11+}-n_{11+}|$ before squaring. This may avoid over-estimation of significance in borderline cases.

2. The small difference between m_{11+} and n_{11+} is indicative of a very good fit to the independence model in this case. However, situations arise when there are associations in opposite directions in individual 2×2 tables that tend to cancel out marked differences between individual m_{11s} and n_{11s}. We consider the test for equality of odds ratios in this example on p. 272.

3. The asymptotic test using M^2 is not always satisfactory for small counts if there are only a few tables.

We now investigate asymptotic procedures that allow us to test the hypothesis H_0 : *all* θ_s *have the same (unspecified) value* θ against the alternative H_1 : *at least two* θ_s *are unequal.* The test we develop has some similarities to the Mantel–Haenszel procedure. We first estimate the common value θ and form a statistic not unlike M^2; however, some modification is required since M^2 has only 1 degree of freedom and effectively we lose this when we estimate θ from the data. Several statistics have been suggested and we use one proposed by Breslow and Day (1980). The common odds ratio is estimated by

$$\theta^* = \frac{\Sigma_s n_{11s} n_{22s}/N_s}{\Sigma_s n_{12s} n_{21s}/N_s}. \tag{10.15}$$

We now determine m_{11s}, the expected value of the count in cell (1,1) with association specified by the common odds ratio θ^*. This is a positive root of the equation (10.16) not exceeding n_{1+s} or n_{+1s}, i.e. a root of

$$\frac{x(N_s - n_{1+s} - n_{+1s} + x)}{(n_{1+s} - x)(n_{+1s} - x)} = \theta^* \tag{10.16}$$

Such a root is generally unique and, denoting it by m_{11s}, Breslow and Day proposed the statistic

$$M_{\mathrm{BD}}{}^2 = \Sigma_s\{(m_{11s} - n_{11s})^2/[\mathrm{Var}(n_{11s})]\} \tag{10.17}$$

where

$$\mathrm{Var}(n_{11s}) = [(1/m_{11s}) + (1/m_{12s}) + (1/m_{21s}) + (1/m_{22s})]^{-1} \tag{10.18}$$

and the m_{ijs} are the expected cell counts given m_{11s} and the observed marginal totals. Asymptotically, under H_0 the statistic $M_{\mathrm{BD}}{}^2$ has a chi-squared distribution with $k-1$ degrees of freedom.

If we accept H_0 interest moves to confidence intervals for the common value θ which is estimated by θ^*. Asymptotically, the distribution of $\phi^* = \log\theta^*$ approaches normality more rapidly than that of θ^*. $\mathrm{E}(\phi^*) = \phi = \log\theta$, and several estimates have been proposed for $\mathrm{Var}(\phi^*)$. That recommended in STATXACT was proposed by Robins, Breslow and Greenland (1986) and works well for small k and large counts or large k even if some of the 2×2 tables are fairly sparse. The expression for this estimate is rather complicated, but we give it for completeness. For each stratum s, $s = 1, 2, \ldots, k$, we define

$$a_s = (n_{11s} + n_{22s})/N_s \quad b_s = (n_{12s} + n_{21s})/N_s$$

$$c_s = (n_{11s} n_{22s})/N_s \quad d_s = (n_{12s} n_{21s})/N_s \quad c_+ = \Sigma_s c_s \quad d_+ = \Sigma_s d_s$$

$$P = [\Sigma_s(a_s c_s)]/(2c_+{}^2) \quad Q = [\Sigma_s(a_s d_s + b_s c_s)]/(2c_+ d_+) \quad R = [\Sigma_s(b_s d_s)]/(2d_+{}^2)$$

and estimate $\text{Var}(\phi^*)$ as

$$\text{Var}(\phi^*)=P+Q+R \tag{10.19}$$

An approximate 95% confidence interval for ϕ is

$$\phi^* \pm 1.96\sqrt{\text{Var}(\phi^*)} \tag{10.20}$$

For all but small data sets (when asymptotic procedures may not be appropriate) a computer program is highly desirable for tests and estimation, but we indicate salient features of the computation in Example 10.9.

Example 10.9

The problem. For the three 2×2 tables in Table 10.7, determine whether the hypothesis of equality of odds ratio is acceptable and, if it is, obtain an approximate 95% confidence interval for the common odds ratio θ.

Formulation and assumptions. Solution requires the following steps: (1) estimate a common odds ratio using (10.15); (2) calculate $M_{BD}{}^2$ using (10.17) and use an asymptotic test for significance; (3) if we accept a common odds ratio, compute a 95% confidence interval for ϕ using (10.20) and back-transform to an interval for θ by taking the natural antilogarithm.

Procedure. We sketch the main steps in the computation. The reader should verify all numerical values (Exercise 10.7). Using (10.15) gives

$$\theta^* = \frac{(42\times9)/86+(23\times7)/50+(41\times12)/86}{(33\times2)/86+(18\times2)/50+(29\times4)/86} = 4.702$$

The expected value m_{111} is the relevant root x of (10.16) which here is

$$\frac{x(86-75-44+x)}{(75-x)(44-x)} = 4.702$$

This gives $m_{111}=41.68$, (the other root, $x=100.54$, exceeds both row and column marginal totals). Similarly, we find $m_{112}=23.07$ and $m_{113}=41.25$. Using (10.18) gives

$$\text{Var}(n_{111})=(1/41.68+1/2.32+1/33.32+1/8.68)^{-1}=1.6660$$

and similarly, $\text{Var}(n_{112})=1.3181$, $\text{Var}(n_{113})=2.4550$, whence

$$M_{BD}{}^2=0.32^2/1.6660+0.07^2/1.3181+0.25^2/2.455=0.09$$

Assuming an asymptotic χ^2 distribution with $k-1=2$ degrees of freedom, the result is clearly not significant so we accept the hypothesis of equal odds ratios.

To obtain the relevant confidence interval we note that $\phi^*=\log\theta^*=$ 1.5480; tedious but otherwise straightforward computation of P, Q, R in (10.19) gives $\text{Var}(\phi^*)=0.1846$, whence a 95% confidence interval for ϕ is

$$1.5480\pm1.96\sqrt{0.1846}$$

i.e. the relevant interval is (0.7059, 2.3901). Taking natural antilogarithms the interval for θ is (2.0256, 10.9145).

Conclusion. We accept the hypothesis of a common odds ratio, our point estimate being $\theta=4.702$ and an approximate 95% confidence interval being (2.026, 10.915).

Comments. The confidence interval looks wide and is not symmetric about the point estimate. Remember that θ has a very skew distribution, the value 1 corresponding to independence, while association with θ in the range (0,1) is matched by an equal but opposite association with θ in the range $(1, \infty)$.

In Example 10.8 we indicated that the classic Mantel–Haenszel test for independence may lead to acceptance of H_0 when there is evidence of opposite associations in some of the 2×2 tables. If there is a suspicion that this may be so, it is appropriate to test first for homogeneity of the odds ratios using ${M_{BD}}^2$ as the test statistic. If homogneity is accepted then M^2 given by (10.14) may be used to test whether $\theta=1$ is an acceptable value. Alternatively we may obtain a confidence interval for θ using the method in Example 10.9. If this interval includes $\theta=1$ the hypothesis of independence is acceptable. Doing this for the data in Table 10.6, we find ${M_{BD}}^2=6.497$ and with 3 degrees of freedom $\Pr(\chi^2\geq6.497)=0.0898$, so we accept equality of odds ratio if testing at the 5% level. The point estimator of the odds ratio is $\theta^*=1.0334$ and, using the method in Example 10.9, the 95% confidence interval is (0.4036, 2.6462).

10.3.2. Exact tests for odds ratios

Zelen (1971) and Gart (1970) have given respectively, exact permutation tests for homogeneity of the common odds ratio and a method for obtaining a confidence interval if a common odds ratio is accepted. STATXACT gives programs for both procedures. Unlike many other nonparametric tests the asymptotic procedures described in Section 10.3.1 are not based on asymptotic properties of the statistics used in the exact test. As computer facilties such as those provided by STATXACT are needed to apply the exact methods, we only sketch their nature.

Zelen's test for homogenity is based on a property of the hypergeometric probabilities arising in the Fisher exact test for 2×2 tables. Subject to the condition that any n_{11s} sum over s to the observed marginal total n_{11+}, i.e. $\Sigma_s n_{11s}=n_{11+}$, the product of the hypergeometric probabilities over all k observed 2×2 tables is a maximum when the observations strongly support a common odds ratio θ, irrespective of the value of θ, and this product decreases steadily as that support weakens. Under the hypothesis that all odds ratios are equal, the actual probability, P, of observing any particular outcomes in which the cell (1, 1) entries sum to the observed n_{11+} is given by the product of the k hypergeometric probabilities for those outcomes divided by the sum of all such products over all possible cell (1, 1) values consistent with marginal totals for the relevant 2×2 table and which sum to n_{11+}. If we denote the probability for our observed outcome by P_0 the corresponding tail probability has size determined by the sum of the probabilities associated with all outcomes for which $P \leq P_0$.

If this test is applied to the data in Table 10.6, STATXACT gives the probability of the observed outcome as 0.008213 and the relevant tail probability as 0.0689. This compares with the probability 0.0898 for the corresponding probability using the Breslow–Day asymptotic test.

The procedure for estimating the common odds ratio or obtaining a confidence interval if H_0 is accepted by the Zelen test takes a different approach. We noted in the asymptotic theory that if we estimate expected values under an assumption that the common odds ratio takes a particular value θ^*, say, then in general, if m_{11s} is the expected value in cell (1, 1) of table s, $\Sigma_s m_{11s} \neq n_{11+}$. Large discrepancies indicate that θ^* is an unsatisfactory value for the common odds ratio. Gart's procedure for obtaining a confidence interval for a common odds ratio θ is in essence to compute a statistic T which is the sum of the observed n_{11s} and to consider its position in the distribution of all possible values the statistic might take if each n_{11s} is replaced by any possible value consistent with the marginal totals in the sth table, i.e. the values considered are no longer constrained to sum to n_{11+}. We may find values θ^-, θ^+, that give values of T for which we just accept these as hypothetical values of the common odds ratio in a test at the $100(\frac{1}{2}\alpha)\%$ significance level, and these provide $100(1-\alpha)\%$ confidence limits for θ. Some modifications are needed to these arguments if the data indicate acceptance of either a zero or infinite common odds ratio. Details are given in the STATXACT manual, Section 6.3.

For the data in Example 10.9 the exact 95% confidence interval given by Gart's method is (1.9149, 12.306), compared to the asymptotic interval (2.0256, 10.9145) obtained in Example 10.9. The latter depends on the validity of certain variance estimates, involves a back-tranformation from logarithms and assumes a reasonable rate of convergence to normality. From the practical viewpoint the difference between the asymptotic and exact permutation theory result should not cause alarm, bearing in mind there are also questions of discontinuity in the permutation distribution. The STATXACT manual suggests ways of allowing for these.

10.4 FIELDS OF APPLICATION

Tests of association may be relevant in any situation where independence is rejected for contingency tables. In three-way (or higher-dimension) tables especially, the nature of associations is often of special interest.

Drug testing

Situations like that in Examples 10.4 and 10.5 for side-effects in drugs are common in clinical trials. In such examples the nature of association between dose levels and side-effects is the main interest in the study. Many such associations are adequately described by an appropriate linear-by-linear association model with suitable choice of row and column scores.

Medicine

In medical research it is often felt that physiological abnormalities may produce undesirable responses and that the seriousness of these may increase with severity of the disorder. An example of such a trend occurred with the spleen size/blood platelet count data considered in Example 9.8. Similarly, responses to environmental factors such as different levels of a known carcinogen nearly always show an ordered response and again linear-by-linear association models may be appropriate for analysis of such associations.

Administration of justice

There has been considerable interest in recent years in difference, in criminal trends, and the way the courts treat offenders, in different ethnic groups. Agresti (1984, p. 32) considers the analysis of counts of death penalty verdicts based on ethnic grouping of victims and defendants, using data given by Radelet (1981).

Sociology

Many studies have been made of association between socio-economic background and educational or career achievement; also between socio-economic background and attitudes towards social problems. Log-linear models are often relevant to studies of such associations.

10.5 SUMMARY

The likelihood ratio statistic, G^2, may be partitioned into independent single-degree-of-freedom components (Section 10.1.1) to study various aspects of association. The partitioning rules are not simple and in general do not lead to unique partitions. Choice of an appropriate partitioning is governed by the questions to be answered.

The general **log-linear** model is the contingency table analogue of the additive linear model in the analysis of variance. If separate interaction terms are introduced for each cell, this leads to a saturated model in which expected counts equal observed counts. Restricted models such as that for **linear-by-linear association** (Section 10.2.2) are of more interest. This model is applicable if there are indications of monotonic association across rows and columns for ordinal categories. Many standard nonparametric tests have exact permutation distributions obtainable by allocating appropriate scores in a linear-by-linear association model.

Binomial responses (Section 10.3) with 2 levels of one explanatory variable and k levels of a further explanatory variable (often called a covariate) often lead to different patterns of association at different levels of the covariate (Section 10.3). The **Mantel–Haenszel test** (Section 10.3.1) may be used to test for overall independence. This has been extended to test for a common odds ratio across strata defined by levels of a covariate, and, when the hypothesis of a common odds ratio is accepted, to test hypotheses about its value or to obtain a confidence interval for it.

EXERCISES

10.1 Verify the value of G^2 and each component thereof in Example 10.1.

10.2 Present cross-section tables of the data in Table 10.1 to show
(i) separately for each cholesterol level, presence or absence of CHD at each blood-pressure level;
(ii) separately for each blood-pressure level, presence or absence of CHD for each cholesterol level.

10.3 Use a Pearson chi-squared test or a likelihood ratio test to determine whether either of the 2×2 sub-tables in Table 10.3 indicates association.

10.4 Verify the value G^2=2.30 obtained in Example 10.2.

10.5 Confirm the result stated in the comments on Example 10.5 that with column scores 1, 2, 20, 200 the linear-by-linear association test indicates significance at the 1.01% level.

10.6 Confirm, by using an asymptotic test based on $\phi=\log\theta$, that for each of the 2×2 sub-tables in Table 10.3 we reject the hypothesis of no association.

10.7 Verify the numerical results stated in Example 10.9.

10.8 O'Muircheartaigh and Sheil (1983) give the following data for numbers of players with scores (i) par or better (P_1), (ii) over par (P_2), for low- and high-handicap golfers under two wind conditions, W_1 and W_2. Are the data adequately described by a first-order interaction model?

	Low handicap		*High handicap*	
	P_1	P_2	P_1	P_2
W_1	9	35	1	49
W_2	37	51	12	115

Use at least two-different methods of analysis and compare your results. Also obtain a 95% confidence interval for the common odds ratio if you accept the hypothesis that they are the same for each 2×2 sub-table.

10.9 Howarth and Curthoys (1987) give numbers of male and female students in English and Scottish universities in the years 1900–1901 and 1910–1911. Are proportions in the sexes independent between countries for each year? Is a first-order interaction model (i) necessary and (ii) sufficient to explain the observations?

	1900–1901		*1910-1911*	
	M	F	M	F
England	11755	2090	16038	3579
Scotland	4432	719	5137	1599

10.10 Considering only the data for patients with coronary heart disease (CHD) in Table 10.1 and assuming the blood pressure levels and cholesterol levels are both in increasing orders, use an appropriate test to decide if there is evidence of an association between blood pressure and cholesterol levels. If you were told that the blood pressure levels were I=normal or below, II= 1–10% above normal, III=11–20% above normal, IV=21–50% above normal, and V= over 50% above normal, and that the cholesterol levels are A=normal or less, B=up to 50% above normal, C=over 50% above normal and including some patients as much as 200% above normal, how might you take this information into account in your analysis?

10.11 In an English Parliamentary electoral constituency a random sample of 400 voters are classified by age and political affiliation as follows:

Political affiliation	*Age group*			
	Under 30	31–40	41–55	over 55
Conservative	31	32	39	34
Liberal Democrat	16	19	25	31
Labour	36	27	58	52

Is there evidence of an association between poltical affiliations and age? It is generally accepted that the Conservative, Liberal Democrat and Labour Parties represent an ordering of right, middle and left in the political spectrum.

10.12 Agresti (1984) quotes the following data on cross-classification of attitudes towards abortion and amount of schooling based on the US General Social Survey, 1972. Test these data for evidence of association between attitudes and educational background.

	Attitude towards abortion		
Schooling	Disapprove	Neutral	Approve
Less than high school	209	101	237
High school	151	126	426
More than high school	16	21	138

11

Robustness

11.1 THE COMPUTER AND ROBUSTNESS

This chapter introduces modern techniques with applications ranging from location tests to advanced regression problems. All require computer programs for realistic application, together with a broad appreciation of basic statistical theory to make best use of them and to avoid pitfalls that go with sophistication in statistics and applied science generally. We indicate some principles by numerically trivial examples. An experimenter with limited statistical training should be aware of these methods, but seek expert advice about application to particular problems.

11.1.1 A special kind of robustness

In a broad sense robustness implies that an analysis does not depend too critically upon specific assumptions. Considerations of Pitman efficiency indicate that a wrong assumption may make a procedure less efficient than we would hope, and further that some procedures may be robust against certain types of departure from a model, but not against others; e.g. a procedure that is robust against 'long-tail' departures from an assumed distribution may be sensitive to skewness. Since about 1970 there has been considerable interest in procedures that are robust against 'small' departures from assumptions, where 'small' may mean either that a small proportion of the observations may be grossly in error or that an appreciable proportion of the observations are subject to small perturbations. Example 2.8 provided a demonstration of the latter in the context of rounding. Outliers illustrate the first situation.

There are many tests for outliers but some of these themselves lack robustness; they may, for example, be notoriously bad at detecting more than one outlier in the same tail; others tend to miss a pair of outliers in opposite tails. A simple and reasonably robust test is to classify any observed x_0 as an outlier if

$$|x_0 - \text{med}(x_i)|/\{\text{med}[|x_i - \text{med}(x_i)|]\} > 5 \qquad (11.1)$$

where $\text{med}(x_i)$ is the median of all observations in a sample. The denominator in (11.1) is a measure of spread called the **median absolute deviation**, often abbreviate to MAD. The choice of 5 as the critical value is somewhat arbitrary, motivated by the reasoning that if our observations, apart from the outlier, have an approximately normal distribution, it picks up as outliers any observations more than about 3 standard deviations from the mean.

Example 11.1

The problem. Use (11.1) to detect any outliers in the data set

8.9 6.2 7.2 5.4 3.7 2.8 17.2 13.7 6.9

Formulation and assumptions. It is easiest to determine the median and MAD after ordering the observations. We first test the observation furthest from the median, using (11.1). If this is not an outlier we stop. If it is we test the next extreme observation (in either tail), and proceed until we find an observation that is not an outlier.

Procedure. The ordered observations are

2.8 3.7 5.4 6.2 6.9 7.2 8.9 13.7 17.2

The median is 6.9; absolute deviations from the median are $|2.8-6.9|=4.1$, and similarly 3.2, 1.5, 0.7, 0.0, 0.3, 2.0, 6.8, 10.3. Ordering these, we easily find that their median, the MAD, is 2.0. Setting $x_0=17.2$ in (11.1) the left-hand side is $(17.2-6.9)/2=5.15$; this classifies 17.2 as an outlier. It is easily verified that no other observation satisfies the outlier criterion (11.1).

Conclusion. The observation 17.2 is an outlier.

Comment. Having decided 17.2 is an outlier our problems are not over. If all checks show it is a correct observation (not a mistake), it suggests our sample is not from a normal distribution but gives little indication whether the distribution is skew or long-tailed symmetric.

For a number of reasons an observation may be genuine even if atypical. Huber (1977) suggests that it is not uncommon for up to 10% of experimental observations to come from a different (and often longer-tailed) distribution than the bulk of the observations, yet both have the same mean. This may occur, for example, if several people make the observations, if some are made using different equipment or with diverse sources of raw material or at appreciably different times (all factors that may introduce different, but often unnoticed, sources of variability).

Mixtures of observations from normal distributions, most from one with mean μ and standard deviation σ, but a few from a distribution with mean μ and standard deviation 3σ, may be unexpected and the mix undetected.

Computers, useful as they are, have compounded the problem of gross errors that go undetected. e.g. those caused by hitting two keys simultaneously when entering data, recording perhaps a correct 3.6 as 34.6. A visual or statistical check may detect such an error, but in large data banks with thousands, or perhaps millions, of data items, mistakes slip through.

Departures from normality due to grouping or rounding may be less severe in effect, but as we saw in Example 2.8 they cannot be ignored entirely; they are examples of small perturbations of most of the data.

Even with basically reasonable models, we may find one or two gross departures or a large number of small perturbations, so we seek estimation procedures that are almost optimal if we have no such upsets and are little affected if we do have a few disturbances.

One such class of estimators are the Huber–Hempel *m*-estimators, the '*m*' because they are like optimal maximum likelihood estimators when these are appropriate and are little disturbed by gross departures of a few, or small perturbations in many, observations.

11.1.2 Motivation for *m*-estimators for location

Using *m*-estimators in non-trivial problems requires adequate computer programs. We demonstrate the basics for estimating the mean of a symmetric distribution. In this case the maximum likelihood estimator is the sample mean; in particular, maximum likelihood estimation of the mean of a normal distribution, given n observations $x_1, x_2, x_3, \ldots, x_n$, is equivalent to least squares estimation when we choose our estimator $\hat{\mu}$ as the value of μ that minimizes

$$U(x, \mu)=\Sigma_i(x_i-\mu)^2$$

In words, $\hat{\mu}$ is the μ value that minimizes the sum of squares of deviations of the observed x_i from μ. Mathematically $(x-\mu)^2$ is referred to as a **distance function** as it provides a measure of the distance (the square) of a data point x from μ. Other possible distance functions include the absolute distance $|x-\mu|$. While $U(x, \mu)$ is minimized by setting μ equal to $\bar{x}$, the sample mean, $V(x, \mu)=\Sigma_i|x_i-\mu|$ is minimized by setting μ equal to $\mu^*=\text{med}(x_i)$, the sample median. For samples from a normal distribution we prefer $\hat{\mu}$ to μ^* because it has a smaller variance for a given n.

However, even one outlier has more effect on $\hat{\mu}$ derived from U than it has on μ^* derived from V. Indeed the influence of any x_i upon $\hat{\mu}$ is linearly dependent on the distance of x_i from μ, i.e. an outlier lying 6 standard deviations from the mean has twice the influence of one lying 3 standard deviations from the mean. We do not prove this intuitively reasonable result, which follows from a study of the **influence function** – an advanced mathematical concept. It is easily illustrated by an example.

Example 11.2

Consider the 5 observations 0, 2, 2, 3, 5. Estimation of μ by the sample mean gives $\hat{\mu}=2.4$. If we made a mistake and recorded the data as 0, 2, 2, 3, 15 we find $\hat{\mu}=4.4$, while if we record it as 0, 2, 2, 3, 45 (the result of hitting adjacent keys simultaneously?) we find $\hat{\mu}=10.4$. This demonstrates the linear effect; altering the entry from 5 to 15 (by 10) increases the sample mean by 2, i.e. from 2.4 to 4.4. We would expect an increase from 15 to 45 (i.e. by 30 or 3×10) to increase the sample mean by $3\times2=6$. This is what happens, the increase being from 4.4 to $4.4+6=10.4$.

If we estimate μ by minimizing $V(x, \mu)$ we choose the sample median which is 2 in all 3 cases. Indeed merely pushing one (or often more) tail entries further into the tail does not alter μ^*.

Our efficient estimator of the mean of a symmetric distribution is sensitive to a small number of aberrant observations in one tail, whereas the less efficient estimator may be unaffected by an odd outlier. Recognizing this, Huber (1972) and others developed m-estimators that have almost all the properties of the best estimators when there are no data aberrations and retain these (almost) when there are outliers, being insensitive to these aberrations.

For location parameters this is achieved by choosing an estimator that behaves like least squares (or an equivalent maximum likelihood estimator) for 'good' observations, but more like the median estimator for outliers in that it reduces their influence by down-weighting them. An advantage of m-estimators is that, as well as being robust, their efficiency is close to that of maximum likelihood estimators, whereas a median estimator, for example, pays for its robustness by loss in efficiency.

11.1.3 Huber's *m*-estimator for location

To describe a simple m-estimator we generalize the idea of a distance function. Some continuity and differentiability properties are needed for a straightforward approach; some distance functions, e.g. $V(x, \mu)$, lack these. We do not discuss such niceties, which can often be surmounted in practice.

To find a minimum of $U(x, \mu)$, regarded as a function of μ, we differentiate U with respect to μ and equate the derivative to zero. This leads to the 'normal' or 'estimating' equation

$$\Sigma_i[-2(x_i-\mu)]=0 \qquad (11.2)$$

with solution $\hat{\mu}=(\Sigma_i x_i)/n=\bar{x}$, the sample mean.

We define a distance function more generally as a function $d(t)$, such that for all t, (i) $d(t)\geq0$, (ii) $d(t)=d(-t)$ and (iii) the derivative $\psi(t)=d'(t)$ exists for all t, and is a non-decreasing function of t. Condition (iii) is relaxed for some m-estimators.

For estimating the mean, μ, of a symmetric distribution, Huber (1972) proposed a function, $d(t)$, where for some fixed $k>0$,

$$\begin{aligned} d(t) &= \tfrac{1}{2}t^2 && \text{if } |t|\leq k \\ &= k|t|-\tfrac{1}{2}k^2 && \text{if } |t|>k. \end{aligned}$$

Clearly d is a distance function with derivatives of the form

$$\begin{aligned} \psi(t) &= -k && \text{if } t<-k \\ &= t && \text{if } |t|\leq k \\ &= k && \text{if } t>k \end{aligned}$$

We estimate μ by a value μ^* that satisfies the normal equation

$$\Sigma_i\ \psi(x_i-\mu)=0 \qquad (11.3)$$

If we allow $k\to\infty$, then for all x_i, $\psi(x_i-\mu)=x_i-\mu$ and our estimator is the sample mean. For finite k, the form of $d(t)$ shows that for $|x_i-\mu|\leq k$ we minimize a $d(t)$ equivalent to that in least squares, while if $|x_i-\mu|>k$ we minimize a linear function of absolute differences.

We have to choose k and then decide how to solve (11.3). In general, we require an iterative solution and for any chosen k we proceed by writing (11.3) in the form

$$\Sigma_i \frac{\psi(x_i-\mu)}{(x_i-\mu)}(x_i-\mu)=0 \qquad (11.4)$$

Writing $w_i=\psi(x_i-\mu)/(x_i-\mu)$, (11.4) becomes $\Sigma_i w_i(x_i-\mu)=0$, which has the solution

$$\mu^*=\Sigma_i(w_i x_i)/(\Sigma_i\ w_i) \qquad (11.5)$$

which is a weighted mean of the x_i. However, the weights are functions of μ^*. Thus we must start with an estimate μ_0 of μ^* and with this calculate initial weights w_{i0} and use these in (11.5) to calculate a new estimate μ_1. This in turn is used to form new weights w_{i1}, say, and the cycle continues until convergence, which is usually achieved after a few iterations. We discuss in more detail the choice of k in Section 11.1.4 but it suffices to choose k such that the interval [med$(x_i)-k$, med$(x_i)+k$] embraces between about 70% and 90% of the observations. For moderate or large samples calculation of an m-estimator is tedious without the requisite computer program, but we illustrate the method for small data sets.

Example 11.3

The problem. Obtain the Huber m-estimator for the three data sets:

(i) 0 2 2 3 5 (ii) 0 2 2 3 15 (iii) 0 2 2 3 45

Formulation and assumptions. The median of all three sets is 2. If we choose $k=2$, the interval (0, 4) in all sets covers 4 from 5 observations. We use the iterative procedure described above to obtain μ^* for each set.

Procedure. For data set (i) we choose the sample median, med$(x_i)=2$ as our initial estimate μ_0. Since $\psi(x_i-\mu_0)=x_i-\mu_0$ if $|x_i-\mu_0|<k$ it follows that for x_i satisfying this condition $w_{i0}=1$. Thus when $k=2$ and $\mu_0=2$, the observations 0, 2, 2, 3 all have weight 1. For $x=5$, $x_i-\mu_0=5-2=3$, which is greater than $k=2$. For this observation $\psi(x_5-\mu_0)=k=2$, whence $w_{5,0}=2/3$. Using (11.5) we get a new estimate

$$\mu_1=[1\times0+1\times2+1\times2+1\times3+(2/3)\times5]/[1+1+1+1+(2/3)]=2.21$$

The cycle is repeated with $\mu_1=2.21$. Now only values between 0.21 and 4.21 have weight 1, so observations 0 and 5 are down-weighted, Clearly 0 has the weight 2/2.21 and 5 has the weight $2/(5-2.21)=2/2.79$. Calculating the weighted mean with these weights gives a new estimate $\mu_2=2.29$. A further iteration gives $\mu_3=2.32$, a value maintained at the next iteration.

Detailed calculations for data sets (ii) and (iii) are left as an exercise (Exercise 11.1). It will be found that the value $x=15$ in (ii) is down-weighted drastically, indeed if $\mu_0=2$, $w_{0,5}=2/13$ and this weight changes only slightly in subsequent iterations. Finally, we find $\mu^*=$ 2.33 for this set and for (iii) the Huber estimate is again $\mu^*=2.33$.

Conclusion. The Huber estimates of μ are respectively 2.32, 2.33, 2.33.

Comments. The estimates are virtually the same for each set, whereas in Example 11.2 the samples means found were 2.4, 4.4, 10.4. The sample median in each case was 2, so that is another location estimator unaffected by an outlier. The median has lower Pitman efficiency than the sample mean if we assume normality; in general, the median also has lower Pitman efficiency than the Huber estimator under normality.

11.1.4 Choice of *k* or other distance functions

Generally speaking, $k>0$ is chosen to down-weigh outliers. A function called the influence function confirms (as we might expect) that the smaller we choose k, the less the influence of 'way-out' observations. However, if we choose k too small and give full weight to only a few observations near μ our estimator may be appreciably influenced by rounding or grouping effects in these observations.

A further difficulty with the Huber estimator and arbitrary k concerns the effect of a change in scale. It is well known that if we replace all x_i by $y_i=a+bx_i$, where a, b are constants, then the least squares estimate of the mean of the Y population is $a+b\hat{\mu}$. A similar relationship holds for sample medians, but this useful property does not hold for the m-estimator described in Section 11.1.3. This desirable property is recovered by a simple modification of the function ψ, replacing $\psi(x-\mu)$ by $\psi[(x-\mu)/s]$, where s is some measure of spread. Experience has shown an appropriate one to be the median absolute deviation (MAD) divided by 0.6745. This divisor makes s a consistent estimator of σ if the underlying distribution is normal. The weights in (11.5) are modified by using $(x-\mu)/s$ in the denominator also. With this modification (which can be looked upon as standardization of the data) choosing $k=1.5$ gives an estimator with desirable properties for down-weighting outliers without overemphasizing the influence of moderate grouping or rounding. Since at any stage, j, say, of the iteration k is now relevant to the values of $(x_i-\mu_j)/s$ where μ_j is the jth iterative estimator of μ, we must recalculate $(x_i-\mu_j)/s$ for each sample value at each stage of the iteration. There is no change in s or k between iterations. For the data sets in Example 11.3 (see Exercise 11.2) the estimate μ^* using this approach is 2.34 in each case.

Hampel, in Andrews *et al.* (1972), and others, have suggested alternative distance functions that completely ignore (i.e. give zero weight to) very extreme observations. Effectively, Huber's distance function limits the influence of an extreme value to what it would be if it were at a distance k from μ^*. Hampel's function gives even less influence to moderately extreme values and zero weight to very extreme ones. Maximum likelihood estimation is a special case of m-estimators, and when estimating μ from a sample from a normal distribution it corresponds to setting $k=\infty$ in the Huber estimator.

Andrews (1974) developed m-estimators for regression problems using yet another distance function. In multiple regression problems outliers are notoriously difficult to spot without an elaborate analysis, and m-estimators have a useful role to play, despite some convergence problems. Solutions are again achieved by what is effectively iterated weighted least squares: obtaining starting values that give reasonable guarantees of convergence may be a large part of the computational effort. Nonparametric methods like those of Agee asnd Turner (1979) may prove useful for this purpose.

For a fuller discussion of m-estimators and influence functions see Randles and Wolfe (1979, Section 7.4), or Huber (1972; 1977) for more mathematical detail of the method and properties. Problems of outliers in regression have given birth to the subject of regression diagnostics described by Cook and Weisberg (1982) and Atkinson (1986).

11.1.5 Trimmed means

A simple method for dealing with outliers or long tails in location problems is by calculating a **trimmed mean**. We omit the top $\alpha\%$ and bottom $\alpha\%$ of the observations and estimate the population mean using the remaining 'central' observations. Common trimming levels are the top and bottom 10% (extreme deciles), or the first and third quartiles; in this latter case the trimmed mean is sometimes called the interquartile mean. There are variants on the trimming procedure. One known as **Winsorization** shrinks extreme observations inward to the value of the largest remaining observation, thus reducing their influence without ignoring them entirely. Winsorization and the Huber m-estimator both down-weigh extreme observations. The Huber estimator down-weighs observations more than a pre-determined distance from the current estimator of the mean, while trimming or Windorization down-weighs a predetermined proportion of the extreme observations. Theoretical and Monte Carlo comparisions have been made between results of trimming and using m-estimators. One may conjure up specific examples where one approach fares better than the other, but it is probably true by and large, if one is dealing with unknown types of data aberration, that the Huber-type estimators provide a more efficient safety net than trimmed means for location estimates.

11.2 THE JACKKNIFE

High-speed low-cost computer calculation has revolutionized attitudes towards obtaining approximate numerical solutions to intractable mathematical problems. Estimation of bias may be tackled this way. **Mean bias** arises when the mean value of an estimator (remember, the value of an estimator varies from sample to sample) is not equal to the parameter it is estimating. Sometimes an easy-to-make analytic adjustment to an estimator eliminates bias. A simple example is provided by the maximum likelihood estimator of the variance σ^2 of a normal distribution based on a sample of n observations. The maximum likelihood estimator is $s^2 = \Sigma_i(x_i - \bar{x})^2/n$ and $E(s^2) = (n-1)\sigma^2/n$. Bias is removed by the simple expedient of replacing n by $n-1$ in the denominator of s^2. It is perhaps less well known that even after making this adjustment and taking the positive square root, s, to estimate the standard deviation σ, we are using a biased estimator for σ. Fortunately, the bias is of little practical consequence. There are many situations where bias virtually defies an analytical study but we wish to make an estimate of how great it is and hopefully to reduce it.

Quenouille (1949) introduced a nonparametric method for estimating bias that is now known as the **jackknife**. It also provides an estimate of the variance of a parameter estimate. In many applications the jackknife has proved less successful than a simpler concept, the **bootstrap**, which we discuss in Section 11.3. Both techniques use the computer's ability to handle routine and repetitive calculations speedily.

11.2.1 The jackknife and bias

The jackknife may be used to estimate bias for any estimator that is the sample analogue of the population characteristic it is estimating, e.g. the sample mean as an estimator of the population mean, the sample variance as an estimator of the population variance, the sample median as an estimator of the population median, and in the case of bivariate samples the Pearson product moment correlation coefficient as an estimate of the population correlation coefficient or the ratio of sample means $\bar{x}/\bar{y}$ as an estimator of the population mean ratio μ_x/μ_y. In all these cases, except that of the sample mean as estimator of the population mean, the estimates are, in general, biased.

Motivation for the jackknife comes from the estimator $s^2=\Sigma_i(x_i-\bar{x})^2/n$ given above. If we write $E(s^2)=\sigma^2-\sigma^2/n$, the term $-\sigma^2/n$ represents the bias in s^2 as an estimator of σ^2. In this case the jackknife eliminates the bias. In other cases, cited above, the bias is more complex and analytically intractable; the jackknife merely reduces it.

So that our discussion covers any parameter estimate of interest (mean, variance, median, correlation coefficient, ratio of means, etc.) we refer to the parameter being estimated as θ; to the estimator which is the sample analogue of θ as $\hat{\theta}$; to the jackknife estimator described below as $\tilde{\theta}$.

To obtain the jackknife estimator we calculate $\hat{\theta}$ for all n observations. We repeat this calculation omitting x_1, obtaining an estimator which we denote by $\hat{\theta}_{(1)}$. This step is repeated a further $n-1$ times, omitting in turn $x_2, x_3, \ldots, x_n$. The estimator associated with omitting x_i is denoted by $\hat{\theta}_{(i)}$. For bivariate data we omit one observation point (x_i, y_i) at a time. We denote the mean of the $\hat{\theta}_{(i)}$ by $\hat{\theta}_{(.)}$, i.e. $\hat{\theta}_{(.)}=\Sigma_i\hat{\theta}_{(i)}/n$. The jackknife estimator $\tilde{\theta}$ of θ is

$$\tilde{\theta}=n\hat{\theta}-(n-1)\hat{\theta}_{(.)} \tag{11.6}$$

and the jackknife estimate of bias is

$$B=(n-1)(\hat{\theta}_{(.)}-\hat{\theta}) \tag{11.7}$$

It is well known that the sample mean is an unbiased estimator of the population mean μ for any distribution, whereas we have already commented on the bias of s^2 defined above as an estimator of the population variance σ^2. We show for a trivial numerical example that the jackknife estimator of the mean is identical with the sample mean and that the jackknife estimator of variance is the usual unbiased estimator with the divisor n replaced by $n-1$. Simple but tedious algebra shows these findings to be generally true for estimation of μ and σ^2.

Example 11.4

Determine jackknife estimators for the mean and variance of the data set 1, 2, 7, 10.

We easily confirm $\bar{x}=5$, $s^2=13.5$. To calculate the jackknife mean we need the mean of the samples 2, 7, 10 and 1, 7, 10 and 1, 2, 10 and 1, 2, 7 and these are 19/3, 6, 13/3, 10/3. The mean of all four is 60/12=5, whence (11.6) gives $\tilde{\theta}=4\times5-3\times5=5$. Also (11.7) gives $B=0$, consistent with the sample mean being unbiased.

For variance we again calculate the sample variances omitting one point at a time, obtaining the four values 98/9, 14, 146/9 and 62/9. The mean of these four values is 12, whence (11.6) gives $\tilde{\theta}=4\times13.5-3\times12=18$. It is easily verified that this equals the unbiased estimator obtained by replacing $n=4$ by $n-1=3$ in s^2. The bias estimate given by (11.7) is -4.5, equal to the true bias.

These well-known results about bias in the mean and variance estimators were the motivation for jackknife estimation. In mathematical jargon, the jackknife is said to eliminate bias of order $1/n$. Unfortunately, many estimators such as the sample correlation coefficient for the population correlation coefficient exhibit a more complicated form of bias. The jackknife estimator does not eliminate such bias, but it usually reduces it.

Broadly speaking, reduction of bias in estimators is desirable, but there are complications, for sometimes a biased estimator has a smaller mean square error than an unbiased estimator, where the **mean square error** is defined for any estimator θ^* of a parameter θ as $\mathrm{E}[(\theta^*-\theta)^2]$. Clearly, this is a measure of how close our estimator is to the true value and it may be broken into additive components representing variance and bias. An alternative estimator that reduces the bias component may have an increased variance component, the overall effect being to increase mean square error.

11.2.2 Jackknife estimates of variance

Tukey (1958) proposed a jackknife estimator for the variance of any parameter estimator $\hat{\theta}$ which is the sample equivalent of the population parameter θ. We write this estimator

$$\mathrm{Var}^*(\hat{\theta})=(n-1)\Sigma_i(\hat{\theta}_{(i)}-\hat{\theta}_{(.)})^2/n \qquad (11.8)$$

The expression $(\hat{\theta}_{(i)}-\hat{\theta}_{(.)})^2$ in (11.8) is of the form $(x_i-\bar{x})^2$ and tedious but elementary algebra shows that if $\hat{\theta}=\bar{x}$ this reduces to the usual unbiased estimator for variance of the mean. We verify this for some simple numerical data.

Example 11.5

For the data 1, 2, 7, 10 in Example 11.4 we found $\tilde{\sigma}^2=18$ was the unbiased estimator of σ^2, whence the usual estimated variance of the sample mean is $\tilde{\sigma}^2/n=18/4=4.5$. To calculate the jackknife estimator (11.8) we require the four jackknife means computed in Example 11.4, viz. 19/3, 6, 13/3, 10/3 and their mean 5, whence, substituting in (11.8), $\mathrm{Var}^*(\bar{x})=4.5$.

This equivalence was the motivation behind (11.8), and although the estimator often gives reasonable approximations for the variance of estimators for other parameters, hopes that use of standard deviations based on $\sqrt{[\text{Var}^*(\hat{\theta})]}$ might lead to acceptable confidence intervals based on a t-distribution with $n-1$ degrees of freedom (as it does exactly for the mean of samples from a normal distribution) have proved unsatisfactory in many cases. Sometimes this is because the distribution of the estimator is not symmetric. However, when exact theory is intractable, jackknife estimates of variance or standard deviation may prove useful.

The more desirable properties of jackknife estimators stem from the fact that, generally speaking, the systematic selection of samples of $n-1$ from the original sample of n will give broadly similar estimates to those from the full sample but will make some allowance for a particularly influential point.

The sample cumulative distribution function $S(x)$ introduced in Section 3.3.1 may be regarded as the nonparametric maximum likelihood estimator of any distribution function $F(x)$. We obtain it by associating with each sample value x_i in a sample of n a probability $p_i=1/n$. There is a strong intuitive appeal in using as an estimate of any parameter in $F(x)$ the corresponding 'parameter' for $S(x)$, e.g. the sample mean as an estimate of the population mean or the sample (Pearson) correlation coefficient as the estimate of the population correlation coefficient. Jackknifing makes use of a very specific type of secondary sampling from our original sample of n. We now explore a further 'sampling from a sample' technique.

11.3 THE BOOTSTRAP

Although conceptually simpler than the jackknife, bootstrap estimation of e.g. the variance or standard deviation of a parameter estimator, usually involves more computation. When jackknifing we essentially calculate the full sample estimate of the parameter and n further estimates omitting one data point each time, a total of $n+1$ estimates.

For bootstrap estimation, even for small n, we may calculate anything from 100 to 1000 or more estimates of a parameter, so implementation is very much a job for a computer.

The motivation for the bootstrap again hinges on the concept of $S(x)$ as defined by (3.3) in Section 3.3.1 for a sample of n, as a nonparametric maximum-likelihood estimator of $F(x)$, and the use of the sample analogue of the population parameter (or other characteristic such as the median) we want to estimate is again appealing and often corresponds to the maximum likelihood estimator in a parametric analysis as well as in the nonparametric situation.

We may be interested in the standard deviation (or some other property) of our estimator. Given only our sample data, this is our sole source of information about that property. However, the standard deviation or other property of an estimator $\hat{\theta}$ is in general a function of F, the population cumulative distribution function. The bootstrap estimator estimates such properties by the same function of S, the sample cumulative distribution function, if indeed we know what the relevant estimating function is. The difficulty is that except in a few cases we do not know what the relevant function is. We do of course know the estimator of the standard deviation of the sample mean when we use it as an estimator of the population mean; it is $s/\sqrt{n}$, where $s^2=\Sigma_i(x_i-\bar{x})^2/n$; note that this is the maximum likelihood estimator of the variance of the sample mean rather than the unbiased estimator with divisor $n-1$ in s^2.

So far the reader might be excused for thinking the bootstrap estimator is a lot of fuss about nothing. But what happens if we have a sample from some unknown distribution, and we proceed to estimate the first quartile by taking the sample first quartile as our estimator and want some measure of its reliability such as its standard deviation? There is not much theory to help us find an analytic solution. This is where what is usually understood by **bootstrap estimation** comes to our aid. We use a Monte Carlo simulation to estimate the bootstrap estimator. We take repeated samples of n from our original sample of n with replacement, i.e. in such a sample (often called a **bootstrap sample**) a particular value x_i may appear more than once, just once, or not at all. The algorithm used to obtain an estimate of the bootstrap standard deviation of any parameter estimator $\hat{\theta}$ (or any other quantity of interest) is as follows:

1. We obtain from our sample $x_1, x_2, \ldots, x_n$ an estimate $\hat{\theta}$ of θ.
2. We obtain a sample (a bootstrap sample) of n values from this sample where each of the x_i has probability $1/n$ of occurring as any member of the sample (i.e. we sample with replacement). From this sample we obtain an estimate $\tilde{\theta}_1$ of $\hat{\theta}$ by the same method as we used to obtain $\hat{\theta}$ itself.
3. Step 2 is repeated a fixed number of times, B, say, giving a sequence of estimates, the bootstrap estimates, $\tilde{\theta}_j$, $j=1, 2, \ldots, B$.
4. The bootstrap estimates are used to calculate any quantities of interest, e.g. the bootstrap standard deviation of $\hat{\theta}$ which is computed as

$$\sigma_B(\hat{\theta})=\sqrt{\{\Sigma_j[(\tilde{\theta}_j-\tilde{\theta}_{(.)})^2]/(B-1)\}} \tag{11.9}$$

where $\tilde{\theta}_{(.)}$ is the mean of the $\tilde{\theta}_j$.

If we allow $B\to\infty$ this estimator will tend to the true bootstrap estimator of the standard deviation. However, as Efron (1982, Chapter 5) points out, in many cases the Monte Carlo estimate is reasonably close to the true bootstrap estimate of standard deviation for values of B as low as $B=100$. In practice, any improvement due to increasing B may not be important since (especially for small samples) there may be considerable differences between the best sample estimates and the true standard deviation based upon the population distribution. This applies even with parametric estimation where the sample standard deviation of the mean $s/\sqrt{n}$ may be very different from the true standard deviation $\sigma/\sqrt{n}$.

Effron (1982, Chapter 10) discusses several ways for obtaining approximate confidence intervals using bootstrap estimators. In a few cases an analytic approach is possible. One particular example he gives is a bootstrap estimator for a confidence interval for a population median based upon the sample median. This turns out to be close to the confidence intervals we obtained for the median in our discussion of the one-sample median test in Section 2.1.2.

We recommend Efron (1982) as a very readable summary of the (then) current position about the use of the jackknife and bootstrap and other important resampling methods. A large number of papers detailing applications to specific problems and also developing the theory have appeared since 1982. The former tend to be of prime interest to workers in specific fields (e.g. regression, analysis of survival data, reliability) and the latter are often statistically and mathematically sophisticated. We conclude this discussion with an illustrative numerical example which, although trivial, gives some flavour of how the method works and indicates the need for caution in interpretation. For this particular example we are able to compare our bootstrap result with certain other approximations arrived at on a theoretical basis. We mix informal discussion of the bootstrap approach with solving a specific problem.

Example 11.6

The problem. Explore the use of the bootstrap and other methods to study the behaviour of the sample 10% trimmed mean as an estimator of location in the population from which the following sample was obtained

0 6 7 8 9 11 13 15 19 40

Discussion of the problem. Since there are 10 observations we trim off 10%(=1) from each tail, i.e. the observations 0, 40. Our estimator of the population trimmed mean is the sample trimmed mean $\hat{\theta}$, i.e. $\hat{\theta}=88/8=11$. We base our discussion on a Monte Carlo sample of 100 bootstrap estimators $\tilde{\theta}_j$. Indeed the first of these samples consisted of the observations

15 13 13 40 40 15 13 0 19 15

Characteristically for a bootstrap sample, not all values in the original sample appear while others are repeated, i.e. 13 and 15, each occur 3 times and 40 twice. To compute $\tilde{\theta}_1$ we reject 0 and *one* of the 40s and compute the mean of the remaining 8 values, giving $\tilde{\theta}_1=17.875$. Another typical sample, in fact the 11th taken, was

13 11 7 19 0 13 15 8 7 8

giving, after rejection of 0 and 19, $\tilde{\theta}_{11}$=82/8=10.25.

An inspection of the original sample suggests that the prime reason for using the trimmed mean is to down-weigh the observation 40. The implication in so doing is that we regard it as in some way atypical or perhaps that if we really believe we are sampling from a fairly skew or long-tailed distribution we prefer a location estimator for the 'bulk' of the population, virtually ignoring the long tail. The median (which may be regarded as a 50% trimmed mean!) is a competitor to our 10% trimmed mean in this situation. The bootstrap procedure indicates a certain skewness in the distribution of the $\tilde{\theta}$. This arises because repetitions of the value 40 in the bootstrap sample are not down-weighted in the bootstrap estimator to the extent they are in the original sample, for if the value 40 occurs twice it is only deleted once using a 10% trim. The 100 bootstap estimators $\tilde{\theta}_j$ generated for this example had a markedly skew distribution (reflecting to some extent the skewness of the initial sample values). Indeed the maximum and minimum $\tilde{\theta}_j$ were 23.13 and 6.88 respectively. The mean $\tilde{\theta}_{(.)}$=11.93, and (11.9) gives $\sigma_B(\hat{\theta})$=3.61. Note that this is close to the bootstrap standard deviation of the sample mean $\bar{x}$ calculated from all 10 sample values (and using the divisor 10 instead of 9 for the variance), which turns out to be 3.26 (the usual, equivalent to the jackknife, estimator with divisor 9 is 3.44). An approximate formula for the estimated variance for a trimmed mean of the original sample given by Efron (1982, p. 15) leads to an estimate of the standard deviation of 1.90.

It would appear, therefore, that in this particular example bootstrap estimation undoes some of the robustness that one attempts to build in by trimming. This is perhaps not surprising when one realizes that with very modest trimming of only one observation from each tail of the sample, the trimming will be less effective than it was in the original sample as soon as there is a repeat of an extreme value. This problem will be less acute in large samples than it is in small ones. It may also be overcome in trimming problems by more sophisticated approaches to trimming such as the use of an adaptive trimmed mean, a concept discussed by Carroll (1979). Indeed the problem is reduced, and a smaller bootstrap estimate of the standard error of the trimmed mean obtained, if we use a 20% trim, i.e. delete the two most extreme observations in each tail.

One of several proposals for a nonparametric confidence interval for a parameter θ, based on a Monte Carlo estimate of the bootstrap distribution of $\hat{\theta}$, is to determine the interval by arranging the $\tilde{\theta}_j$ in ascending order and taking the range specified by the central $100(1-\alpha)\%$ of the $\tilde{\theta}_j$ as a $100(1-\alpha)\%$ confidence interval. For example, if $B=100$ and we require an estimated 90% interval we would determine the relevant limits by rejecting the 5 smallest and 5 largest $\tilde{\theta}_j$. For $B=100$ for the 10% trimmed mean in this example our Monte Carlo simulation gave 90% confidence limits of

7.50 and 18.62, i.e. an estimated 90% confidence interval of (7.50, 18.62). This is asymmetric about $\hat{\theta} = 11.0$.

The normal theory t-distribution based 90% confidence interval for the original sample of 10 is (6.50, 19.11). This, of course, is symmetrical about $\bar{x} = 12.8$ and is not a satisfactory interval in the light of evidence of asymmetry in the data.

A 90% confidence interval based on the trimmed mean estimated variance given by Efron (1982, p. 15) and assuming this estimator has an approximate t-distribution with (again approximately) 9 degrees of freedom is (7.51, 14.88). Recognizing the influence of the asymmetry of the distribution of bootstrap estimators, Efron (1982, p. 82) proposes what he calls a 'bias-corrected' percentile confidence interval which in fact rejects a differing proportion of the $\tilde{\theta}_j$ in each tail. We omit detail, but this leads to a 90% bootstrap confidence interval of (6.88, 15.00), close to that derived using the trimmed mean estimated variance.

The above very brief description of bootstrap estimation, with many 'ifs' and 'buts', may sound derogartory. That is not the intention; bootstrapping is a valuable technique when there is no clear analytic theory to obtain a measure of accuracy of an estimator. The simplicity of the concept, allied to the complexity of the questions it may be called upon to answer, means it is an approach to be used with careful assessment of what it may achieve. Our simple illustrative example indicates both its potential and pitfalls.

11.3.1 Cross-validation

Other resampling schemes in wide use include **cross-validation.** The original use was to split samples randomly into two equal portions and to perform a number of alternative possible analyses (e.g. fitting regressions with different numbers of parameters, with and without data transformations, etc.) until one obtained a satisfactory fit with the first half-sample. The method of analysis that best suited this half-sample was 'cross-validated' by applying it to the remaining data. Usually, the model does not fit quite so well to the latter because it was 'tailored' to fit the first half-sample; however, if the fit is reasonable this gives support for the model.

Modern computers have led to extensions; in particular, it is now often used to detect aberrant points or individual observations in a manner that superficially resembles jackknifing. The data are analysed repeatedly omitting one (or sometimes a group) of points, in turn, and the effect on estimates is studied. The resemblance to jackknifing is superficial in that this form of cross-validation is not concerned with bias or other aspects of estimating parameters, but simply with determining whether some individual or groups of observations have a large influence on estimates.

11.4 FIELDS OF APPLICATION

Rather than suggesting specific applications we indicate here only the types of statistical problems where methods in this chapter have met with success. Most of these problems arise in a wide range of applications.

Correlation coefficients

We pointed out in Section 11.2 that the sample correlation coefficient, r, is generally a biased estimator of its population counterpart, ρ. Only in the case of the normal and one or two other distributions is the nature of this bias determined theoretically. In most cases little is known about the precision of r as an estimator of ρ. Efron (1982) studied the correlation coefficient between performance measures for a sample of 15 law students, The sample estimate of the correlation coefficient was $r=0.776$ and the jackknife estimate of bias was -0.007. Assuming normality, the bias would be -0.011. In a series of experiments each with 200 samples of 14 from a bivariate normal distribution with $\rho=0.5$, Efron found that the average estimate of standard deviation of jackknife estimates of ρ for all 200 trials was 0.223 with a standard deviation of 0.085, while for the bootstrap with $B=512$ it was 0.206 with a standard deviation of 0.063. The true value of the standard deviation of a sample estimate of ρ was, in this case, 0.218.

Ratio estimates

An estimate, $\bar{x}/\bar{y}$, of the ratio of population means is widely used in sampling. Jackknifing is a convenient way to estimate the often considerable bias as well as the variance of such estimators. A simple numerical example is given in Exercise 11.7.

Regression

All of m-estimators, the jackknife and bootstrap procedures have been widely used in regression to deal with outliers and to provide alternative approaches to some of those developed in Chapter 8. The classic paper on use of m-estimators in regression is that of Andrews (1974).

Adaptive trimmed means

Adaptive trimming of means is a procedure whereby analyses are carried out with, say, 5%, 10% and 20% trimming of observations in each tail. One seeks a degree of trimming that gives a location estimate with the smallest variance. This variance may be estimated by the jackknife or bootstrap. For this problem in experiments where the true standard deviation is known, Efron (1982) found that the bootstrap with $B=200$ performed better than the jackknife for samples of 10 or 20 from several different distributions.

11.5 SUMMARY

Outliers may often be detected by using (11.1), but many other tests exist.

Huber and other ***m*-estimators** behave in a similar manner to maximum likelihood estimators for 'well-behaved data' and reduce the influence of outliers. A well-known m-estimator for location due to Huber is given in Section 11.1.3. Iterative solutions of the normal equation are usually required and standardization is necessary to preserve solutions under linear transformation of the data.

Trimmed means (Section 11.1.5), with certain available modifications, provide robust estimates of location that are closely related to m-estimators.

The jackknife (Section 11.2) is a resampling procedure that provides a partial or sometimes complete correction for bias in sample estimators that are the equivalent of the parent distribution parameter. It also provides an estimator of the variance or standard deviation of the estimator.

Bootstrapping (Section 11.3) is a powerful resampling technique for obtaining the estimated standard deviation of, and also approximate confidence intervals for, certain parameter estimates. It is particularly useful when the mathematics of such estimation problems is intractable.

Cross-validation (Section 11.3.1) is a useful technique for exploring how well data fit a proposed model.

EXERCISES

11.1 Complete the computation of the m-estimators in Example 11.3.

11.2 Using the standardization factor based on the MAD with $k=1.5$, recalculate $\hat{\mu}$ for the data sets in Example 11.3.

11.3 Determine the jackknife estimates of mean and variance based on the sample -6, 0, 1, 2, 3. Verify that the sample variance is a biased estimator of σ^2 and determine the bias.

11.4 Smith and Naylor (1987) give the following data for strengths of 15 cm lengths of glass fibre and suggest that the two shortest may be outliers. Does the statistic (11.1) confirm this?

0.37	0.40	0.70	0.75	0.80	0.81	0.83	0.86	0.92	0.92	0.94
0.95	0.98	1.03	1.06	1.06	1.08	1.09	1.10	1.10	1.13	1.14
1.15	1.17	1.20	1.20	1.21	1.22	1.25	1.28	1.28	1.29	1.29
1.30	1,35	1.35	1.37	1.37	1.38	1.40	1.40	1.42	1.43	1.51
1.53	1.61									

11.5 Calculate the mean of the data in Exercise 11.4 and also the trimmed means trimming the top and bottom (i) deciles and (ii) quartiles. Explain any differences between these location estimates. Which do you prefer?

11.6 A sample of two gives the values 0, 1. The sample mean $\hat{\theta} = ½$ is the analogue of the population mean and has bootstrap variance 0.125. Obtain 100 bootstrap estimates of the mean and use these to compute the estimated bootstrap variance using (11.9). Is your estimate reasonably close to the true bootstrap variance?

11.7 A sample of paired observations of X, Y are

$$X = 1 \quad 2 \quad 3 \quad 5$$
$$Y = 3 \quad 7 \quad 9 \quad 7$$

Use the jackknife to estimate the bias in $\bar{x}/\bar{y}$ as an estimator of the ratio of population means.

11.8 Use the data in Exercise 11.7 and the jackknife to estimate bias in the sample correlation coefficient as an estimate of the population coefficient.

11.9 Use a Monte Carlo sampling technique to obtain B=20 bootstrap samples to estimate the bootstrap standard deviation of the sample median as an estimate of the population median, given the sample

1 2 3 9 11 12 20

12

Other developments

12.1 OTHER CONCEPTS AND METHODS

We have described many commonly used nonparametric concepts, tests and estimation procedures for location, dispersion, matching distributions, trends and association, but our coverage has been far from exhaustive. Alternative straightforward concepts and procedures range from quick and easy tests to techniques that are specific to one type of problem, or to those in a defined field of application. There are many nonparametric methods applicable to sophisticated problems. We have hinted at some of these in earlier chapters, e.g. at analysis of factorial treatment structures in Section 6.4.2 and also the use of splines in curve fitting in Section 8.2.

In Section 12.2 we list some commonly used elementary procedures not described in this book for testing and estimation, referring to several well-known and widely used texts where a description of each appears, often with examples, as well as additional references (often to source papers) where further detail may be found. Some of the books to which we refer are no longer in print, but they are often available in college or other libraries. Also in this section, we give references to a number of papers dealing with developments from basic techniques.

Section 12.3 provides a brief survey of some modern developments at a more sophisticated level (again with references for further reading), The aim of this chapter is to give a 'feel' for the scope of the subject rather than any in-depth or broad coverage. The fact that the *Journal of the American Statistical Association* published over 40 papers that dealt with very diverse aspects of the theory and application of nonparametric methods in the year 1990 alone is not atypical, and indicates the wide use and applicability of these methods.

12.2 FURTHER WIDELY USED PROCEDURES

We classify some well-known concepts and procedures under broad headings that indicate where each may be relevant. These are not hard and fast divisions, for, as emphasized throughout this book, many procedures initially introduced for a particular problem are closely related to tests used to answer what at first sight appear very different questions; e.g., the Jonckheere–Terpstra procedure was first introduced as a test for an ordered treatment effect (location shift) but is also a useful measure of association between counts equivalent to Kendall's tau for ordered categorical data. To avoid reptition we use codes given in Table 12.1 to refer to various texts (details of which are given in the references). In the lists of concepts and procedures the relevant text entry is followed by a number which indicates the page on which a description commences, e.g. NW 119 refers to page 119 of Neave and Worthington (1988).

Table 12.1. Codes for text references to additional methods.

Text authors	*Code*
Agresti (1990)	A
Bradley (1968)	B
Conover (1980)	C
Daniel (1990)	D
Kendall and Gibbons (1990)	KG
Leach (1979)	Lch
Lehmann (1975)	Leh
Marascuilo and McSweeney (1977)	MM
Neave and Worthington (1988)	NW
Siegel and Castellan (1988)	SC

Using the codes in Table 12.1 we list some concepts and procedures in various categories:

1. *Location*

Aligned rank test (alternative to Friedman) D 267, Leh 270, MM 401
Bradley–Terry model (paired comparisons) A 370, KG 193
Change point test (change in a process) SC 64
Durbin test (incomplete blocks) C 310, D 284, KG 129, MM 379, 464
Gart test (effect of order) Lch 137
Hollander test (two-sample) D 116
Mood (or Brown–Mood) (asymptotic version of median tests in this book) B 237, MM 290. NW 341
Neave and Worthington match tests NW 273 (and elsewhere in NW)
Rosenbaum test (differences in average performance) NW 144
Savage (exponential scores) test (comparing times to failure) Leh 103
Terry–Hoeffding test (expected normal scores) B 149, MM 280
Tukey's quick test NW 119

2. *Dispersion*

The following are tests for equal dispersion:

Ansari–Bradley test D 103
Freund–Ansari test B 121
Klotz test MM 291
Mood test MM 290
Mood–Westenberg test NW 344
Moses test D 107, SC161

3. *Samples from identical distributions or matching samples to populations*

Birnbaum–Hall test (several samples) C 377
Shapiro–Wilk test (normality) C 363
Wald–Wolfowitz run test (two samples) B 263, D 113, NW 352

4. *Association*

Brown–Mood method (regression) D 431, MW 220
Cramér coefficient C 181, D 403, SC 225
Hotelling–Pabst test (test of Spearman coefficient) B 91, C 255
Kappa (Cohen's kappa) (agreement) Lch 287, SC 284
Lambda statistic (asymmetric association) SC 298
Olmstead–Tukey test D 381, NW 144
Partial rank correlation C 260, D 395, KG Chap. 8, SC 254
Pearson contingency coefficient C 182
Point biserial correlation coefficient D 409
Somers' delta Lch 195, SC 303

5. *Miscellaneous*

Noether's test for cyclic trend B 179
Runs tests for randomness (in general) B Chaps 11, 12

Many basic techniques are undergoing continuing refinement or modification to increase efficiency or to make them applicable to new problems. We illustrate such developments for selected problems.

Sample size and power. How large a sample do we need to attain a given power when testing some H_0 against a specific alternative? While this may be reasonably easy to determine for some parametric tests, the less restrictive assumptions for nonparametric tests make power determination harder. Noether (1987a) discusses criteria for sample sizes needed for tests with given power against H_1 that differ from H_0 by an amount of interest. The criteria are applicable to the sign test, the Wilcoxon signed rank and rank sum tests and Kendall's tau. Noether also gives the Pitman efficiency of the sign test relative to the signed rank test for samples from a number of distributions; broadly, his findings confirm the known result that the sign test has superior efficiency for symmetric distributions with extremely long tails. On the other hand, the sign test has Pitman efficiency only ⅓ that of the signed rank test for samples from a uniform distribution.

Tests for several types of difference. While the Kruskal–Wallis test is essentially one for location differences, Boos (1986) has developed more comprehensive tests for k samples using linear rank statistics that test specifically for location, scale, skewness and kurtosis.

Simple approximate confidence intervals. Markowski and Hettmansperger (1982) consider computationally easy and reasonably simple estimates and confidence intervals for one- and two-sample location problems.

Bivariate means. Dinse (1982) gives a nonparametric analogue of Hotelling's T^2 test for hypotheses about means given samples from a normal bivariate distribution.

Extensions of Kolmogorov–Smirnov type tests. Chandra, Singpurwalla and Stephens (1981) give tables of critical values for a test of the Kolmogorov type for the extreme value and Weibull distributions – distributions that play a key role in studies of life duration or times to failure. Modifications of Smirnov tests applicable to discrete distributions are discussed by Eplett (1982). Saunders and Laud (1980) present a multivariate Kolmogorov goodness-of-fit test.

Symmetry. Tests for symmetry based on a Cramér–von Mises type of statistic proposed by various workers are reviewed by Koziol (1980), and Breth (1982) gives improved confidence bands for a distribution when the centre of symmetry is known or can be esitmated.

Binomial trends. A test for ordered trends in binomial responses was developed by Cochran (1954) and Armitage (1955), and STATXACT provides a program for their test, which is sometimes referred to as the Cochran–Armitage test.

Multivariate trend. Dietz and Killen (1981) use the concept of Kendall's tau to develop a multivariate test for trend and discuss its application to pharmaceutical data.

Nomination sampling. An ingenious paper by Willemain (1980) tackles the problem of estimating the median of a population given not one sample of n observations, but instead only the largest observed value in each of n independent samples. Willemain calls this **nomination** sampling. This may be the only type of information available in historical studies or when only selected cases are included in an experiment. For example, in testing a new and scarce or expensive drug, each of n consultants may be offered sufficient to treat only one patient and asked to give it to the patient on their list in greatest need. Records may be available for the change in level of some blood constituent for each treated patient; we may wish to use these to estimate the mean or median change level of this constituent that might be expected if the drug were used for all patients suffering from the disease.

Minimum scale parameter. Sometimes samples come from several populations for each of which the kth quantile is known to be the same. We may then be interested in selecting the population with the smallest variance, or some other scale measure. Nonparametric methods for this problem are discussed by Gill and Mehta (1989; 1991). A situation where this selection may be of interest is that in which several methods of producing an item all result in the same proportion of defectives (e.g. falling below a minimum permitted weight) and we want to select the production method with the lowest variability in weight of items produced.

Dominance Another problem of interest for two random variables X, Y is evaluation of $\Pr(X<Y)$ or $\Pr(X>Y)$. This is discussed for continuous distributions by Halperin, Gilbert and Lachin (1987) and for categorical data by Simonoff, Hochberg and Reiser (1986).

Minimum response levels. Shirley (1977) proposed a nonparametric test to determine the lowest dose level at which there is a detectable treatment reponse compared with the level for untreated control subjects. An improved version of this test was proposed by Williams (1986), and a version based on a Friedman-type ranking procedure applicable to a randomized block design is given by House (1986).

Small proportion of subjects responding. Johnson, Verill and Moore (1987) present rank tests when treatments do not affect all units in a sample but there is a response (perhaps an increase) for only a proportion of the units. The situation is common in some biochemical contexts where not all subjects respond to a stimulus.

Equality of marginal distributions. Lam and Longnecker (1983) present an adaptation of the Wilcoxon–Mann–Whitney test to test for equality of marginal distributions when sampling from a bivariate population.

Change-over points. The change-over point problem arises when observations are ordered in time and after a fixed time, τ, they have a distribution differing in location from that before τ. The problem is to determine τ. The problem may extend to several changes at different times. The basic problem is discussed by Pettitt (1979; 1981) and, for several change-over times, by Lombard (1987).

Angular distributions. Angular distributions are common in the geophysical sciences and meteorology. At a given weather station are daily wind directions at 6 a.m. correlated with those at noon? Are the directions of two kinds of geological fault correlated? Do prevailing wind directions show a seasonal shift? Lombard (1986) discusses the change-point problem for angular data. Fisher and Lee (1982; 1983) discuss nonparametric methods for assesing angular association based on an analogue for Kendall's tau.

Spatial medians. An ingenious extension of nonparametric location tests to bivariate data was proposed by Brown (1983). He applies angle tests analogous to the sign test for location shifts in a specified direction, or indeed in any direction, to what is termed the **spatial median**. This is the location measure in two dimensions that minimizes the sum of absolute distances to observations. Brown's tests are based on the sums of sines and cosines for polar coordinates associated with the directions of points from a hypothetical spatial median. Brown *et al.* (1992) review various aspects of bivariate sign tests. Multivariate sign tests are discussed by Randles (1989), and Peters and Randles (1990) develop a multivariate signed rank test for one-sample location problems.

12.3 MORE SOPHISTICATED DEVELOPMENTS

In this section we outline one or two recent and more sophisticated developments and applications not dealt with specifically in earlier chapters.

12.3.1 Design and analysis of experiments

We introduced some unsophisticated multiple comparisons tests in Section 6.4.1 and gave references to further work in this area. Some writers, including Fligner (1985), have taken a closer look at established nonparametric methods. Fligner makes a case for replacing Kruskal–Wallis tests by one based on all possible pairwise Wilcoxon–Mann–Whitney tests. While an analogous parametric procedure using pairwise sample t-tests has little to commend it, Fligner shows his procedure has good efficiency relative to Kruskal–Wallis, but for reasons indicated in Chapter 6 care is needed with significance levels as the tests are not independent. Hayter and Stone (1991) examine simultaneous (multiple comparison) confidence intervals for the Jonckheere–Terpstra situation where treatment differences are ordered. Akritas (1990) discusses the use of ranks in the analysis of two-factor factorial experiments in some detail.

Shirley (1981) proposed a nonparametric analysis of covariance based on adjustments to the sample rank means used in the Kruskal–Wallis test, and illustrated the method with a numerical example. Nonparametric analysis of covariance is also considered by Quade (1982) and by Conover and Iman (1982).

Klotz (1980) discusses the effect of missing categorical data on Cochran's test described in Section 6.4.4.

Sequential methods analogous to two-stage sampling for quality control are introduced by Spurrier and Hewett (1980) for a test of independence based on Kendall's tau. A first sample is taken and if the evidence from it is insufficient to reach a decision we take a second sample and examine an appropriate combination of the data sets. Lehmann (1975, pp. 103–4) draws attention to other sequential tests involving ranks. Sequential rank tests in clinical trials where repeated measurements are made are discussed by Lee and DeMets (1992).

Mack and Wolfe (1981) introduced a k-sample test called the **umbrella** test. This tests H_0 : *all* $F_i(x)$ *identical* against $H_1 : F_1(x) \leq F_2(x) \leq \ldots \leq F_u(x) \geq F_{u+1}(x) \geq \ldots \geq F_k(x)$ with at least one strict inequality. The umbrella value is u, where $1 \leq u \leq k$. This extends the idea of ordered location implicit in the Jonckheere–Terpstra test.

Experimental design problems for censored data are mentioned in Section 12.3.3.

12.3.2 Regression and curve fitting

A comprehensive discussion of much of the theory of distribution-free regression is contained in Maritz (1981, Chapters 5 and 6).

In a parametric context the relationship between covariance analysis and multiple regression is well known. Oja (1987) discusses permutation tests in multiple regression and the analysis of covariance.

Motivated by m-estimators, a method of fitting non-linear curves is described by Härdle and Gasser (1984). A more sophisticated approach using kernels (essentially weights) in nonparametric regression is that of Gasser, Müller and Mammitzsch (1985). Hall and Hart (1990) use bootstrap methods to test for differences between means expressed as very general regression functions. McKean, Sheather and Hettmansperger (1990) describe regression diagnostics for rank-based methods, an approach discussed in earler papers by McKean and co-workers.

The nonparametric calibration (or inverse regression) problem in which one uses regression to estimate x corresponding to later observed y is the subject of a paper by Knafl, Sacks, Spiegelman and Ylvisaker (1984).

A completely different approach to nonparametric curve fitting in the nonlinear case uses splines, a sound method for fitting smooth curves to data. Problems of oversmoothing are avoided by using penalty functions. We referred in Chapter 8 to a major paper by Silverman (1985) discussing both theoretical and practical aspects, and an interesting practical application to data on growth of sunflowers receiving nitrogen at different times by Silverman and Wood (1987). In an earlier paper Silverman (1984) describes a cross-validation method (see Section 11.3.1) for choice of smoothing parameters in spline regression. A general review of the use of splines in statistical problems is given in Wegman and Wright (1983).

12.3.3 Censored data

We discussed censored data in two samples in Section 5.1.8, where we pointed out that survival data are often incomplete in that we may not know the time of death of a patient but only be aware that he or she is still alive at a certain instant. This may arise due to loss of contact or termination of the study while some patients are still alive. Such censoring is often called 'right' censoring as it applies to the right or upper-tail of the survival time distribution. 'Left' or lower-tail censoring is also possible. This may occur if one is interested in the time of onset of a tumour, but it is impossible to determine this prescisely. If all that is known is the time of detection, onset can only be said to be earlier. In extreme cases a tumour may be detected only after death. In animal experiments where animals have been exposed to, say, a known carcinogen, which may or may not produce a tumour, and tumours can only be detected after death, some animals that do not die naturally may be sacrificed (i.e. deliberately killed) and examined to see if a tumour is present, an approach that introduces serious ethical questions.

Determination of death rates due to, say, a cancer, is further complicated by competing hazards; e.g. a person with a malignant tumour may die from a cause unconnected with the cancer itself.

All these complexities result in incomplete data, many of which have an element of censoring. Assumptions about times of onset or times of survival cannot always be precise and subjects may only be ranked for such criteria. Then nonparametric methods are virtually the only ones available.

Key papers in this area with a strong reliance on nonparametric methods, especially those based on ranks (or scores that are transformations of ranks), are those by Gehan (1965a; 1965b), Peto and Peto (1972) and Prentice (1978). Peto and Peto's work was extended by Harrington and Fleming (1982).

Different rank tests for survival data are compared and their differences elucidated by a video game in a not too technical paper by Lan and Wittes (1985).

Woolson and Lachenbruck (1983) consider analysis of covariance for censored data, and Hanley and Parnes (1983) consider nonparametric estimation of a multivariate distribution with censoring, while earlier Campbell (1981) considered bivariate estimation.

Dinse (1982) and Dinse and Lagakos (1982) both discuss estimation problems for lifetimes and times to failure for incomplete data, while Emerson (1982) and Brookmeyer and Crowley (1982) discuss confidence intervals for the median of censored data.

Efron (1981) and Robinson (1983) apply the bootstrap to censored data. Pettitt (1983; 1985) discussed aspects of regression with censored data.

Other papers dealing with various refinements of analysing censored data include those by Woolson and Lachenbruch (1980; 1981), Woolson (1981),

Slud, Byar and Green (1984), Michalek and Mihalko (1984), Turnbull and Mitchell (1984), Sandford (1985), Dinse (1986), Albers and Akritas (1987), Davis and Lawrence (1989), Dabrowska (1990), O'Quigley and Prentice (1991), Babu, Rao and Rao (1992); this list is indicative rather than exhaustive.

12.3.4 Nonparametric density estimation

For univariate data the histogram is a simple distribution-free representation of a probability density function. Modern computer power has made possible analogous representations of multivariate probability density functions based on large data sets. Just as 'smoothing techniques' can be used to iron out discontinuities inherent in a one-dimensional histogram, the same can be done in many dimensions. More sophisticated methods of estimation may also be used, including one known as kernel density estimation. The mathematical techniques involved are quite advanced and the subject has links with the use of splines and other approaches in sophisticated regression problems. Cross-validation techniques and bootstrapping have also proved useful.

Numerous applications may be found in the literature. For example, Kasser and Bruce (1969) measured a number of variables on patients with coronary heart disease and on normal patients. These included heartbeat rates at rest and after exercise. A three-dimensional plot of a nonparametrically estimated joint density function of these two variables for diseased and normal patients is given by Izenman (1991) and dramatically illustrates the differences in the pattern for the two groups.

A recent view of the subject, including over 200 references, is that by Izenman (1991). The references include a number of papers by P. Hall and by B.W. Silverman, two leading workers in this rapidly developing field, with applications in exploratory data analysis in areas as diverse as medicine and the interpretation of satellite pictures of the earth and other problems in the field generally known as pattern recognition.

12.3.5 Some miscellaneous applications

It would need an intensive search of the statistical literature and of journals in fields such as economics, psychology, biology, education and medicine to produce anything like a complete list of applications of nonparametric methods, but we demonstrate versatility by mentioning a few diverse applications. For example, Heltshe and Forrester (1983) and Smith and van Belle (1984) discuss use of the bootstrap and jackknife in relation to quadrat sampling for determining abundance of plant species in a locality.

Discrimination and classification problems are of wide interest; the first deals with allocation of individual units to known classes on the basis of observed measurements or categorizations, while classification is concerned with determining whether groups of individuals can be broken down into sub-classes such that measurements or alloted categories are more similar for members of a particular sub-class than they are between members of different sub-classes. Papers dealing with nonparametric methods of discrimination include those by Aitchison and Aitken (1976) and Hall (1981a). Gordon (1981) gives a full account of exploratory analysis in classification problems.

Hall and Titterington (1984) deal with nonparametric estimation of mixture proportions where a sample consists of observations some of which come from one distribution and some from another. The distributions may differ in location or scale. A special case of this type of problem is the 'outlier' situation where some observations are thought to come, for example, from a distribution with larger variance than the other, or from one with a different mean. We may sometimes be content to annul the effect of mixtures by using, say, a robust method or other nonparametric method known to be insensitive to outliers. However, there are occasions where we wish to know the proportion of our observations coming from each distribution. If this is something like 50:50 it makes no sense to regard observations from either distribution as outliers. Further papers on this topic include Hall (1981b) and Titterington (1983).

12.4 THE BAYESIAN APPROACH

Readers familiar with inference theory will know that an important school of inference is the Bayesian school, which regards paprameters not as fixed constants but as having a distribution (known as a **prior distribution**) which may be modified on the basis of experimental evidence.

With this emphasis on parameters, the term 'nonparametric Bayesian inference' may seem a contradiction. However, Bayesian concepts can be used in a nonparametric or distribution-free context. Ferguson (1973) gave an early survey of applications. In a quantal response bio-assay (where subjects either survive or die) one may assign a prior distribution to the effective dose, an approach described by Ramsay (1972) and by Disch (1981).

There is increasing interest in the theory of nonparametric Bayesian inference and one may expect more accounts of practical applications in future. Kestermont (1987) suggests a measure for comparing parametric and nonparametric Bayesian inferences.

12.5 CHOOSING A MODEL

We have made clear that *distribution-free* does not mean *assumption-free*, although the assumptions required are often minimal and clearly acceptable in many experimental situations. However, in other cases it is possible, indeed highly desirable, to specify a more detailed and essentially parametric or distribution-dependent model. In this section we cite just one situation where we may feel uneasy with a nonparametric analysis and indeed be stimulated by these difficulties to seek a more appropriate model to describe our data.

The situation we envisage is common in animal toxicology. In preliminary phases of drug testing a drug may be fed to pregnant female rats at several dose levels and various observations made on their offspring. Typically these will include litter size, numbers stillborn or deformed, and weight at birth, often followed by growth records and recording of abnormalities for some weeks after birth. Thomas E. Bradstreet kindly made available to me a large set of such records that he presented at the ASA statistical meetings in Altlanta, Georgia, in August 1991. For this discussion we confine our attention to records of live and dead offspring at birth. The full records give for each dam the dose level of an anti-hypertensive drug administered prior to and during pregnancy, the sex of each offspring and numbers alive and dead for each sex at birth. The dose levels were 0, 30, 100, 300 mg kg^{-1} (with and without access to an aqueous saline solution in place of water at this highest dose level). Twenty dams were allocated randomly to each dose level. In Table 12.2 we give for each of the dose levels the numbers of litters (a litter being the experimental unit) in which all births were live and the numbers of litters in which there was at least one stillbirth. These do not add to 20 for most dose levels because a few dams were not impregnated, or had to be sacrificed before the end of the pregnancy. Table 12.3 gives for each treatment total numbers of live births and total numbers stillborn. For illustrative simplicity we have combined data for sexes; in a fuller analysis the sexes would almost certainly be considered separately.

A key question in a study of this type is whether there is evidence of an association between dose level and toxicity (as measured by deaths). We might apply either a Jonckheere–Terpstra test (Section 9.3.2) or a linear-by-linear association test (Section 10.2.2) to the data in either Table 12.2 or 12.3. Using a linear-by-linear association test with rank scores, we reject H_0 : *no association between dose and deaths* in a one-tail test at an actual 3.36% level for Table 12.2 and at an actual exact 0.35% level for Table 12.3.

Closer joint inspection of these tables gives a hint of a feature of the original data that is lost if we look only at each table in isolation. While there is only one isolated stillborn offspring at zero dose and at other doses (Table 12.2) less than one half of litters have any stillbirths, examination of both tables makes it evident that if we use the drug, then among those litters in which some are stillborn, the average number stillborn is approximately 2 per litter. However, inspecting the full data, it is evident

Table 12.2 Numbers of rat litters with none or some stillborn members at various dose levels of an antihypertensive drug

Dose (mg kg^{-1})	*None dead*	*At least one dead*
0	18	1
30	14	3
100	14	6
300	17	2
300 + saline	12	7

Table 12.3 Numbers of rats born alive and numbers stillborn at various dose levels of an antihypertensive drug

Dose (mg kg^{-1})	*Alive*	*Stillborn*
0	260	1
30	203	4
100	263	14
300	252	2
300 + saline	248	14

that toxicity levels (as measured by stillbirths) at any applied dose level differ between litters. At the level '300+saline' stillbirths occur in 7 out of 19 litters; numbers stillborn in those 7 litters were 5, 1, 1, 2, 1, 3, 1. Of course, for the remaining 12 litters there are in each case zero stillbirths.

In Section 10.1 we briefly mentioned a parametric model called the logisitic regression model. This is particularly apposite when the proportions responding in some binary fashion (e.g. alive/dead) are binomial with the parameter p increasing with dose. In the situation we are considering here there is evidence that that model needs extending to allow for differing probabilities of responses within each litter at any given dose level. Statisticians dealing with toxicological data have developed parametric models to deal with such data. One such is the beta-binomial model described by Crowder (1978), Williams (1982) and others.

While the crude tests based on Jonckheere–Terpstra or the linear-by-linear model indicate association in this example, parametric models of the beta-binomial type will generally prove more efficient in situations like this.

APPENDIX

A1 RANDOM VARIABLES

In Chapter 1 we recommended books giving elementary accounts of basic ideas. This appendix does not supplant these, but we summarize a few basic concepts that occur frequently and that are especially relevant to nonparametric methods. Readers who find this summary too terse should refer to a recommended text (or another of their choice) for greater detail.

A1.1 Discrete and continuous random variables

A discrete random variable, X, takes only a finite (or countably infinite) set of values. These values are often integers (perhaps zero and a range of positive integers). If x_i is the ith observable value of X then $\Pr(X=x_i)=p_i$ is the value of the **probability mass function** (p.m.f.) of X when $X=x_i$.

The **cumulative distribution function** (c.d.f.) is

$$F(x)=\Pr(X\leq x)$$

If $x_i \leq x < x_{i+1}$ then $F(x)=\Sigma_{r\leq i}\, p_r$.

Example. For a binomial distribution with parameters n, p the p.m.f. is

$$\Pr(X=r)={}^nC_r\, p^r q^{n-r}, \quad r=0, 1, 2, \ldots, n$$

where $q=1-p$, and the c.d.f. is

$$F(x)=\Sigma_{r\leq [x]}\Pr(X=r)=\Sigma_{r\leq [x]}{}^nC_r\, p^r q^{n-r}, \quad 0\leq r\leq n$$

where $[x]$ denotes the integral part of x, i.e. the largest integer not exceeding x, e.g. if $x=3.72$, $[x]=3$ (but note that if $x=-3.72$ then $[x]=-4$). $F(x)$ is a non-decreasing step function with a step of height ${}^nC_r\, p^r q^{n-r}$ at each r from 0 to n inclusive. $F(x)=0$ if $x<0$, $F(0)=q^n$, $F(n)=1$ and $F(x)=1$ if $x>n$.

A useful and easily verified recursive formula for successive values of the p.m.f. of the binomial distribution is

$$\Pr(X=r+1)=\frac{(n-r)p}{(r+1)q}\Pr(X=r)$$

Example. For the Poisson distribution with parameter λ the p.m.f. is

$$\Pr(X=r)=\lambda^r e^{-\lambda}/(r!) \quad r=0, 1, 2, \ldots$$

The c.d.f. is

$$F(x)=\Sigma_{r\leq[x]}\ \lambda^r e^{-\lambda}/(r!),\quad x\geq 0$$

This is a step function with a step of height $\lambda^r e^{-\lambda}/(r!)$ at $x=r$. $F(x)=0$ if $x<0$, $F(0)=e^{-\lambda}$ and $F(\infty)=1$. For the p.m.f. the recursive formula

$$\Pr(X=r+1)=\{\lambda/(r+1)\}\Pr(X=r)$$

is easily verified.

A continuous random variable X may take any value in an interval (a, b), finite or infinite. The **probability density function** (p.d.f.), $f(x)$, is zero for x outside (a, b) and in (a, b), $f(x)\delta x$ represents the probability that X takes a value in the arbitrarily small interval $(x, x+\delta x)$. The **cumulative distribution function** (c.d.f.) is the monotonic non-decreasing function

$$F(x)=\Pr(X\leq x)=\int_{-\infty}^{x} f(x)\mathrm{d}x$$

The equalities $F(-\infty)=0$ and $F(\infty)=1$ always hold. If a and/or b are finite then also $F(a)=0$ and $F(b)=1$; and $F(x)=0$ if $x<a$ and $F(x)=1$ if $x\geq b$. A distribution is said to be **symmetric** about θ if, for all x, $f(\theta-x)=f(\theta+x)$.

Example. The normal distribution with mean μ and standard deviation σ is symmetrically distributed about μ.

A1.2 Means, medians and quantiles

The mean value or **expectation** of X is, for a discrete distribution,

$$E(X)=\Sigma_i p_i x_i$$

For the binomial distribution with parameters n, p, $E(X)=np$. For the Poisson distribution with parameter λ, $E(X)=\lambda$.

For a continuous distribution

$$E(X)=\int_{-\infty}^{\infty} xf(x)\mathrm{d}x$$

For $0<k<1$, the kth **quantile** of a random variable X is a value x_k that satisfies the inequalities $\Pr(X<x_k)\leq k$ and $\Pr(X>x_k)\leq 1-k$. For discrete distributions if x_k is not unique a unique kth quantile is obtained by taking the mean of the largest and smallest x_k satisfying the kth quantile condition. If $k=½$ the corresponding quantile is the **median.** If $k=r/4$, $r=1, 2, 3$, the quantile is the rth **quartile**, while if $k=r/10$, $r=1, 2, \ldots, 9$, the quantile is the rth **decile.** Similarly $k=r/100$ defines the rth **percentile.** The second quartile, 5th decile and 50th percentile all correspond to the median.

Example. For the binomial distribution with $n=5, p=\frac{1}{3}$ the probabilities associated with each X are:

r	0	1	2	3	4	5
$\Pr(X=r)$	32/243	80/243	80/243	40/243	10/243	1/243

It is easily verified that $\Pr(X<2)<\frac{1}{2}$ and $\Pr(X>2)<\frac{1}{2}$. Thus 2 is the unique median. For the binomial distribution with $n=5$, $p=\frac{1}{2}$ the probabilities associated with each X are

r	0	1	2	3	4	5
$\Pr(X=r)$	1/32	5/32	10/32	10/32	5/32	1/32

Clearly $X=2$ and $X=3$ both satisfy the condition for a median, so we choose $x_m=(2+3)/2=2.5$ as a unique median. Figure A1 shows the graph of $F(x)$ for this distribution and illustrates clearly why there is no unique median without a convention such as averaging.

Population means and medians are commonly referred to as measures of location or **location parameters.**

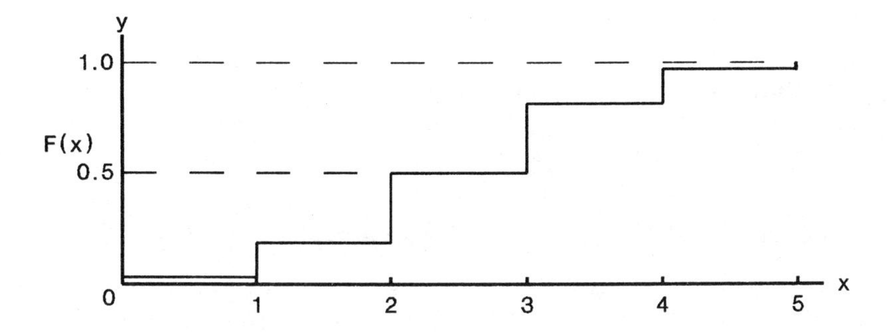

Figure A1 A non-unique median for a binomial variate.

A1.3 Higher moments

The kth moment of X about the point a is

$$E[(X-a)^k]=\Sigma_i\, p_i(x_i-a)^k$$

for a discrete random variable and

$$E[(X-a)^k]=\int_{-\infty}^{\infty}(x-a)^k f(x)\mathrm{d}x$$

for a continuous random variable, where k is a positive integer. The integral may not be convergent for all k. If $a=0$, $E(X^k)$ is the kth moment about the origin and if $a=\mu$, $E[(X-\mu)^k]$ is the kth moment about the mean and is often denoted by μ_k. The second moment about the mean, μ_2, is the variance. For symmetric distributions the mean is the point of symmetry and for odd k, $\mu_k=0$. For any distribution $\mu_1=E(X-\mu)=0$. A commonly used measure of asymmetry or skewness is $\gamma_1=\mu_3/\mu_2^{3/2}$. Another feature of interest is the peakedness of distributions relative to the normal; broadly speaking, this is measured by $\gamma_2=(\mu_4/\mu_2^2)-3$, which has the value zero for the normal distribution. Although not always true, usually if γ_2 is negative the graph of the probability density function $f(x)$ is flatter than that for the normal distribution, while it is more peaked if γ_2 is positive.

A1.4 The sample cumulative distribution function

Given a random sample of n observations $x_1, x_2, \ldots, x_n$, we may arrange them in ascending order, If there is ambiguity we customarily denote the ordered sample values by $x_{(1)}, x_{(2)}, \ldots, x_{(n)}$, where $x_{(1)} \leq x_{(2)} \leq \ldots \leq x_{(n)}$. The **sample** (or **empirical**) **cumulative distribution function** denoted by $S(x)$ is a step fuunction with step height $1/n$ at each unique $x_{(i)}$. In the case of r tied values the step height is r/n at that tied x value. $S(x)$ provides an estimate of the population c.d.f. An example of a sample c.d.f. is given in Figure 3.5.

Sample quantiles are defined with respect to $S(x)$, associating a probability $1/n$ with each sample value. If $n=2m+1$ is odd we have a unique sample median at $x_{(m)}$, whereas if $n=2m$ is even the unique sample median is defined as $\frac{1}{2}[x_{(m)}+x_{(m+1)}]$.

A2 PERMUTATIONS AND COMBINATIONS

Many nonparametric tests require a knowledge of the number of possible orderings of a set of n observations, or of all sub-sets of r of these. These are **permutation** problems. On other occasions we are not interested in different arrangements within any sub-set of r, but only wish to know how many sub-sets of r (each differing by at least one member) we can form from n observations. This is a **combination** problem.

The number of permutations of r items from n is written ${}^nP_r$, and

$${}^nP_r=n(n-1)(n-2)\ldots(n-r+1)$$

Example. If we denote $n=4$ objects by A, B, C, D the possible ordered sets of $r=2$ are AB, AC, AD, BC, BD, CD together with these pairs in reversed order; i.e. 12 in all, consistent with ${}^4P_2=4\times3=12$.

The general form of ${}^nP_r$ arises because we have n possible first choices, each of which can be combined with one of the remaining $n-1$ as a second choice, giving $n(n-1)$ permutations of 2. Continuing this way gives the general form. When $r=n$, ${}^nP_n=n!$, the product of all integers from 1 to n.

In a combination problem we denote the number of groups of r objects from n, each group having at least one different member, by ${}^nC_r$. Clearly for each combination of r items there are ${}^rP_r=r!$ permutations whence ${}^nC_r \times (r!)={}^nP_r$ or ${}^nC_r=({}^nP_r)/(r!)$.

It is not difficult to show that

$$ {}^nC_r=n(n-1)...(n-r+1)/(r!)=(n!)/[r!(n-r)!] $$

and that ${}^nC_r={}^nC_{n-r}$.

A3 SELECTING A RANDOM SAMPLE OF *r* ITEMS FROM *n*

It is easy to select a random sample of r from n items if one has a computer program to list a random set of r numbers between 1 and n (or between 0 and $n-1$). If we number our items 1 to n (or 0 to $n-1$) the program effectively tells us which numbered items are to form our sample of r. Samples may also be obtained using tables of random digits. If we want a sample of 12 items from 30 a useful way to proceed is to number our 30 items 0 to 29. We next select paired random digits from a table. If these lie bewtween 00 (read as 0) and 29 we include the corresponding number in our sample (ignoring any number appearing a second time if we are sampling without replacement). We need not waste all digit pairs over 29. What we do is divide them by 30 and note the remainder, e.g 45/30 gives remainder 15 and 72/30 gives remainder 12, while 60/30 gives remainder 00. These remainders may be used as further sample numbers (again ignoring repeats if sampling without replacement). This is fine for digit pairs 30-59 and 60-89 inclusive, as all remainders between 0 and 29 are equally likely. However, we reject digit pairs between 90 and 99, for if we included remainders when these are divided by 30 we would have a higher overall probability of remainders between 0 and 9 over all digit pairs.

Better use of repeated numbers is possible, but has little advantage for small samples. If sampling with replacement (as in the bootstrap) we retain repeated values, some items appearing twice or more in a sample.

The selection methods generalize to triplets or even larger sets of random digits. For example, if we want a sample of 26 items from a total of 123 we might use all triplets from 000 to 983, taking remainders on division by 123 for numbers between 123 and 983.

A4 THE *t*-DISTRIBUTION AND *t*-TESTS

A4.1 The single-sample test

If $x_1, x_2, \ldots, x_n$ is a sample of n observations from a normal distribution with mean μ and unknown standard deviation σ, to test $H_0 : \mu=\mu_0$ against a one- or two-sided alternative the optimum test statistic is

$$t = \frac{\bar{x} - \mu_0}{s_x/\sqrt{n}} \tag{A4.1}$$

where $(n-1)s_x^2 = \Sigma_i x_i^2 - (\Sigma_i x_i)^2/n$. Under H_0, t has a t-distribution with $n-1$ degrees of freedom. When testing H_0 against $H_1: \mu \neq \mu_0$, large values of $|t|$ indicate significance. Critical values at the 5%, 1% or other levels depend upon the degrees of freedom, $n-1$, and are widely tabulated (see e.g. Neave, 1981, p. 20). For a one-tail test, t must be taken with the relevant sign and the critical value is that corresponding to the required upper- or lower-tail probability. The distribution of t under H_0 is symmetric, so only positive critical values need be tabulated. Many computer programs indicate tail probabilities associated with an observed t when H_0 is true. A $100(1-\alpha)\%$ confidence interval for μ is given by

$$\bar{x} \pm t_{n-1,\alpha}\, s_x/\sqrt{n}$$

where $t_{n-1,\alpha}$ is the critcal t-value for significance at the $100\alpha\%$ level in a two-tail test.

A4.2 Two independent samples

If we have two independent samples of size m, n, i.e.

$$x_1, x_2, \ldots, x_m \text{ and } y_1, y_2, \ldots, y_n$$

from normal distributions with means μ_x, μ_y, both with standard deviation σ, then

$$t = \frac{\bar{x} - \bar{y} - (\mu_x - \mu_y)}{s\sqrt{[1/m + 1/n]}} \tag{A4.2}$$

has a t-distribution with $m+n-2$ degrees of freedom, where $(m+n-2)s^2 = (m-1)s_x^2 + (n-1)s_y^2$, with s_x defined as in Section A4.1 and s_y similarly defined for the second sample. A commonly occurring case is that in which H_0 is $\mu_x = \mu_y$, when the numerator of (A4.2) simplifies to $\bar{x} - \bar{y}$. Critical values for $m+n+2$ degrees of freedom may be obtained from t-tables.

A $100(1-\alpha)\%$ confidence interval for $\mu_x - \mu_y$ has, in an obvious notation, the form

$$\bar{x} - \bar{y} \pm t_{m+n-2,\alpha}\, s[\sqrt{(1/m + 1/n)}]$$

A5 THE CHI-SQUARED AND F-DISTRIBUTIONS

The sum of squares of n independent standard normal variables has a chi-squared distribution with n degrees of freedom. One reason why this distribution is important is that many test statistics in both parametric and nonparametric inference have for large sample sizes approximately a chi-squared distribution under a relevant null hypothesis.

Broadly speaking, high values of chi-squared indicate significance and critical values relevant to one tail of the distribution are usually appropriate. These are tabulated at the 5% and 1% level for a wide range of degrees of freedom by Neave (1981, p. 21) who gives examples of how to carry out appropriate tests.

The F-distribution in its simplest form is used to test $H_0 : \sigma_x^2 = \sigma_y^2$ when given independent samples of size m, n from normal populations. The test statistic against a two-sided alternative is the larger of $F_1 = s_x^2/s_y^2$ and its reciprocal $F_2 = s_y^2/s_x^2$, where s_x^2, s_y^2 are as defined in Section A4. Under H_0 the relevant statistic has an F-distribution with $m-1$, $n-1$ degrees of freedom if s_x^2 is in the numerator and $n-1$, $m-1$ degrees of freedom if s_y^2 is in the numerator. High values of F_1 or F_2, as appropriate, indicate significance. Critical values for both one- and two-tail tests are given by Neave (1981, pp. 22–5) for a wide range of numerator and denominator degrees of freedom.

In most analysis of variance type situations a one-tail test is appropriate: the numerator in the quotient of estimates of variance is always the one with the higher expected 'population' variance under H_1. High values above critical levels appropriate to a one-tail test indicate rejection of H_0. Examples are given by Neave. In most nonparametric tests in this book that use F, one-tail critical values are appropriate, and degrees of freedom for the numerator are given first. Note that the one-tail critical region values for the chi-squared or F-statistic may be relevant even when the alternative hypothesis is essentially two-sided; e.g. in a test of independence against that of association in a general $r \times c$ contingency table.

A6 LEAST SQUARES REGRESSION

Given n paired observations $(x_1, y_1), (x_2, y_2), \ldots, (x_n, y_n)$ the straight line of best fit given by least squares is appropriate if we may write our model

$$y_i = \alpha + \beta x_i + \epsilon_i$$

where the ϵ_i are unknown 'errors' independently and identically (and often assumed normally) distributed with mean 0 and (generally unknown) variance σ^2.

Least squares estimates a, b, of α, β are obtained by choosing a, b to minimize

$$U = \Sigma_i e_i^2 = \Sigma_i (y_i - a - bx_i)^2, \text{ where } e_i = y_i - a - bx_i$$

Equating partial derivatives of U with respect to a, b to zero gives the 'normal' or estimating equations

$$\Sigma_i e_i = \Sigma_i (y_i - a - bx_i) = 0$$
$$\Sigma_i x_i e_i = \Sigma_i x_i (y_i - a - bx_i) = 0$$

Solving for a, b we find

$$b = c_{xy}/c_{xx} \text{ and } a = \bar{y} - b\bar{x}$$

where

$$c_{xy} = \Sigma_i x_i y_i - (\Sigma_i x_i)(\Sigma_i y_i)/n, \quad c_{xx} = \Sigma_i x_i^2 - (\Sigma_i x_i)^2/n$$

A7 BADENSCALLIE DATA SETS

Several examples in this book use data on age at death of male members of four Scottish clans as recorded in the burial ground at Badenscallie in the Coigach district of Wester Ross, Scotland. The data were collected in June 1987. Clan names have been changed but the records are as complete as possible for four real clans. There are a few missing values because names or dates were unreadable on a few headstones, and indeed some headstones appeared to be missing, but these were few in number. Minor spelling variations, especially those of M', Mc, Mac, were ignored. Ages are given for complete years, e.g. 0 means before first birthday, 79 means on or after 79th but before 80th birthday, according to the information given on the tombstone. Data have been put in ascending order within each clan.

McAlpha (59 members):

0 0 1 2 3 9 14 22 23 29 33 41 41 42 44 52 56 57
58 58 60 62 63 64 65 69 72 72 73 74 74 75 75 75 77 77
78 78 79 79 80 81 81 81 81 82 82 83 84 84 85 86 87 87
88 90 92 93 95

McBeta (24 members):

0 19 22 30 31 37 55 56 66 66 67 67 68 71 73 75 75 78
79 82 83 83 88 96

McGamma (21 members):

13 13 22 26 33 33 59 72 72 72 77 78 78 80 81 82 85 85
85 86 88

McDelta (13 members):

1 11 13 13 16 34 65 68 74 77 83 83 87

Table I Cumulative binomial probabilities, $p=0.5$, $6 \leq n \leq 20$

$\Pr(X \leq r) = \Sigma_{i \leq r} \, {}^{n}C_{i} \, (0.5)^{n}$ is recorded for given n, r.

n \ r	0	1	2	3	4	5	6	7	8	9	10	11	12	13	14	15	16	17	18	19	20
6	0.016	0.109	0.344	0.656	0.891	0.984	1.00														
7	0.008	0.062	0.227	0.500	0.773	0.938	0.992	1.00													
8	0.004	0.035	0.144	0.363	0.637	0.856	0.965	0.998	1.00												
9	0.002	0.020	0.090	0.254	0.500	0.746	0.910	0.980	0.996	1.00											
10	0.001	0.011	0.055	0.172	0.377	0.623	0.828	0.945	0.989	0.999	1.00										
11	0.001	0.006	0.033	0.113	0.274	0.500	0.726	0.887	0.967	0.994	0.999	1.00									
12	0.000	0.003	0.019	0.073	0.194	0.387	0.613	0.806	0.927	0.981	0.997	1.00	1.00								
13	0.000	0.002	0.011	0.046	0.133	0.291	0.500	0.710	0.867	0.954	0.989	0.998	1.00	1.00							
14	0.000	0.001	0.006	0.029	0.090	0.212	0.395	0.605	0.788	0.910	0.971	0.994	0.999	1.00	1.00						
15	0.000	0.000	0.004	0.018	0.059	0.151	0.304	0.500	0.696	0.849	0.941	0.982	0.996	1.00	1.00	1.00					
16	0.000	0.000	0.002	0.011	0.038	0.105	0.227	0.402	0.598	0.773	0.895	0.962	0.989	0.998	1.00	1.00	1.00				
17	0.000	0.000	0.001	0.006	0.024	0.072	0.166	0.314	0.500	0.686	0.834	0.928	0.976	0.994	0.999	1.00	1.00	1.00			
18	0.000	0.000	0.001	0.004	0.015	0.048	0.119	0.240	0.407	0.593	0.760	0.881	0.952	0.985	0.996	0.999	1.00	1.00	1.00		
19	0.000	0.000	0.000	0.002	0.010	0.032	0.084	0.180	0.324	0.500	0.676	0.820	0.916	0.968	0.990	0.998	1.00	1.00	1.00	1.00	
20	0.000	0.000	0.000	0.001	0.006	0.021	0.058	0.132	0.252	0.412	0.588	0.748	0.868	0.942	0.979	0.994	0.999	1.00	1.00	1.00	1.00

Table II Critical values for the Wilcoxon signed rank test

Values for the maximum of the lesser of S_- or S_+ for significance at nominal 5% and 1% levels for one- and two-tail tests for $6 \leq n \leq 20$.

n	6	7	8	9	10	11	12	13	14	15	16	17	18	19	20
One-tail test															
5% level	2	3	5	8	10	13	17	21	25	30	35	41	47	53	60
1% level	*	0	1	3	5	7	9	12	15	19	23	27	32	37	43
Two-tail test															
5% level	0	2	3	5	8	10	13	17	21	25	29	34	40	46	52
1% level	*	*	0	1	3	5	7	9	12	15	19	23	27	32	37

*Sample too small for test at this level.

Table III Critical values for $|F(x)-S(x)|$ for Kolmogorov's test.
Maximum values of $|F(x)-S(x)|$ for significance at 5% and 1% levels for one- and two-tail tests, sample sizes $5 \le n \le 40$. An approximation is given for $n > 40$.

n	5	6	7	8	9	10	11	12	13	14	15	16	17	18	19	20	21	22	23
One-tail test																			
5% level	0.509	0.468	0.436	0.410	0.388	0.369	0.352	0.338	0.326	0.314	0.304	0.295	0.286	0.279	0.271	0.265	0.259	0.253	0.248
1% level	0.627	0.577	0.538	0.507	0.480	0.457	0.437	0.419	0.404	0.390	0.377	0.366	0.355	0.346	0.337	0.329	0.321	0.314	0.307
Two-tail test																			
5% level	0.563	0.519	0.483	0.454	0.430	0.409	0.391	0.375	0.361	0.349	0.338	0.327	0.318	0.309	0.301	0.294	0.287	0.281	0.275
1% level	0.669	0.617	0.576	0.542	0.513	0.489	0.468	0.449	0.433	0.418	0.404	0.392	0.381	0.371	0.361	0.352	0.344	0.337	0.330

n	24	25	26	27	28	29	30	31	32	33	34	35	36	37	38	39	40	$n>40$
One-tail test																		
5% level	0.242	0.238	0.233	0.229	0.225	0.221	0.218	0.214	0.211	0.208	0.205	0.202	0.199	0.197	0.194	0.192	0.189	$1.22/\sqrt{n}$
1% level	0.301	0.295	0.290	0.284	0.279	0.275	0.270	0.266	0.262	0.258	0.254	0.251	0.247	0.244	0.241	0.238	0.235	$1.52/\sqrt{n}$
Two-tail test																		
5% level	0.269	0.264	0.259	0.254	0.250	0.246	0.242	0.238	0.234	0.231	0.227	0.224	0.221	0.218	0.215	0.213	0.210	$1.36/\sqrt{n}$
1% level	0.323	0.317	0.311	0.305	0.300	0.295	0.290	0.285	0.281	0.277	0.273	0.269	0.265	0.262	0.258	0.255	0.252	$1.63/\sqrt{n}$

Adapted from Miller (1956) by permission of the publishers of the *Journal of the American Statistical Association.*

Table IV Critical values for significance for Lilliefors' test statistic for normality.
Values are the minimum for significance at 5% and 1% levels in a two-tail test for $6 \le n \le 20$, $n=25, 30$. Linear interpolation may be used for $20<n<30$. An approximation is given for $n>30$.

n	6	7	8	9	10	11	12	13	14	15	16	17	18	19	20	25	30	$n>30$
5% level	0.319	0.300	0.285	0.271	0.258	0.249	0.242	0.234	0.227	0.220	0.213	0,206	0.200	0.195	0.190	0.173	0.161	$0.886/\sqrt{n}$
1% level	0.364	0.348	0.331	0.311	0.294	0.284	0.275	0.268	0.261	0.257	0.250	0.245	0.239	0.235	0.231	0.203	0.187	$1.031/\sqrt{n}$

Adapted from Lilliefors (967) by permission of the publishers of the *Journal of the American Statistical Association.*

Table V Critical values for the Wilcoxon—Mann—Whitney U statistic (equal sample sizes)
The maximum value of the lesser of U_m, U_n for significance in a one- or two-tail test for equal sample sizes $5 \le m=n \le 20$ is given for nominal 5% and 1% significance levels.

$n=m$	5	6	7	8	9	10	11	12	13	14	15	16	17	18	19	20
One-tail test																
5% level	4	7	11	15	21	27	34	42	51	61	72	83	96	109	123	138
1% level	1	3	6	9	14	19	25	31	39	47	56	66	77	88	101	114
Two-tail test																
5% level	2	5	8	13	17	23	30	37	45	55	64	75	87	99	113	127
1% level	0	2	4	7	11	16	21	27	34	42	51	60	70	81	93	105

Table VI Critical values for the Wilcoxon–Mann–Whitney U statistic (unequal sample sizes)

The maximum of the lesser of U_m, U_n for significance in one- or two-tail tests is given for unequal sample sizes between 5 and 20. Values above and to the right of the diagonals apply to a nominal 5% level and values below and to the left apply to a nominal 1% level.

One-tail test

m \ n	5	6	7	8	9	10	11	12	13	14	15	16	17	18	19	20
5		5	6	8	9	11	12	13	15	16	18	19	20	22	23	25
6	2		8	10	12	14	16	17	19	21	23	25	26	28	30	32
7	3	4		13	15	17	19	21	24	26	28	30	33	35	37	39
8	4	6	7		18	20	23	26	28	31	33	36	39	41	44	47
9	5	7	9	11		24	27	30	33	36	39	42	45	48	51	54
10	6	8	11	13	16		31	34	37	41	44	48	51	55	58	62
11	7	9	12	15	18	22		38	42	46	50	54	57	61	65	69
12	8	11	14	17	21	24	28		47	51	55	60	64	68	72	77
13	9	12	16	20	23	27	31	35		56	61	65	70	75	80	84
14	10	13	17	22	26	30	34	38	43		66	71	77	82	87	92
15	11	15	19	24	28	33	37	42	47	51		77	83	88	94	100
16	12	16	21	26	31	36	41	46	51	56	61		89	95	101	107
17	13	18	23	28	33	38	44	49	55	60	66	71		102	109	115
18	14	19	24	30	36	41	47	53	59	65	70	76	82		116	123
19	15	20	26	32	38	44	50	56	63	69	75	82	88	94		130
20	16	22	28	34	40	47	53	60	67	73	80	87	93	100	107	

Two-tail test

m \ n	5	6	7	8	9	10	11	12	13	14	15	16	17	18	19	20
5		3	5	6	7	8	9	11	12	13	14	15	17	18	19	20
6	1		6	8	10	11	13	14	16	17	19	21	22	24	25	27
7	1	3		10	12	14	16	18	20	22	24	26	28	30	32	34
8	2	4	6		15	17	19	22	24	26	29	31	34	36	38	41
9	3	5	7	9		20	23	26	28	31	34	37	39	42	45	48
10	4	6	9	11	13		26	29	33	36	39	42	45	48	52	55
11	5	7	10	13	16	18		33	37	40	44	47	51	55	58	62
12	6	9	12	15	18	21	24		41	45	49	53	57	61	65	69
13	7	10	13	17	20	24	27	31		50	54	59	63	67	72	76
14	7	11	15	18	22	26	30	34	38		59	64	69	74	78	83
15	8	12	16	20	24	29	33	37	42	46		70	75	80	85	90
16	9	13	18	22	27	31	36	41	45	50	55		81	86	92	98
17	10	15	19	24	29	34	39	44	49	54	60	65		93	99	105
18	11	16	21	26	31	37	42	47	53	58	64	70	75		106	112
19	12	17	22	28	33	39	45	51	57	63	69	74	81	87		119
20	13	18	24	30	36	42	48	54	60	67	73	79	86	92	99	

Table VII Critical values for significance in the Smirnov test (equal sample sizes)
Minimum value of $|S(x)-S(y)|$ for significance in one- and two-tail tests at nominal 5% and 1% levels for equal sample sizes $5 \le m = n \le 20$. Here and in Table VIII discontinuities often result in large differences between nominal and actual levels; this explains discontinuities in the general downward trend in critical values as sample size increases.

$m=n$	5	6	7	8	9	10	11	12	13	14	15	16	17	18	19	20
One-tail test																
5% level	0.800	0.833	0.714	0.625	0.667	0.600	0.545	0.500	0.538	0.500	0.467	0.438	0.471	0.444	0.421	0.400
1% level	1.00	1.00	0.857	0.750	0.778	0.700	0.727	0.667	0.615	0.571	0.600	0.563	0.529	0.555	0.526	0.500
Two-tail test																
5% level	1.00	0.833	0.857	0.750	0.667	0.700	0.636	0.583	0.538	0.571	0.533	0.500	0.471	0.500	0.474	0.450
1% level	1.00	1.00	0.857	0.875	0.778	0.800	0.727	0.667	0.692	0.643	0.600	0.625	0.583	0.555	0.526	0.550

Table VIII Critical values for significance in the Smirnov test (unequal sample sizes)
Minimum values of $|S(x)—Sy)|$ for significance in one- and two-tail tests for unequal sample sizes between 5 and 20. Entries above and to right of diagonals are for testing at 5% level; entries below and to left are for testing at 1% level. See also text for Table VII.

One-tail test

n / *m*	5	6	7	8	9	10	11	12	13	14	15	16	17	18	19	20
5		0.800	0.714	0.675	0.667	0.700	0.636	0.600	0.615	0.600	0.667	0.600	0.588	0.578	0.589	0.600
6	1.00		0.667	0.625	0.611	0.600	0.576	0.667	0.590	0.571	0.567	0.563	0.649	0.511	0.561	0.550
7	0.857	0.833		0.607	0.571	0.571	0.571	0.547	0.549	0.571	0.533	0.526	0.513	0.516	0.519	0.514
8	0.875	0.833	0.750		0.556	0.550	0.545	0.500	0.519	0.517	0.500	0.563	0.500	0.500	0.487	0.500
9	0.800	0.778	0.746	0.750		0.556	0.525	0.528	0.504	0.500	0.511	0.479	0.484	0.500	0.468	0.467
10	0.800	0.733	0.714	0.700	0.678		0.518	0.500	0.492	0.486	0.500	0.475	0.465	0.456	0.447	0.500
11	0.800	0.742	0.714	0.693	0.636	0.627		0.485	0.469	0.474	0.461	0.455	0.455	0.444	0.440	0.436
12	0.800	0.750	0.690	0.667	0.639	0.617	0.583		0.445	0.464	0.467	0.458	0.441	0.444	0.434	0.433
13	0.769	0.692	0.692	0.644	0.624	0.600	0.601	0.590		0.428	0.446	0.438	0.434	0.423	0.421	0.415
14	0.729	0.714	0.714	0.643	0.635	0.600	0.584	0.560	0.560		0,438	0.428	0.420	0.413	0.414	0.407
15	0.800	0.700	0.667	0.625	0.622	0.600	0.576	0.567	0.574	0.529		0.421	0.412	0.411	0.400	0.417
16	0.738	0.688	0.652	0.688	0.604	0.588	0.568	0.562	0.538	0.536	0.500		0.401	0.403	0.395	0.400
17	0.741	0.667	0.647	0.625	0.601	0.582	0.556	0.549	0.534	0.525	0.514	0.511		0.386	0.390	0.383
18	0.722	0.722	0.659	0.611	0.611	0.578	0.545	0.556	0.526	0.519	0.511	0.493	0.490		0.389	0.378
19	0.737	0.675	0.647	0.612	0.579	0.547	0.545	0.535	0.526	0.508	0.498	0.497	0.489	0.468		0.379
20	0.750	0.667	0.650	0.625	0.578	0.600	0.536	0.533	0.519	0.507	0.500	0.488	0.479	0.472	0.450	

Table VIII (continued)

Two-tail test

n m	5	6	7	8	9	10	11	12	13	14	15	16	17	18	19	20
5		0.800	0.800	0.750	0.778	0.800	0.709	0.717	0.692	0.657	0.733	0.800	0.647	0.667	0.642	0.650
6	1.00		0.714	0.708	0.722	0.667	0.652	0.667	0.667	0.643	0.533	0.625	0.667	0.667	0.614	0.600
7	1.00	0.857		0.714	0.667	0.657	0.623	0.631	0.615	0.643	0.590	0.571	0.571	0.571	0.571	0.564
8	0.875	0.833	0.857		0.639	0.600	0.602	0.625	0.596	0.571	0.588	0.625	0.556	0.611	0.539	0.550
9	0.889	0.883	0.778	0.764		0.589	0.596	0.583	0.556	0.556	0.556	0.542	0.536	0.556	0.520	0.517
10	0.900	0.800	0.757	0.750	0.700		0.545	0.550	0.569	0.529	0.533	0.525	0.524	0.511	0.495	0.550
11	0.818	0.818	0.766	0.727	0.707	0.700		0.545	0.524	0.532	0.509	0.506	0.497	0.490	0.488	0.486
12	0.833	0.833	0.714	0.708	0.694	0.667	0.652		0.519	0.512	0.517	0.500	0.490	0.500	0.474	0.483
13	0.800	0.769	0.714	0.692	0.667	0.646	0.636	0.608		0.489	0.492	0.486	0.475	0.470	0.462	0.462
14	0.800	0.762	0.786	0.679	0.667	0.643	0.623	0.619	0.571		0.467	0.473	0.467	0.460	0.455	0.450
15	0.800	0.767	0.714	0.675	0.667	0.667	0.618	0.600	0.590	0.586		0.475	0.455	0.456	0.446	0.450
16	0.800	0.750	0.688	0.688	0.653	0.625	0.602	0.604	0.582	0.563	0.554		0.456	0.444	0.437	0.429
17	0.800	0.716	0.706	0.647	0.647	0.624	0.588	0.583	0.576	0.563	0.557	0.526		0.435	0.437	0.429
18	0.778	0.778	0.690	0.653	0.667	0.600	0.596	0.583	0.585	0.556	0.544	0.535	0.536		0.415	0.422
19	0.747	0.728	0.684	0.645	0.626	0.595	0.584	0.570	0.559	0.556	0.533	0.526	0.514	0.515		0.421
20	0.800	0.733	0.664	0.650	0.617	0.650	0.577	0.583	0.550	0.543	0.533	0.525	0.515	0.506	0.492	

Table IX Critical values for Kendall's tau

Values are minimum of $|t|$ for significance at nominal 5% and 1% levels in one- and two-tail tests, $5 \leq n \leq 20$

n	5	6	7	8	9	10	11	12	13	14	15	16	17	18	19	20
One-tail test																
5% level	0.800	0.733	0.619	0.571	0.500	0.467	0.418	0.394	0.359	0.363	0.333	0.317	0.309	0.294	0.287	0.274
1% level	1.000	0.867	0.810	0.714	0.667	0.600	0.564	0.545	0.513	0.473	0.467	0.433	0.426	0.412	0.392	0.379
Two-tail test																
5% level	1.000	0.867	0.714	0.643	0.556	0.511	0.491	0.455	0.436	0.407	0.390	0.383	0.368	0.346	0.333	0.326
1% level	*	1.000	0.905	0.786	0.722	0.644	0.600	0.576	0.564	0.516	0.505	0.483	0.471	0.451	0.439	0.421

*Sample too small for test at this level.

Table X Critical values for the Spearman rank correlation coefficient

Values are given for the minimum $|r_S|$ for $5 \leq n \leq 20$ for significance at nominal 5% and 1% levels in one- and two-tail tests. The approximation for $n > 20$ is moderately accurate for $21 \leq n \leq 30$ and good for $n > 30$.

n	5	6	7	8	9	10	11	12	13	14	15	16	17	18	19	20	$n>20$
One-tail test																	
5% level	0.900	0.829	0.714	0.643	0.600	0.564	0.536	0.503	0.484	0.464	0.446	0.429	0.414	0.401	0.391	0.380	$1.64/\sqrt{(n-1)}$
1% level	1.00	0.943	0.893	0.833	0.783	0.746	0.709	0.678	0.648	0.626	0.604	0.585	0.566	0.550	0.535	0.522	$2.33/\sqrt{(n-1)}$
Two-tail test																	
5% level	1.00	0.886	0.786	0.738	0.700	0.648	0.618	0.587	0.560	0.538	0.521	0.503	0.488	0.474	0.460	0.447	$1.96/\sqrt{(n-1)}$
1% level	*	1.00	0.929	0.881	0.833	0.794	0.755	0.727	0.703	0.679	0.657	0.635	0.618	0.600	0.584	0.570	$2.58/\sqrt{(n-1)}$

*Sample too small for test at this level.

References

Adichie, J.N. (1967) Estimates of regression parameters based on rank tests. *Ann. Math. Statist.*, **38**, 894–904.

Agee, W.S. and Turner, R.H. (1979) Applications of robust regression to trajectory data reduction. In *Robustness in Statistics* (eds. R. L. Learner and G.N. Wilkinson). London: Academic Press.

Agresti, A. (1984) *Analysis of Ordinal Categorical Data*. New York: John Wiley & Sons.

Agresti, A. (1990) *Categorical Data Analysis*. New York: John Wiley & Sons.

Agresti, A. (1992) A survey of exact inferences for contingency tables. *Statistical Science*, **7**, 131–53.

Aitchison, J. and Aitken, C.G.G. (1976) Multivariate binary discrimination by the kernel method. *Biometrika*, **63**, 413–20.

Aitchison, J.W. and Heal, D.W. (1987) World patterns of fuel consumption; towards diversity and a low cost energy future. *Geography*, **72**, 235–9.

Akritas, M.G. (1990) The rank transformation in some two factor designs. *J. Amer. Statist. Assoc.*, **85**, 73–8.

Albers, W. and Akritas, M.G. (1987) Combined rank tests for the two sample problem with randomly censored data. *J. Amer. Statist. Assoc.*, **82**, 645–55.

Anderson, O.D. (1992) Accuracy and precision. *Teaching Statistics*, **14**, (3), 2–5.

Andrews, D.F. (1974) A robust method for multiple linear regression. *Technometrics*, **16**, 523–31.

Andrews, D.F., Bickel, P.J., Hampel, F.R., Huber, P.S., Rogers, W.H. and Tukey, J.W. (1972) *Robust Estimates of Location. Survey and Advances*. Princeton, NJ: Princeton University Press.

Anon (1991) *1991 Britannica Book of the Year*. Chicago: Encyclopaedia Britannica Inc.

Ansari, A.R. and Bradley, R.A. (1960) Rank sum tests for dispersion. *Ann. Math. Statist.*, **31**, 1174–89.

Arbuthnot, J. (1710) An argument for Divine Providence, taken from the constant regularity observ'd in the births of both sexes. *Phil. Trans. Roy. Soc.*, **27**, 186–90.

Armitage, P. (1955) Tests for linear trends in proportions and frequencies. *Biometrics*, **11**, 375–86.

Atkinson, A.C. (1986) *Plots, Transformations and Regression. An Introduction to Graphical Methods of Diagnostic Regression Analysis*. Oxford: Clarendon Press.

Babu, G.J., Rao, C.R. and Rao, M.B. (1992) Nonparametric estimation of specific occurrence exposure rates in risks and survival analysis. *J. Amer. Statist. Assoc.*, **87**, 84–9.

Bahadur, R.R. (1967) Rates of convergence of estimates and test statistics. *Ann. Math. Statist.*, **38**, 303–24.

Bardsley, P. and Chambers, R.L. (1984) Multipurpose estimation from unbalanced samples. *Applied Statistics*, **33**, 290–9.

Bartlett, M.S. (1935) Contingency table interactions. *J. Roy. Statist. Soc., Suppt.*, **2**, 248–52.

Bassendine, M.F., Collins, J.D., Stephenson, J., Saunders, P and James, O.W.F. (1985) Platelet associated immunoglobulins in primary biliary cirrhosis: a cause of Thrombocytopenia? *Gut*, **26**, 1074–9.

Berry, D.A. (1987) Logarithmic transformations in ANOVA. *Biometrics*, **43**, 439–56.

Biggins, J.D., Loynes, R.M. and Walker, A.N. (1987) Combining examination results. *Brit. J. Math. and Statist. Psychol.*, **39**, 150–67.

Bishop, Y.M.M., Fienberg, S.E. and Holland, P. (1975) *Discrete Multivariate Analysis: Theory and Practrice.* Cambridge, Mass: MIT Press.
Blomqvist, N. (1950) On a measure of dependence between two random variables. *Ann. Math. Statist.*, **21**, 593–600.
Blomqvist, N. (1951) Some tests based on dichotomization. *Ann. Math. Statist.*, **22**, 362–71.
Bodmer, W.F. (1985) Understanding statistics. *J. Roy. Statist. Soc. A*, **148**, 69–81.
Boos, D.D. (1986) Comparing K populations with linear rank statistics. *J. Amer. Statist. Assoc.*, **81**, 1018–25.
Bowker, A.H. (1948) A test for symmetry in contingency tables. *J. Amer. Statist. Assoc.*, **43**, 572–4.
Bowman, K.O. and Shenton, L.R. (1975) Omnibus test contours for departures from normality based on $\sqrt{b_1}$ and b_2. *Biometrika*, **62**, 243–50.
Bradley, J.V. (1968) *Distribution-free Statistical Tests.* Englewood Cliffs, NJ: Prentice Hall.
Breslow, N.E. and Day, N.E. (1980) *The Analysis of Case–Control Studies.* Lyon: IARC Scientific Publications, No. 32.
Breth, M. (1982) Nonparametric estimation for a symmetric distribution. *Biometrika*, **69**, 625–34.
Brookmeyer, R. and Crowley, J. (1982) A confidence interval for the median survival time. *Biometrics*, **38**, 29–41.
Brown, B.M. (1980) Median estimation in simple linear regression. *Austral. J. Statist.*, **22**, 154–66.
Brown, B.M. (1983) Statistical use of the spatial median. *J. Roy. Statist. Soc. B*, **45**, 25–30.
Brown, B.M. and Maritz, J.S. (1982) Distribution-free methods in regression. *Austral. J. Statist.*, **24**, 318–31.
Brown, B.M., Hettmansperger, T.P., Nyblom, J. and Oja, H. (1992) On certain bivariate sign tests and medians. *J. Amer. Statist. Assoc.*, **87**, 127–35.
Campbell, G. (1981) Nonparametric bivariate estimation with randomly censored data. *Biometrika*, **68**, 417–22.
Carroll, R. (1979) On estimating variance of robust estimators when the errors are asymmetric. *J. Amer. Statist. Assoc.*, **74**, 674–9.
Carter, E.M. and Hubert, J.J. (1985) Analysis of parallel line assays with multivariate responses. *Biometrics*, **41**, 703–10.
Chandra, M., Singpurwalla, N.D. and Stephens, M.A. (1981) Kolmogorov statistics for test of fit for the extreme value and Weibull distributions. *J. Amer. Statist. Assoc.*, **76**, 729–31.
Chatfield, C. (1983) *Statistics for Technology.* London: Chapman & Hall.
Christensen R.A. (1990) *Log linear models.* New York: Springer-Verlag.
Cochran, W.G. (1950) The comparison of percentages in matched samples. *Biometrika*, **37**, 256–66.
Cochran, W.G. (1954) Some methods for strengthening the common χ^2 tests. *Biometrics*, **10**, 417–54.
Cohen, A. (1983) Seasonal daily effect on the number of births in Israel. *Applied Statistics*, **32**, 228–35.
Conover, W.J. (1980) *Practical Nonparametric Statistics.* 2nd edn. New York: John Wiley & Sons.
Conover, W.J. and Iman, R.L. (1976) On some alternative procedures using ranks for the analysis of experimental designs. *Communications in Statistics*, **A5**, 1349–68.
Conover, W.J. and Iman, R.L. (1982) Analysis of covariance using the rank transformation. *Biometrics*, **38**, 715–24.
Cook, R.D. and Weisberg, S. (1982) *Residuals and Influence in Regression.* London: Chapman & Hall.

Cox, D.R. and Snell, E.J. (1989) *Analysis of Binary Data*. 2nd edn. London: Chapman & Hall.

Cox, D.R. and Stuart, A (1955) Some quick tests for trend in location and dispersion. *Biometrika*, **42**, 80–95.

Crowder, M.J. (1978) Beta-binomial ANOVA for proportions. *Applied Statistics*, **27**, 34–7.

Dabrowska, D.M. (1990) Signed rank tests for censored matched pairs. *J. Amer. Statist. Assoc.*, **85**, 478–85.

Daniel, W.W. (1990) *Applied Nonparametric Statistics*. 2nd. edn Boston: PWS-Kent Publishing Company.

Dansie, B.R. (1986) Normal order statistics as permutation probability models. *Applied Statistics*, **35**, 269–75.

Davis, T.P. and Lawrance A.J. (1989) The likelihood for competing risk survival analysis. *Scand. J. Statist.*, **16**, 23–28.

de Kroon, J. and van der Laan, P. (1981) Distribution-free test procedures in two-way layouts: a concept of rank interaction. *Statistica Neerlandica*, **35**, 189–213.

Dietz, E.J. and Killen, T.J. (1981) A nonparametric multivariate test for trend with pharmaceutical applications. *J. Amer. Statist. Assoc.*, **76**, 169–74.

Dinse, G.E. (1982) Nonparametric estimation for partially-complete time and type of failure data. *Biometrics*, **38**, 417–31.

Dinse, G.E. (1986) Nonparametric prevalence and mortality estimators for animal experiments with incomplete cause-of-death data. *J. Amer. Statist. Assoc.*, **81**, 328–36.

Dinse, G.E. and Lagakos, S.W. (1982) Nonparametric estimation of lifetime and disease onset distribution from incomplete observations. *Biometrics*, **38**, 921–32.

Disch, D. (1981) Bayesian nonparametric inference for effective doses in a quantal-response experiment. *Biometrics*, **37**, 713–22.

Dobson, A.J. (1990) *An Introduction to Generalized Linear Models*. London: Chapman & Hall.

Durbin, J. (1987) Statistics and statistical science. *J. Roy. Statist. Soc. A*, **150**, 177–91.

Efron, B. (1981) Censored data and the bootstrap. *J. Amer. Statist. Assoc.*. **76**, 312–19.

Efron, B. (1982) *The Jackknife, the Bootstrap and other Resampling Plans*. Philadelphia: Society for Industrial and Applied Mathematics.

Emerson, J.D. (1982) Nonparametric confidence intervals for the median in the presence of right censoring. *Biometrics*, **38**, 17–27.

Eplett, W.J.R. (1982) The distribution of Smirnov type two-sample rank tests for discontinuous distribution functions. *J. Roy. Statist. Soc. B*, **44**, 361–9.

Everitt, B.S. (1992) *The Analysis of Contingency Tables*. 2nd. edn London: Chapman & Hall.

Ferguson, T.S. (1973) A Bayesian analysis of some nonparametric problems. *Annals of Statistics*, **1**, 209–30.

Fienberg, S.E. (1980) *The Analysis of Cross Classified Categorical Data*. 2nd edn. Cambridge, Mass: MIT Press.

Fisher, N.I. and Lee, A.J. (1982) Nonparametric measures of angular-angular association. *Biometrika*, **69**, 315–21,

Fisher, N.I. and Lee, A.J. (1983) A correlation coefficient for circular data. *Biometrika*, **70**, 327–32.

Fisher, R.A. (1935) *The Design of Experiments*. Edinburgh: Oliver & Boyd.

Fisher, R.A. and Yates, F. (1957) *Statistical Tables for Biological, Agricultural and Medical Research*. 5th edn. Edinburgh: Oliver & Boyd.

Fligner, M.A. (1985) Pairwise versus joint ranking. Another look at the Kruskal–Wallis statistic. *Biometrika*, **72**, 705–9.

Freeman, G.H. and Halton, J.H. (1951) Note on an exact treatment of contingency, goodness of fit and other problems of significance. *Biometrika*, **38**, 141–9.

Friedman, M. (1937) The use of ranks to avoid the assumptions of normality implicit in the analysis of variance. *J. Amer. Statist. Assoc.*, **32**, 675–701.

Gart, J. (1970) Point and interval estimation of the common odds ratio in the combination of 2×2 tables with fixed marginals. *Biometrika*, **57**, 471–5.

Gasser, T., Müller, H.G. and Mammitzsch, V. (1985) Kernels for nonparametric curve estimation. *J. Roy. Statist. Soc. B*, **47**, 238–52.

Gat, J.R. and Nissenbaum, A. (1976) Limnology and ecology of the Dead Sea. *Nat. Geog. Soc. Res. Reports – 1976 Projects*, 413–18.

Geffen, G., Bradshaw, J.L. and Nettleton, N.C. (1973) Attention and hemispheric differences in reaction time during simultaneous audio-visual tasks. *Quart. J. Expt. Psychology*, **25**, 404–12.

Gehan, E.A. (1965a) A generalized Wilcoxon test for comparing arbitrarily singly censored samples. *Biometrika*, **52**, 203–23.

Gehan, E.A. (1965b) A generalized two-sample Wilcoxon test for doubly censored data. *Biometrika*, **52**, 650–3.

Gibbons, J.D. (1985) *Nonparametric Statistical Inference*. 2nd edn. New York: Marcel Dekker.

Gideon, R.A. and Hollister, R.A. (1987) A rank correlation coefficient resistant to outliers. *J. Amer. Statist. Assoc.*, **82**, 656–66.

Gill, A.N. and Mehta, G.P. (1989) Selection procedures for scale parameters using two-sample statistics. *Sankhya*, Series B, **51**, 149–57.

Gill, A.N. and Mehta, G.P. (1991) Selection procedures for scalar parameters using two-sample U-statistics. *Austral. J. Statist.*, **33**, 347–62.

Gordon, A.D. (1981) *Classification*. London: Chapman & Hall.

Graubard, B.I. and Korn, E.L. (1987) Choice of column scores for testing independence in ordered $2 \times k$ contingency tables. *Biometrics*, **43**, 471–6.

Groggel, D.J. and Skillings, J.H. (1986) Distribution-free tests for the main effects in multifactor designs. *American Statistician*, **40**, 99–102.

Grizzle, J.E., Starmer, C.F. and Koch, G.G. (1969). Analysis of categorical data by linear models. *Biometrics*, **25**, 489–504.

Hall, P. (1981a) On nonparametric binary discrimination. *Biometrika*, **68**, 287–94.

Hall, P. (1981b) On the nonparametric estimation of mixture proportions. *J. Roy. Statist. Soc. B*, **43**, 147–56.

Hall, P. and Titterington, D.M. (1984) Efficient nonparametric estimation of mixture proportions. *J. Roy. Statist. Soc. B*, **46**, 465–73.

Hall, P. and Hart, J.D. (1990) Bootstrap tests for differences between means in nonparametric regression. *J. Amer. Statist. Assoc.*, **85**, 1039–49.

Halperin, M., Gilbert, P.R. and Lachin, J.M. (1987). Distribution free confidence intervals for $\Pr(X_1 < X_2)$. *Biometrics*, **43**, 71–80.

Hanley, J.A. and Parnes, M.N. (1983) Nonparametric estimation of a multivariate distribution in the presence of censoring. *Biometrics*, **39**, 129–39.

Härdle, W. and Gasser, T. (1984) Robust nonparametric function fitting. *J. Roy. Statist. Soc. B*, **46**, 42–51.

Harrington, D.P. and Fleming, T.R. (1982) A class of rank test procedures for censored survival data. *Biometrika*, **69**, 553–66.

Hayter, A.J. and Stone, G. (1991) Distribution-free multiple comparisons for monotonic ordered treatment effects. *Austral. J. Statist.*, **33**, 335–46.

Heltshe, J.F. and Forrester. N.E. (1983) Estimating species richness using the jackknife procedure. *Biometrics*, **39**, 1–11.

Hill, N.S. and Padmanabhan, A.L. (1984) Robust comparison of two regression lines and biomedical applications. *Biometrics*, **40**, 985–94.

Hinkley, D.V. (1989) Modified profile likelihood in transformed linear models. *Applied Statistics*, **38**, 495–506.

Hollander, M. and Wolfe, D.A. (1973) *Nonparametric Statistical Methods*. New York: John Wiley & Sons.

Hora, S.C. and Conover, W.J. (1984) The F-statistic in the two way layout with rank score transformed data. *J. Amer. Statist. Assoc.*, **79**, 668–73.

House, D.E. (1986) A nonparametric version of Williams' test for a randomized block design. *Biometrics*, **42**, 187–90.

Howarth, J. and Curthoys, M. (1987) The political economy of women's higher education in late nineteenth and early twentieth century Britain. *Historical Research*, **60**, 208–31.

Huber, P.J. (1972) Robust statistics: a review. *Ann. Math. Statist.*, **43**, 1041–67.

Huber, P. J. (1977) *Robust Statistical Procedures*. Philadelphia: Society for Industrial and Applied Mathematics.

Hussain, S.S. and Sprent, P. (1983) Nonparametric regression. *J. Roy. Statist. Soc. A*, **146**, 182–91.

Iman, R.L. (1974) A power study of rank transformation for the two-way classification model when interaction may be present. *Canadian J. Statist.*, **2**, 227–39.

Iman, R.L. and Davenport, J.M. (1980) Approximations of the critical region of the Friedman statistic. *Communications in Statistics*, **A9**, 571–95.

Iman, R.L., Hora, S.C. and Conover, W.J. (1984) Comparison of asymptotically distribution-free procedures for the analysis of complete blocks. *J. Amer. Statist. Assoc.*, **79**, 674–85.

Izenman, A.J. (1991) Recent developments in nonparametric density estimation. *J. Amer. Statist. Assoc.*, **86**, 205–24.

Jaeckel, L.A. (1972) Estimating regression equations by minimizing the dispersion of residuals. *Ann. Math. Statist.*, **43**, 1449–58.

Jarque, C.M. and Bera, A.K. (1987) A test for normality of observations and regression residuals. *International Statist. Rev.*, **55**, 163–72.

Jarrett, R.G. (1979) A note on the intervals between coal mining disasters. *Biometrika*, **66**, 191–3.

Jonckheere, A.R. (1954) A distribution free k-sample test against ordered alternatives. *Biometrika*, **41**, 133–45.

Johnson, R.A., Verrill, S. and Moore, D.H. (1987) Two-sample rank tests for detecting changes that occur in a small proportion of the treated population. *Biometrics*, **43**, 641–55.

Kasser, I.S. and Bruce, R.A. (1969) Comparative effects of aging and coronary heart disease on submaximal and maximal exercise. *Circulation*, **39**, 759–74.

Katti, S.K. (1965) Multivariate covariance analysis. *Biometrics*, **21**, 957–74.

Kendall, M.G. (1938) A new measure of rank correlation. *Biometrika*, **30**, 81–93.

Kendall, M.G. (1970) *Rank Correlation Methods*. 4th edn. London: Charles Griffin & Co.

Kendall, M.G. and Gibbons, J.D. (1990) *Rank Correlation Methods*. 5th edn. London: Edward Arnold.

Kestermont, M.-P. (1987) The Kolmogorov distance as comparison measure between parametric and nonparametric Bayesian inference. *The Statistician*, **36**, 259–64.

Kildea, D.G. (1978) Estimators for regression methods – the Brown–Mood approach. Ph. D. Thesis, Latrobe University, Melbourne, Australia.

Kimber, A.C. (1987) When is a χ^2 not a χ^2? *Teaching Statistics*, **9**, 74–7.

Kimber, A.C. (1990) Exploratory data analysis for possibly censored data from skewed distributions. *Applied Statistics*, **39**, 21–30.

Kimura, D.K. and Chikuni, S. (1987) Mixtures of empirical distributions: an iterative application of the age-length key. *Biometrics*, **43**, 23–35.

Klotz, J. (1962) Nonparametric tests for scale. *Ann. Math. Statist.*, **33**, 498–512.

Klotz, J. (1980) A modified Cochran–Friedman test with missing observations and ordered categorical data. *Biometrics*, **36**, 665–70.

Knafl, G., Sacks, J., Spiegelman, C. and Ylvisaker, D. (1984) Nonparametric calibration. *Technometrics*, **26**, 233–41.

Knapp, T.R. (1982) The birthday problem: some empirical data and some approximations. *Teaching Statistics*, **4**, 10–14.

Kolmogorov, A.N. (1933) Sulla determinazione empirica di una legge di distribuzione. *G. Inst. Ital. Attuari*, **4**, 83–91.

Kolmogorov, A.N. (1941) Confidence limits for an unknown distribution function. *Ann. Math. Statist.*, **12**, 461–3.

Kotz, S. and Johnson, N.L. (eds) (1983) *Encyclopedia of Statistical Sciences*. New York: John Wiley & Sons.

Koziol, J.A. (1980) On a Cramér–von Mises-type statistic for testing symmetry. *J. Amer. Statist. Soc.*, **75**, 161–7.

Krauth, J. (1988) *Distribution-free Statistics – An Application-oriented Approach*. Amsterdam: Elsevier.

Kruskal, W.H. and Wallis, W.A. (1952) Use of ranks in one-criterion variance analysis. *J. Amer. Statist. Assoc.*, **47**, 583–621.

Lam, F.C. and Longnecker, M.T. (1983) A modified Wilcoxon rank sum test for paired data. *Biometrika*, **70**, 510–13.

Lan, K.K.G. and Wittes, J.T. (1985) Rank tests for survival analysis: a comparison by analogy with games. *Biometrics*, **41**, 1063–9.

Leach, C. (1979) *Introduction to Statistics. A Nonparametric Approach for the Social Sciences*. Chichester: John Wiley & Sons.

Lee, J.W. and DeMets, D.L. (1992) Sequential rank tests with repeated measurements in clinical trials. *J. Amer. Statist. Assoc.*, **87**, 136–42.

Lehmann, E.L. (1975) *Nonparametrics: Statistical Methods Based on Ranks*. San Francisco: Holden Day, Inc.

Lilliefors, H.W. (1967) On the Kolmogorov–Smirnov test for normality with mean and variance unknown. *J. Amer. Statist. Assoc.*, **62**, 399–402.

Lindsey, J.C., Herzberg, A.M. and Watts, D.G. (1987) A method of cluster analysis based on projections and quantile–quantile plots. *Biometrics*, **43**, 327–41.

Lombard, F. (1986) The change-point problem for angular data. A nonparametric approach. *Technometrics*, **28**, 391–7.

Lombard, F. (1987) Rank tests for change-point problems. *Biometrika*, **74**, 615–24.

Lubischew, A.A. (1962) On the use of discriminant functions in taxonomy. *Biometrics*, **18**, 455–77.

McCullagh, P. and Nelder, J.A. (1989) *Generalized Linear Models*. 2nd edn. London: Chapman & Hall.

Mack, G.A. and Skillings, J.H. (1980) A Friedman type rank test for main effects in a two-factor ANOVA. *J. Amer. Statist. Assoc.*, **75**, 947–51.

Mack, G.A. and Wolfe D.A. (1981) *K*-sample rank test for umbrella alternatives. *J. Amer. Statist. Assoc.*, **76**, 175–81.

McKean, J.W., Sheather, S.J., and Hettsmansperger, T.P. (1990) Regression diagnostics for rank based methods. *J. Amer. Statist. Assoc.*, **85**, 1018–28.

Manly, B.F.J. (1991) *Randomization and Monte Carlo Methods in Biology*. London: Chapman & Hall.

Mann, H.B. and Whitney, D.R. (1947) On a test of whether one of two random variables is stochastically larger than the other. *Ann. Math. Statist.*, **18**, 50–60.

Mantel, N. and Haenszel, W. (1959) Statistical aspects of the analysis of data from retrospective studies of disease. *J. Nat. Cancer Res. Inst.*, **22**, 719–48.

Marascuilo, L.A. and McSweeney, M. (1977) *Nonparametric and Distribution-free Methods for the Social Sciences*. Monterey, Calif: Brooks/Cole Publishing Company.

Marascuilo, L.A. and Serlin, R.C. (1979) Tests and contrasts for comparing change parameters for a multiple sample McNemar data model. *Brit. J. Math. and Statist. Psychol.*, **32**, 105–12.

Maritz, J.S. (1979) On Theil's method in distribution-free regression. *Austral. J. Statist.*, **21**, 30–5.

Maritz, J.S. (1981) *Distribution-free Statistical Methods*. London: Chapman & Hall.

Markowski, E.P. and Hettmansperger, T.P. (1982) Inferences based on simple rank step scores statistics for the location model. *J. Amer. Statist. Assoc.*, **77**, 901–7.

Marriott, F.H.C. (ed.) (1990). *A Dictionary of Statistical Terms. ISI*, 5th. edn. Harlow: Longman Scientific & Technical.

Mattingley, P.F. (1987) Pattern of horse devolution and tractor diffusion in Illinois, 1920–82. *The Professional Geographer*, **39**, 298–309.

Mehta, C.R. and Patel, N.R. (1983) A network algorithm for performing Fisher's exact test in $r \times c$ contingency tables. *J. Amer. Statist. Assoc.*, **78**. 427–34.

Mehta, C.R. and Patel, N.R. (1986) A hybrid algorithm for Fisher's exact test on unordered $r \times c$ contingency tables. *Communications in Statistics*, **15**, 387–403.

Mehta, C.R. Patel, N.R. and Gray, R. (1985) On computing an exact commmon odds ratio in several 2×2 contingency tables. *J. Amer. Statist Assoc.*, **80**, 969–73.

Mehta, C.R., Patel, N.R. and Senchaudhuri, P. (1988). Importance sampling for estimating exact probabilities in permutational inference. *J. Amer. Statist. Assoc.*, **83**, 999–1005.

Mehta, C.R., Patel, N.R. and Tsiatis, A.A. (1984). Exact significance tests to establish treatment equivalence for ordered categorical data. *Biometrics*, **40**, 819–25.

Michalek, J.E. and Mihalko, D. (1984) Linear rank procedures on litter-match models. *Biometrics*, **40**, 487–91.

Miller, L.H. (1956) Table of percentage points of Kolmogorov statistics. *J. Amer. Statist. Assoc.*, **51**, 111–21.

Mood, A.M. (1954) On the asymptotic efficiency of certain nonparametric two-sample tests. *Ann. Math. Statist.*, **25**, 514–22.

Moses, L.E. (1963) Rank tests for dispersion. *Ann. Math. Statist.*, **34**, 973–83.

Neave, H.R. (1981) *Elementary Statistical Tables*. London: George Allen & Unwin Ltd.

Neave, H.R. and Worthington, P.L. (1988) *Distribution-free tests*. London: Unwin Hyman.

Noether, G.E. (1984) Nonparametrics: the early years – impressions and recollections. *The American Statistician*, **38**, 173–8.

Noether, G.E. (1987a) Sample size determination for some common nonparametric tests. *J. Amer. Statist. Assoc.*, **82**, 645–7.

Noether, G.E. (1987b) Mental random numbers: perceived and real randomness. *Teaching Statistics*, **9**, 68–70.

Noether, G.E. (1991) *Introduction to Statistics: The Nonparametric Way*. New York: Springer-Verlag.

Oja, H. (1987) On permutation tests in multiple regression and analysis of covariance problems. *Austral. J. Statist.*, **29**, 91–100.

Olkin, I. and Yitzhaki, S. (1992) Gini regression analysis. *International Statist. Rev.*, **60**, 185–96.

O'Muircheartaigh, I.G. and Sheil, J. (1983) Fore or five? The indexing of a golf course. *Applied Statistics*, **32**, 287–92.

O'Quigley, J. and Prentice, R.L. (1991). Nonparametric tests of association between survival times and continuously measured covariates: the logit-rank and associated procedures. *Biometrics*, **47**, 117–27.

Page, E.B. (1963) Ordered hypotheses for multiple treatments: a significance test for linear ranks. *J. Amer. Statist. Assoc.*, **58**, 216–30.

Paul, S.R. (1979) Models and estimation procedures for for the calibration of examiners. *Brit. J. Math. and Statist. Psychol.*, **32**, 242–51.

Pearce, S.C. (1965) *Biological Statistics – An Introduction*, New York: McGraw-Hill.

Pearson, J.C.G. and Sprent, P. (1968) Trends in hearing loss associated with age or exposure to noise. *Applied Statistics*, **17**, 205–15.

Pearson, K. (1900) On a criterion that a given system of deviations from the probable in the case of a correlated system of variables is such that it can reasonably be supposed to have arisen in random sampling. *Phil. Mag.* (5), **50**, 157–75.

Peters, D. and Randles, R.H. (1990) A multivariate signed rank test for the one-sample location problem. *J. Amer. Statist. Assoc.*, **85**, 552–7.

Peto, R. and Peto, J. (1972) Asymptotically efficient rank-invariant test procedures. *J. Roy. Statist. Soc. A*, **135**, 185–206.

Pettitt, A.N. (1979) A nonparametric approach to the change-point problem. *Applied Statistics*, **28**, 126–35.

Pettitt, A.N. (1981) Posterior probabilites for a change-point using ranks. *Biometrika*, **68**, 443–50.

Pettitt, A.N. (1983) Approximate methods using ranks for regression with censored data. *Biometrika*, **70**, 121–32.

Pettitt, A.N. (1985) Re-weighted least squares estimation with censored and grouped data: an application of the EM algorithm. *J. Roy. Statist. Soc. B*, **47**, 253–60.

Pitman, E.J.G. (1937a) Significance tests that may be applied to samples from any population. *J. Roy. Statist. Soc., Suppt.*, **4**, 119–130.

Pitman, E.J.G. (1937b) Significance tests that may be applied to samples from any population, II: The correlation coefficient test. *J. Roy, Statist, Soc., Suppt.*, 225–32.

Pitman, E.J.G. (1938) Significance tests that may be applied to samples from any population. III. The analysis of variance test. *Biometrika*, **29**, 322–35.

Pitman, E.J.G. (1948) Mimeographed lecture notes on nonparametric statistics. Columbia University.

Plackett, R.L. (1981) *The Analysis of Categorical Data*. 2nd edn. London: Charles Griffin & Co.

Prentice R.L. (1978) Linear rank tests with right censored data. *Biometrika*, **65**, 167–79.

Quade, D. (1982) Nonparametric analysis of covariance by matching. *Biometrics*, **38**, 597–611.

Quenouille, M. H. (1949) Approximate tests of correlation in time series. *J. Roy. Statist. Soc. B*, **11**, 18–84.

Radelet, M. (1981) Racial characteristics and the imposition of the death penalty. *Amer. Sociol. Rev.*, **46**, 918–27.

Ramsay, F.L. (1972) A Bayesian approach to bioassay. *Biometrics*, **28**, 841–58.

Randles, R.H. (1989) A distribution-free multivariate sign test based on interdirections. *J. Amer. Statist. Assoc.*, **84**, 1045–50.

Randles, R.H., Fligner, M.A., Policello, G.E. and Wolfe, D.A, (1980) An asymptotically distribution-free test for symmetry versus asymmetry. *J. Amer. Statist. Assoc.*, **75**, 168–72.

Randles, R.H. and Wolfe, D.A. (1979) *Introduction to the Theory of Nonparametric Statistics*. New York: John Wiley & Sons.

Rees, D.G. (1987) *Foundations of Statistics*. London: Chapman & Hall.

Robins, J., Breslow, N. and Greenland, S. (1986) Estimation of the Mantel–Haenszel variance consistent in both sparse data and large-strata limiting models. *Biometrics*, **42**, 311–23.

Robinson, J.A. (1983) Bootstrap confidence intervals in location-scale models with progressive censoring. *Technometrics*, **25**, 179–87.

Rogerson, P.A. (1987) Changes in US national mobility levels. *The Professional Geographer*, **39**, 344–51.

Rosenthal, I. and Ferguson, T.S. (1965) An asymptotic distribution-free multiple comparison method with application to the problem of *n* rankings of *m* objects. *Brit. J. Math. and Statist. Psychol.*, **18**, 243–54.

Ross, G.J.S., (1990) *Nonlinear Estimation*, New York: Springer-Verlag.

Rowntree, D. (1981) *Statistics Without Tears*. London: Penguin Books.

Sandford, M.D. (1985) Nonparametric one-sided confidence intervals for an unknown distribution function using censored data. *Technometrics*, **27**, 41–48.

Saunders, R. and Laud, P. (1980) The multidimensional Kolmogorov goodness of fit test. *Biometrika*, **67**, 237.

Scheirer, C.J., Ray, W.S. and Hare, N. (1976) The analysis of ranked data derived from completely randomized factorial designs. *Biometrics*, **32**, 429–434.

Scholz, F.W. (1978) Weighted median regression estimates. *Annals of Statistics*, **6**, 603–9.

Scott, A.J., Smith, T.M.F. and Jones, R.G. (1977) The application of times series methods to the analysis of repeated surveys. *International Statist. Rev.*, **45**, 13–28.

Sen, P.K. (1968) Estimates of the regression coefficient based on Kendall's tau. *J. Amer. Statist. Assoc.*, **63**, 1379–89.

Shirley, E. (1977) A nonparametric equivalent of Williams' test for contrasting existing dose levels of a treatment. *Biometrics*, **33**, 386–9.

Shirley, E.A.C. (1981) A distribution-free method for the analysis of covariance based on ranked data. *Applied Statistics*, **30**, 158–62.

Shirley, E.A.C. (1987) Applications of ranking methods to multiple comparison procedures and factorial experiments. *Applied Statistics*, **36**, 205–13.

Siegel, S. and Castellan, N.J. (1988) *Nonparametric Statistics for the Behavioral Sciences*. 2nd edn. New York:McGraw-Hill.

Sievers, G.L. (1978) Weighted rank statistics for simple linear regression. *J. Amer. Statist. Assoc.*, **73**, 628–31.

Silverman, B.W. (1984) A fast and efficient cross-validation method for smoothing parameter choice in spline regression. *J. Amer. Statist. Assoc.*, **79**, 584–9.

Silverman, B.W. (1985) Some aspects of the spline smoothing approach to nonparametric regression curve fitting. *J. Roy. Statist. Soc. B*, **47**, 1–52.

Silverman, B.W. and Wood, J.T. (1987) The nonparametric estimation of branching curves. *J. Amer. Statist. Assoc.*, **82**, 551–8.

Simonoff, J.S., Hochberg, Y. and Reiser, B. (1986) Alternative estimation procedures for Pr($X<Y$) in categorical data. *Biometrics*, **42**, 897–907.

Singer, B. (1979) Distribution-free methods for nonparametric problems. A classified and selected bibliography. *Brit. J. Math. and Statist. Psychol.*, **32**, 1–60.

Slud, E.V., Byar, D.P. and Green, S.B. (1984) A comparison of reflected versus test-based confidence intervals for the median survival time based on censored data. *Biometrics*, **40**, 587–600.

Smirnov, N.V. (1939) On the estimation of discrepancy between empirical curves of distribution for two independent samples. (In Russian.) *Bulletin Moscow University*, **2**, 3–16.

Smirnov, N.V. (1948) Tables for estimating the goodness of fit of empirical distributions. *Ann. Math. Statist.*, **19**, 279–81.

Smith, E.P. and van Belle, G. (1984) Nonparametric estimation of species richness. *Biometrics*, **40**, 119–29.

Smith, R.L. and Naylor, J.C. (1987) A comparison of likelihood and Bayesian estimators of the three-parameter Weibull distribution. *Applied Statistics*, **36**, 358–69.

Snee, R.D. (1985) Graphical display of results of three treatment randomized block experiments. *Applied Statistics*, **34**, 71–7.

Spearman, C. (1904) The proof and measurement of association between two things. *Amer. J. Psychol.*, **15**, 72–101.

Spurrier, J.D. and Hewett, J.E. (1980) Two-stage test of independence using Kendall's statistic. *Biometrics*, **36**, 517–22.

Stuart, A. (1955) A test for homogeneity of the marginal distributions in a two-way classification. *Biometrika*, **42**, 412–16.

Stuart, A. (1957) The comparison of frequencies in matched samples. *Brit. J. Statist. Psychol.*, **10**, 29–32.

Swed, F.S. and Eisenhart, C. (1943) Tables for testing randomness of grouping in a series of alternatives. *Ann. Math. Statist.*, **14**, 83–6.

Sweeting, T.J. (1982) A Bayesian analysis of some pharmacological data using a random coefficient regression model. *Applied Statistics*, **31**, 205–13.

Terpstra, T.J. (1952) The asymptotic normality and consistency of Kendall's test against trend when ties are present in one ranking. *Indag. Math.*, **14**, 327–33.

Theil, H. (1950) A rank invariant method of linear and polynomial regression analysis, I, II, III. *Proc. Kon. Nederl. Akad. Wetensch. A*, **53**, 386–92, 521–5, 1397–1412.

Thomas, G.E. (1989) A note on correcting for ties with Spearman's ρ. *J. Statist. Comp. and Simulation*, **31**, 37–40.

Thomas, G.E. and Kiwanga, G.E. (1991) Use of ranking and scoring methods in the analysis of ordered categorical data from factorial experiments. *Technical report series, STAT-91-11*. University of Waterloo, Canada.

Titterington. D.M. (1983) Minimum distance nonparametric estimation of mixture proportions. *J. Roy. Statist. Soc., B*, **45**, 37–46.

Tukey, J.W. (1958) Bias and confidence in not quite large samples. Abstract. *Ann. Math. Statist.*, **29**, 614.

Turnbull, B.W. and Mitchell, T.J. (1984) Nonparametric estimation of the distribution of time to onset for specific diseases in survival/sacrifice experiments. *Biometrics*, **40**, 41–50.

Upton, G.J.G. (1992) Fisher's exact test. *J. Roy. Statist. Soc., A*, **155**, 395–402.

van der Waerden, B.L. (1952) Order tests for the two sample problem and their power. *Proc. Kon. Nederl. Akad. Wetensch. A*, **55**, 453–8.

van der Waerden, B.L. (1952) Order tests for the two sample problem and their power. *Proc. Kon. Nederl. Akad. Wetensch. A*, **56**, 303–16.

Wahrendorf, J., Becher, H. and Brown, C.C. (1987) Bootstrap comparisons of non-nested generalized linear models: applications in survival analysis and epidemiology. *Applied Statistics*, **36**, 72–81.

Wegman, E.J. and Wright, I.W. (1983) Splines in statistics. *J. Amer. Statist. Assoc.*, **78**, 351–65.

Welch, B.L. (1937). On the *z*-test in randomized blocks and latin squares. *Biometrika*, **29**, 21–52.

Wilcoxon, F. (1945) Individual comparisons by ranking methods. *Biometrics*, **1**, 80–3.

Willemain, T.R. (1980) Estimating the population median by nomination sampling. *J. Amer. Statist. Assoc.*, **75**, 908–11.

Williams, D.A. (1982) Extra-binomial variation in logisitic linear models. *Applied Statistics*, **31**, 144–8.

Williams, D.A. (1986) A note on Shirley's nonparametric test for comparing several dose levels with a zero-dose control. *Biometrics*, **42**, 183–6.

Woolson, R.F. (1981) Rank tests and a one-sample log rank test for comparing observed survival data to a standard population. *Biometrics*, **37**, 687–96.

Woolson, R.F. and Lachenbruch, P.A. (1980) Rank tests for censored matched pairs. *Biometrika*, **67**, 597–606.

Woolson, R.F. and Lachenbruch, P.A. (1981) Rank tests for censored randomized block designs. *Biometrika*, **68**, 427–35.

Woolson, R.F. and Lachenbruch P.A. (1983) Rank analysis of covariance with right-censored data. *Biometrics*. **39**, 727–33.

Yates, F. (1984) Tests of significance for 2×2 contingency tables. *J. Roy. Statist. Soc. A*, **147**, 426–463.

Zelen, M. (1971) The analysis of several 2×2 contingency tables. *Biometrika*, **58**, 129–37.

Solutions to odd-numbered exercises

Many exercises in this book are open ended in that they require, as do most practical statistical problems, not simply numerical outcomes (a statistic, significance level, confidence interval, etc.) but an interpretation of such numerical values in the context of the real-world problem that gave rise to the original data. The brief summaries below are not solutions in this sense, but rather a guide to indicate appropriate calculations. In many cases there are alternative tests or estimation procedures that could well have been used; these will often, but not invariably, lead to similar conclusions. Bearing such points in mind it is hoped this section will alleviate any difficulties with the odd-numbered exercises.

Chapter 1

1.1 2.55. **1.3** Open interval (156, 454); see Example 2.2 for refinement of limits. **1.5** $S=13$; $\Pr(S \le 13)=7/126=0.056$. Do not reject H_0 at nominal 5% level but note that exact tail probability only just exceeds 0.05. **1.7** Open interval (0, 454); due to discontinuity this is an exact 98.1% interval.

Chapter 2

2.1 Accept H_0; (effective sample size 10; 4 plus observed). **2.3** Due to discontinuities distribution appears flatter than normal distribution. **2.5** Set I: (1, 5); exact limits reject 1 but include 5, i.e. interval is $1<\theta\le 5$. Set II: (3, 5); $3\le\theta<5$). **2.7** Accept H_0; Pr(3 or less plus)=0.073. **2.9** For sign test effective sample size is 14 with 4 plus; Pr(4 or less plus)=0.090 so accept H_0. With symmetry assumption use Wilcoxon (asymptotic statistic is $Z=-1.576$). Accept H_0. **2.11** Asymptotic sign test gives $Z=0.64$ so do not reject H_0. Asymptotic Wilcoxon test gives $Z=$ 0.775, so again do not reject H_0; note that observations 210, 141, 117, 111 cast doubt upon symmetry assumption. **2.13** Cox–Stuart trend test gives 6 plus, 1 minus; relevant tail probability is 0.062, so result just fails to attain significance at nominal 5% level in a one-tail test. **2.15** Asymmetry suggests sign test limits are appropriate. For $n=16$, reject H_0 if 3 or fewer plus or minus. Appropriate 95% interval is (11, 36). **2.17** 7 plus, 1 minus. If one-tail test considered appropriate, evidence of a monotonic trend at 3.5% level. **2.19** Normal approximation gives $Z=6.77$; strong evidence of pessimism.

Chapter 3

3.1 $S_-=10$. $\Pr(S_-\leq 10)=0.0469$; accept H_0 as relevant two-tail probability is $2\times 0.0469=0.0938$. **3.3** No. Lilliefors' statistic is -0.259, almost indicating significance at a nominal 1% level. **3.5** Using modified van der Waerden scores $S_-=5.70$; $Z=-2.10$; $\Pr(Z\leq -2.10)=0.018$. Permutation test tail probability is 0.017; reject H_0. **3.7** Allowing for ties Wilcoxon asymptotic test gives $Z=-2.57$. $\Pr(|Z|\geq 2.57)=0.0102$. Van der Waerden scores (using (3.1) to allow for ties) give $Z=-2.80$. Clearly a strong case for rejection of H_0. **3.9** No. Lilliefors' statistic is 0.30, significant at 1% level.

Chapter 4.

4.3 Reject H_0 at 4.5% level. Wilcoxon 95% and 99% confidence intervals are (0.05, 0.85) and (−0.20, 1.05) respectively. Corresponding normal theory limits are (0.03, 0.85) and (−0.14, 1.02). **4.5** Sign test, 4 M equivalent to 4 minus. Significant at 4.8% level in two-tail test. **4.7** Asymptotic McNemar test $Z=1.42$; accept H_0 : *no change in attitudes*. **4.9** $Z=-3.14$. Reject H_0 : *75% prefer new formula* at 0.1% level in two-tail test. **4.11** Exact Wilcoxon test gives two-tail probability 0.125 so do not reject H_0. (Asymptotically, $Z=-1.669$, $\Pr(|Z|\geq 1.669)=0.095$). **4.13** Exact Wilcoxon two-tail probability 0.0313. Reject H_0 at 3.1% level. **4.15** Exact Wilcoxon two-tail probability 0.2786 (asymptotic test probability is 0.2559). Accept H_0. 95% confidence interval for difference is (−2.95, 0.95).

Chapter 5

5.1 $P=0.0029$. **5.5** Variance reduced from 1932 to 1929. **5.7** Asymptotic WMW $Z=-1.62$. $\Pr(|Z|\geq 1.62)=0.1049$. Permutation test two-tail probability is 0.1106. Do not reject hypothesis of equal variance (close to significance if one-tail test considered appropriate). **5.9** $U=13$ corresponding to a two-tail probability 0.0549, just failing to reach significance at 5% level. The test appropriate if distributions identical apart from shift in mean or if one distribution dominates. **5.11** $U=275$ for men and $U=116$ for women. Asymptotic $Z=-2.16$, $\Pr(|Z|\geq 2.16)=0.034$. Men show greater anxiety. **5.13** Smirnov statistic is 0.417; not significant at nominal 5% level. **5.15** (i) $U=35$; not significant; (ii) squared rank test or parametric F-test indicate variances differ; (iii) Smirnov test fails to distinguish but Cramér–von Mises test indicates difference at nominal 5% level; (iv) Lilliefors' test does not reject normality. **5.17** Yes. $U=12$. Significant at exact 1.06% level in two-tail test. **5.19** $U=17$, reject H_0 at 0.1% level using asymptotic test. For 95% confidence limits reject 62 largest and smallest differences giving interval (5, 13). Hodges–Lehmann estimator is 9. The 95% normal theory confidence interval is (4.99, 13.03). **5.21** $Z=-2.848$. Significant at 0.44% level in two-tail test. Tied ranks have virtually no influence, reducing denominator of Z from 116.7 to 116.1.

Chapter 6

6.1 F=6.78 with 2, 15 degrees of freedom. Significant at 1% level, similar to result for Kruskal–Wallis test. **6.3** Z=2.83, indicating significance at 0.23% level. **6.7** Over all 4 groups F=2.73 with 4, 5 degress of freedom. Not significant. The analysis of variance is sensitive to heterogeneity of variance between groups. **6.9** Jonckheere–Terpstra U=90. Exact one-tail probability is 0.0568 (asymptiotic probability is also 0.0568) so not quite significant at 5% level in one-tail test. **6.11** Friedman T=6.47 and T_1=2.27. Accept H_0 : *no consistent preference.* **6.15** (i) T_1=1.42, not significant. (ii) T_1=5.28. Significant at 1% level. **6.17** Not significant, T=4.70 using (6.2). **6.19** Multiple comparison test on ranks indicates numbers significantly higher than controls for gibberellic acid, indole acetic acid and adenine sulphate. **6.21** U=119.0 for Jonckheere–Terpstra test, giving Z=−4.114. Significant at a level greater than 0.1%.

Chapter 7

7.5 Do not reject H_0 : *no trend.* **7.7** r_s=0, t=1/21. Strong evidence against association. **7.9** r_s=0.90, t_b=0.78; former significant at 5% level in a two tail test. **7.11** (i) Friedman T_1=6.09. Significant at 0.1% level. Clearly examiners show agreement but this is not consistent across all candidates. (ii) T_1=5.24. Significant at 0.1% level; in particular examiner 5 consistently awards higher marks.

Chapter 8

8.3 y=−13.6+2.42x. Graph shows nonlinearity. With only a small data set it is a good idea to plot a scatter diagram to save the trouble of fitting a possibly inappropriate line. **8.5** The fitted x, y are:

x	4	5	6	7	8	9	10	11	12	13
y	40	45	51	55	60	65	67	68	71	74

x	14	15	16	17	18	19	20
y	75.33	76.67	78	82	83	85	89

8.7 Equation is y=5.74+0.0064x; 95% confidence interval is (0.0032, 0.0095). **8.9** Mann–Whitney U=5. Exact one-tail probability is 0.0754. Accept H_0 : *slopes equal.*

Chapter 9

9.9 Pearson $X^2=25.34$ (10 degrees of freedom). Significant at 0.47% level. A Kruskal–Wallis test just fails to reach significance at the 5% level, indicating the test may be insensitive to some types of association when there are many ties. **9.11** Overwhelming evidence that poor make relatively less use of services. Asymptotic Jonckheere–Terpstra, Pearson X^2, and Fisher exact tests all indicate significance at nominal 0.1% level. **9.13** Highly significant indication that manufacturer's claim not justified. (Hint: under H_0 the probability that 0, 1, 2, 3, 4 parts survive has a binomial distribution with $p=0.95$.) **9.15** Yes. (From data we expect 36.8 positive responses per 100 interviewed under H_0.) **9.19** McNemar $X^2=23$. Very highly significant indication of change in attitudes. **9.21** If choices random, expect 33.33 for each group. $X^2=38.14$ with 23 degrees of freedom. Significant at nominal 5% level. First choices are 248 for A, 181 for B, 198 for C and 173 for D. Expected number for each is 200. $X^2=16.99$ with 3 degrees of freedom. Significant at 0.1% level. **9.23** $X^2=40.48$; asymptotically $\Pr(X^2 \geq 40.48)<0.00005$; $G^2=13.43$; asymptotically $\Pr(G^2 \geq 13.43)=0.266$. In both cases the permutation distribution exact probability is 0.0182. Different patterns in columns 11 and 12 relative to the rest of the table dominate in determing X^2 and G^2 values; the former is more susceptible to small expected frequencies in row 1. There is really no justification for using asymptotic results with these low expected frequencies.

Chapter 10

10.3 Both show evidence of association significant at a nominal 0.1% level. **10.9** First-order interaction model does not suffice. One way of establishing this is to compute the Breslow–Day statistic (10.17). This has the value 54.11 with 1 degree of freedom, providing overwhelming evidence that there is not a common odds ratio. Another approach would be that used in Example 10.2. **10.11** As parties are 'ordered' we may use a linear-by-linear association test with rank scores. This gives an asymptotic $Z=1.28$ so we do not reject H_0. Jonckheere–Terpstra test is a possible alternative. One might query whether any association need be monotonic with age.

Chapter 11

11.3 Mean zero; unbiased. Variance 12.5; bias of sample estimate is -2.5. **11.5** Mean 1.130, 10% trimmed mean 1.114, 25% trimmed mean 1.163. Little difference between these; in absence of any evidence of an outlier probably best not to trim. **11.7** 0.0018.

Index

Topics marked with an asterisk are not dealt with in detail but appropriate references are given on the page(s) indicated.